RNases in Health and Diseases

FOUNDATIONS AND FRONTIERS IN ENZYMOLOGY Series

Munishwar Nath Gupta, Series Editor

Series Website: https://www.elsevier.com/books-and-journals/book-series/books-series-landing-page-request-foundations-and-frontiers-in-enzymology

Foundations and Frontiers in Enzymology

RNases in Health and Diseases

From Housekeeping Enzymes to Moonlight Proteins

Edited by

ESTER BOIX

Department of Biochemistry and Molecular Biology, Universitat Autònoma de Barcelona, Cerdanyola del Vallès, Spain

ELSEVIER

ACADEMIC PRESS

An imprint of Elsevier

Academic Press is an imprint of Elsevier
125 London Wall, London EC2Y 5AS, United Kingdom
50 Hampshire Street, 5th Floor, Cambridge, MA 02139, United States

ISBN: 978-0-443-27312-4

For Information on all Academic Press publications
visit our website at https://www.elsevier.com/books-and-journals

Publisher: Megan Ball
Acquisitions Editor: Ali Afzal-Khan
Editorial Project Manager: Shreya Jain
Production Project Manager: Karthick Jeganathan
Cover Designer: Gopalakrishnan Venkatraman

Working together to grow libraries in developing countries
www.elsevier.com • www.bookaid.org

Typeset by MPS Limited, Chennai, India

Contents

List of contributors

Brian Becknell
The Abigail Wexner Research Institute at Nationwide Children's Hospital, Columbus, OH, United States

Ester Boix
Department of Biochemistry and Molecular Biology, Universitat Autònoma de Barcelona, Cerdanyola del Vallès, Spain

Raúl Anguita Contreras
Department of Biochemistry and Molecular Biology, Faculty of Biosciences, Universitat Autònoma de Barcelona, Cerdanyola del Vallès, Spain

Hanna Cortado
The Abigail Wexner Research Institute at Nationwide Children's Hospital, Columbus, OH, United States

Rosanna Culurciello
Department of Biology, University of Naples Federico II, Naples, Italy

Giovanni Gotte
Biological Chemistry Section of the Department of Neuroscience, Biomedicine and Movement, University of Verona, Verona, Italy

Xincheng Kang
Department of Biochemistry and Molecular Biology, Universitat Autònoma de Barcelona, Cerdanyola del Vallès, Spain

Jiarui Li
Department of Biochemistry and Molecular Biology, Autonomous University of Barcelona, Cerdanyola del Vallès, Spain

Mohammed Moussaoui Keribii
Department of Biochemistry and Molecular Biology, Universitat Autònoma de Barcelona, Cerdanyola del Vallès, Spain

Mahesh Narayan
Department of Chemistry and Biochemistry, University of Texas at El Paso, El Paso, TX, United States

Elio Pizzo
Department of Biology, University of Naples Federico II, Naples, Italy

Guillem Prats-Ejarque
Department of Biochemistry and Molecular Biology, Autonomous University of Barcelona, Cerdanyola del Vallès, Spain; ALBA synchrotron Light Source, Cerdanyola del Vallès, Spain

Laura Diaz Satizabal
Department of Biochemistry and Molecular Biology, Universitat Autònoma de Barcelona, Cerdanyola del Vallès, Spain

Jinghao Sheng
Zhejiang University School of Medicine, Hangzhou, Zhejiang, P.R. China

About main contributing authors

Mahesh Narayan is a full professor in chemistry, biochemistry, and bioinformatics at the University of Texas, United States. His main area of interest is in protein structure-function and particularly in protein misfolding and neuroscience.

Elio Pizzo is a full professor in biochemistry at the Università di Napoli Federico II, Italy. He is currently working on the structural-function characterization and evolutionary studies of ancestral vertebrate RNases.

Giovanni Gotte is a full professor in biochemistry at Università degli Studi di Verona, Italy. His work focuses on protein oligomerization pathways within the RNase A superfamily.

Jinghao Sheng is a professor at the School of Medicine at Zhejiang University, Hangzhou, China. One of his main topics of interest is human RNase 5/angiogenin role in chronic digestive diseases.

Brian Becknell is a pediatric nephrologist and physician scientist at Nationwide Children's Hospital and research director of the Kidney and Urinary Tract Center at the Abigail Wexner Research Institute, Ohio, United States. He is studying the role of antimicrobial RNases in urinary tract infections.

Ester Boix is a full professor in biochemistry and molecular biology at the Universitat Autònoma de Barcelona, Spain. She is currently working on the mechanism of action of extracellular RNases and the processing of non-coding RNA.

Preface

The present project aims to serve as a companion book for both students and researchers who decide to start working with ribonucleases and are willing to acquire an overall background from the field. This book covers from basic science to methodological tools and therapeutic applications.

The book also wants to introduce new concepts in the field. Ribonucleases are not considered anymore as simple enzymes composed merely of a scaffold and a catalytic active site. They exert a diversity of biological activities, some of them just recently discovered. As proteins at the front line of immune surveillance, they need to rapidly adapt to the cell environment. They can work as RNA scavengers, cell sentinels, and alarmins, generate intercellular signaling molecules, and can even control the tissue-specific translation. Based on their multifaceted nature, we have named them moonlight proteins. As RNases keep writing their evolving story, we invite the readers to join us as travel companions in this exciting journey.

Acknowledgments

I would like to dedicate this book to Prof. Richard Youle, who was my mentor during my postdoctoral stay at the *National Institutes of Health* in Bethesda in 1995. I still have such good memories from that period of my life where I learned how to enjoy Open Science, without barriers or boundaries, and felt part of a truly engaged team.

I also want to thank all of the book chapter authors, who kindly accepted to participate in this project and contributed to making it real.

SECTION I

Ribonucleases structure–function

CHAPTER 1

Overview: ribonucleases, a family of multifaceted proteins

Ester Boix
Department of Biochemistry and Molecular Biology, Universitat Autònoma de Barcelona, Cerdanyola del Vallès, Spain

1.1 Introduction

Ribonucleases (RNases) are back on stage. A matter of study by enzymologists at the origin of the biochemistry discipline, RNases are now attracting renewed attention upon the discovery of signaling RNAs.

RNases are nuclease enzymes that cleave phosphodiester bonds in ribonucleotide substrates. We can further classify RNases based on their exo or endo cleavage selectivity, single-stranded or double-stranded RNA substrate preference, or their subcellular location or function (Aravind and Koonin, 2001). This book aims to introduce readers to the RNase field by taking the RNaseA superfamily as a reference (Fig. 1.1). Here, in this chapter, we aim to provide an overall perspective of this protein family and to persuade readers of the many sides and roles these small proteins can achieve.

RNaseA is a secretory endonuclease enzyme that targets single-stranded RNA. As mentioned above, RNaseA was the subject of pioneer biochemistry methodologies during the 20th century and was the first protein where the 3D folding was related to its primary sequence. The protein was purified in large quantities from bovine pancreas provided by slaughterhouses; upon ion exchange chromatography, a main nonglycosylated peak fraction was named RNaseA to differentiate it from the glycosylated B form (Raines, 1998). Despite the protein being one of the first enzymes to be crystallized, the identification of its catalytic mechanism even preceded the solving of its 3D structure, representing the first example of characterization of protein catalytic site functional groups, where two His imidazole groups were ascribed to a general acid–base catalysis (Cuchillo et al., 2011). Following, secondary binding sites were identified, which aligned at the 5' and 3' ends of the cleaved phosphodiester bond and contributed to the proper substrate anchoring at the protein active site groove (Fig. 1.2).

1.2 Exploring the substrate-binding site architecture

Chapter 2 reviews current knowledge of the RNaseA catalytic mechanism, together with the differences and specificities in substrate recognition characteristic to other

RNases in Health and Diseases
DOI: https://doi.org/10.1016/B978-0-443-27312-4.00007-9

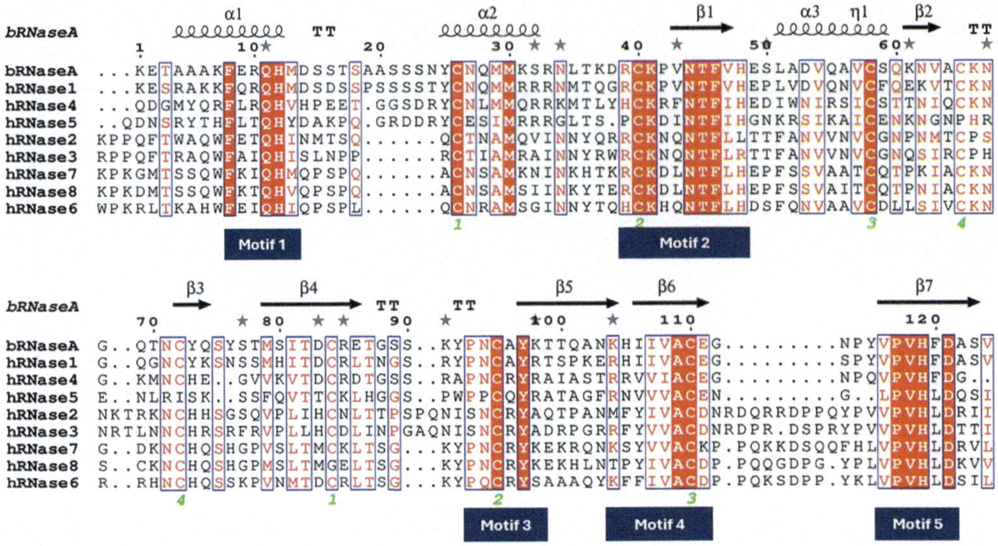

Figure 1.1 Sequence alignment of human canonical RNases that belong to the RNaseA superfamily. The alignment was performed with *ClustalW* and depicted with the *Espript* Server. We can identify five conserved motifs within the family : motif 1 is related to antimicrobial activity; motif 2 encompasses the conserved family signature (CKXXNTF); motif 3 is related to recognition by a putative tyrosine kinase receptor; motif 4 is related to secondary base binding site (B2), and motif 5 to primary base binding site (B1). Last, the catalytic triad (H12, H119, and K41 in RNaseA) is associated with motifs 1, 2, and 5. The spatial locations of motifs in the 3D structure are illustrated in Fig. 1.4.

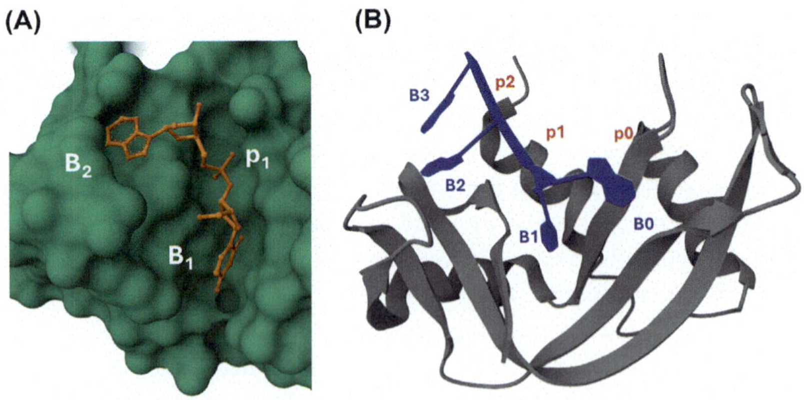

Figure 1.2 Schematic drawing of RNaseA binding to nucleotide substrates. (A) RNaseA binding mode to a dinucleotide (3D structure of protein-d(CpA) complex; PDB ID: 1RPG); p1 being the main phosphate-binding site where cleavage takes place, and B1 and B2 are the main and secondary base binding sites; (B) RNaseA in complex with d(ApTpApA) (PDB ID: 1RCN); base and phosphate subsites are indicated in blue and orange respectively.

family members (Boix et al., 2013). Identification of substate specificity and nucleotide-binding selectivity is particularly challenging when studying a polymeric substrate that is rapidly transforming itself during catalysis (Moussaoui et al., 1996). The RNA substrate interacts with the protein via several subsite cavities for binding to bases (B_n) and phosphates (P_n) (Nogués et al., 1998; Parés et al., 1991). Subsite disposition on the protein surface contributes to RNA alignment and favors an endonuclease activity to release a population of degradation products, which in turn will offer a collection of novel substrate fragments for the next cleavage cycle. This dynamic process represents an additional challenge for enzymologists to achieve an accurate analysis of the catalytic reaction. In **Chapter 3**, we present some methodologies that successfully addressed the quantification of RNaseA catalysis on polymeric RNA substrates.

Another challenge to overcome was the obtention of appropriate enzyme–substrate complexes for RNase structural enzymology studies. Although the RNaseA protein alone could easily be crystallized, analysis of the mechanism of action by early X-ray crystallography studies required the use of nucleotide analogs, which could not totally mimic the protein interactions with its natural substrate (Raines, 1998). NMR and time-resolved crystallography are suitable alternatives to monitor the RNases' catalytic processes. To date, there are scarce PDB depositions of RNase in complex with extended nucleotide analogs. Still, one of the best glimpses we have now for RNaseA recognition at the active site groove of a substrate analog is the 3D complex of the protein with the d(CpA) dinucleotide (Zegers et al., 1994) (Fig. 1.2). A main base interaction site (B1) specific for pyrimidines was identified, whereas the secondary site (B2) was reported to have a preference for purines. Using kinetic and chemical modification approaches, additional protein substrate-binding sites at the 5' and 3' of the cleaved phosphodiester bond (p1) were identified and named as p_0 and p_2 (Nogués et al., 1998). The 3D complex of a tetramer at the active site, deposited in 1994 by Fontecilla-Camps, outlined the residues contributing to the secondary binding sites (Fontecilla-Camps et al., 1994) (see Fig. 1.2). However, study of RNase-RNA analog complexes is mostly hindered by inherent complexity in RNA structuration, potential posttranscriptional modification, and substrate polymeric type and flexibility. Some of these obstacles have slowly been overcome by novel molecular modeling and structural dynamics AI-based software. Doucet and coworkers ideated a very elegant approach that was able to dissect the protein flexible regions and the network of interactions that vibrate in harmony to adapt to substrate-binding and catalysis processes (Tran et al., 2024).

1.3 A reference family for protein folding and evolutionary studies

Apart from pioneering enzymology work, the small size and stability of these protein families have provided them a privileged place in the world of protein folding research.

Chapter 4 reviews the protein folding pathways of RNaseA. The protein adopts a kidney-shaped $\alpha+\beta$ 3D structure (Fig. 1.2) stabilized by 3–4 disulfide bonds. The presence of disulfide bonds is characteristic of secretory proteins and is advantageous for stabilization, whereas it can complicate the pathways to native fold (Narayan, 2011). Although statistically the probability of forming the proper disulfide bond is very low, in vivo the process is promoted by an adequate oxidizing environment and the presence of chaperones at the endoplasmic reticulum, which prevent the accumulation of misfolded proteins.

But to understand how structure meets function, to track how diversity within the family reveals the protein adaptation to evolutionary pressure, we must jump to Chapter 5. In **Chapter 5**, we travel backward to explore the family origins. The RNaseA superfamily presents a particular interest as a vertebrate-specific family (Goo and Cho, 2013; Cho et al., 2005). Following a trait characteristic of innate immunity proteins, some genes have undergone a rapid evolutionary pace (Rosenberg, 2008). Interestingly, family homologs are exclusively found in vertebrate tetrapods and teleosts. To find out where all these RNases come from, Pizzo and coworkers studied fish RNases in depth (Pizzo and D'Alessio, 2007). Interestingly, fish RNases not only have catalytic activity but are also endowed with angiogenesis activity, a trait characteristic of RNase5 type members, also named angiogenins (Pizzo et al., 2008, Pizzo et al., 2010).

Structure–functional comparative analysis should guide us to discover both shared innate immune traits for vertebrates and singular properties ascribed to each family subbranch. Based on current known members in humans, we can define eight canonical RNase types (Sorrentino, 2010) (see Fig. 1.1), where RNases2 and 3 evolved from a common ancestor during primate evolution (Rosenberg, 2008) (Fig. 1.3). Note that the

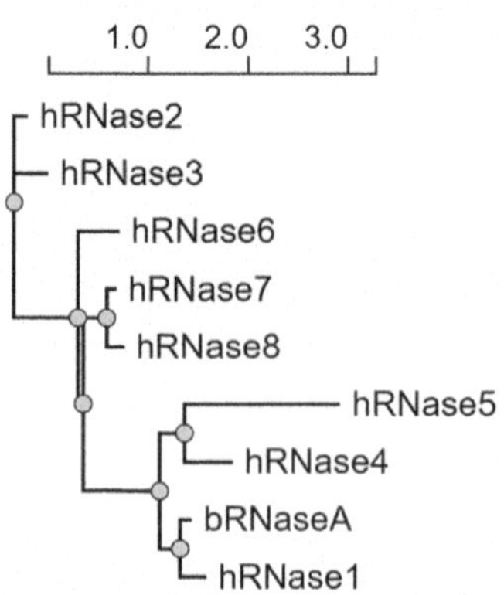

Figure 1.3 Evolutionary tree for RNaseA family. Evolutionary tree drawn by the *NGPhylogeny* server based on the eight canonical human RNases (see alignment in Fig. 1.1). Scale shown by the *Presto* program is arbitrary, based on sequence similarity.

RNase5 type, considered the most ancestral member, shows only three disulfide bonds, out of the four disulfide bonds characteristic of all other family members.

But even more intriguing are noncanonical RNases (RNases9–12), which are classified as noncanonical due to their lack of the conserved family signature (CKXXNTF; see motif2 at Fig. 1.1), together with the absence of the structural determinants for enzymatic activity. RNase9 was originally identified in sperm (Cheng et al., 2009), and RNases9–12 contribute to sperm maturation and participate in heredity by controlling the composition of the small RNA population (Shaffer et al., 2024). RNases' participation in cell identity and interstrain competition has also been described for other RNase families, such as RNaseT2 members associated with pollen incompatibility, or RNases involved in interstrain competition (Igic, 2001).

Regarding the family origins, another hypothesis suggests that the most ancestral family members were endowed with antimicrobial activity. Among RNase types, we find members that behave as antimicrobial proteins (Boix and Nogués, 2007; Pizzo et al., 2008). As a family of secretory proteins, most members are located at the extracellular level and might contribute to protecting biological fluids from infection (Lu et al., 2018). So, we speculate: have these enzymes evolved to protect host organisms from pathogens, to discriminate self from nonself? Are the most ancestral nucleases involved in a simple early defense system, precursor of complex adaptive immunity? (Hernandez, 2025).

Next, in **Chapter 6**, there is an exhaustive overview of all the complexity that can be achieved at the 4th level of protein structuration. Starting from a monomeric unit and thanks to N- and C-terminus domain swapping, a bunch of quaternary structures can be engineered (Noro et al., 2023). Interestingly, RNase motives 1 and 5 (Fig. 1.1) participate in domain swapping, and hybrid active sites can be reconstructed. Gotte and coworkers tell us about many imaginative oligomers engineered in their lab (Gotte et al., 2006) and how to create novel active sites by domain swapping. Interestingly, we find a natural RNase dimer in bovine seminal fluid (Ercole et al., 2009). This protein has enhanced catalytic activity against double-stranded RNA compared to the corresponding monomeric form (Jermann et al., 1995; Opitz et al., 1998).

1.4 Small RNA processing by ribonucleases

Following, we go back from protein quaternary structure to RNA substrate complexity and selective cleavage. **Chapter 7** is focused on non-coding RNA (ncRNA) processing by RNases. Rapid developments in RNA biology are revealing a growing pool of ncRNAs associated with disease propensities (Canestrari and Paroo, 2018). Among ncRNAs, we find miRNAs, lncRNAs, tRNAs, and piRNAs, which can participate in gene expression regulation. Recent results demonstrate that some RNaseA family members can generate functional cleavage products. We can say that RNases work as small RNA tailors and generate a defined singularized cell RNAome. One of the most abundant cellular small ncRNAs is tRNA and its derived fragments (tRFs). RNases1, 2, and 5 have been reported to cleave tRNAs at anticodon loops (Akiyama et al., 2021; Li et al., 2022, 2024; Lu et al., 2022; Su et al., 2019) and generate tRNA halves in response to stress conditions, called stress-responsive fragments (tiRNAs). Changes in tRF population are dependent on cell types (Artamonov et al., 2024). On the other hand, RNA posttranscriptional modifications, such as methylation, can protect ncRNAs from being cleaved by RNases, as demonstrated by the enhanced degradation of hypomethylated tRNAs by angiogenin (Pereira et al., 2021). Cleavage of tRNA by internalized RNases was originally associated with cytotoxic properties (Saxena et al., 2002) and implemented to design novel antitumoral agents (Ardelt et al., 2009).

RNases can act as alarmins: processing of foreign RNA at the endosomal level has been reported to generate immunostimulatory fragments that can activate Toll-like receptors (Ostendorf et al., 2020). A very interesting work by Dalpke and coworkers on host–pathogen interplay suggests that bacteria can protect their own RNAs from cleavage by host RNases by the addition of posttranscriptional modifications (Nunes et al., 2024).

Interestingly, some specific tRFs are abundant at the extracellular level and can be enriched in specific biological fluids, such as sperm (Sharma et al., 2016); where noncanonical RNases were associated with sperm maturation (Shaffer et al., 2024). Some specific tRFs are transported from donor to receptor cells by vesicles and can trigger an immune response in the recipient cells (Pawar et al., 2020). Thus ncRNAs will participate in an information network, interconnecting individual cells and enabling a coordinated response to external stimuli. Together with RNaseA family members, other RNases, such as RNaseT2, RNaseL, or the Schlafen protein (SLFN) family, were reported to participate in the overall tRF population and cellular defense mechanisms (Elder et al., 2024). Recently, tRFs were reported as an additional system to regulate cell translation (Jonkhout et al., 2017). Therefore tRNA fragments (tRFs), also named as tRNA-derived small RNAs, are considered a novel class of ncRNAs with high biological relevance in health and disease and powerful therapeutic applications (Di Fazio and Gullerova, 2023). Very recently, Das and coworkers demonstrated that specific tRFs can work by selective autophagy activation and can be applied to treat hypoxia conditions in the kidney (Li et al., 2025).

1.5 Exploring RNase-based therapies

Next, Section II focuses on the physiological properties of RNases and their potential clinical applications. First, **Chapter 8** introduces the role of RNases as antimicrobial proteins. Human RNaseA superfamily members participate in innate immunity and provide a first line of defense against bacterial, viral, fungal, and parasitic pathogens

(Koczera et al., 2016; Lu et al., 2018). Their antimicrobial mechanism is based on a double-sword action, combining direct disruption of the cell envelope together with enzymatic activity. Regarding antipathogen properties, we also observe some task distribution among RNases. Whereas some members have been reported to fight bacterial infection, such as RNase3 (Boix et al., 2008), RNase6 (Pulido et al., 2016), or RNase7 (Harder et al. 2002), RNase2 is expressed upon viral infection (Li & Boix, 2021; Anguita et al., 2025). We also observe a specific distribution associated with the cell type source. For example, RNase7 is secreted by epithelial cells and is abundant in the skin, with protection against fungal and bacterial skin infections (Harder et al., 2012). On its side, overexpression of RNase3 has been correlated to parasite infections (Adu et al., 2011; Diop et al., 2018; Eriksson et al., 2007). Levels of RNases in blood were proposed as valuable diagnostic biomarkers (Venge et al., 1999; Lu et al., 2018). Besides, we observe differences in the expression of RNases related to different stages of infection. For example, in tuberculosis infection, we can differentiate early/late stages that are associated with distinct protein levels and activities (Anguita et al., 2025). Additionally, we observe how antipathogen activities can be complemented by immunomodulatory properties. The observed pattern suggests that antimicrobial RNases, as other antimicrobial proteins and peptides, might exert a protective role at the initial stages of infection but can damage affected tissues in chronic infections (Lu et al., 2018).

In this context, **Chapter 9** deals with the "bad guy" side of RNases, focusing on disease conditions associated with RNases biological properties. Even before the molecular mechanisms of pro-inflammatory action of RNases started to be unraveled, RNase3 was routinely used in the clinics for asthma prognosis and monitoring (Venge et al., 1999). Levels of RNases have also been associated with sepsis (Martin et al., 2016), and RNase3 has been found among the top six markers linked to sepsis bad prognosis (Li et al., 2024). A study by single-cell transcriptomics and machine learning to discover novel biomarkers proposed RNase2, together with other three top indicators, for early diagnosis of infective endocarditis and sepsis (Yi et al., 2023). In contrast, RNase1 has been reported to attenuate sepsis cardiopathies (Zechendorf et al., 2020). So, whereas RNase3 can work as a potent cytotoxin, promoting in some contexts apoptosis or necrosis, RNase1, abundant in blood, is associated with the removal of bulk RNA released from damaged tissues and proposed as an antiinflammatory agent and cardiac protector.

RNases can participate in cell signaling and act as immune alarmins. Curiously, RNases can act as ligands for tyrosine kinase receptors and promote cell proliferation. Activation of tyrosine kinase receptors has been reported for RNases1, 5, and 7. This action can be associated with tissue healing but can also lead to cancer. Indeed, high levels of RNase1 are indicative of nonresponse to antitumoral treatments in hepatocarcinoma (Li et al., 2024). Another study identifies RNase1 as responsible for the failure of a T-cell antitumoral therapy due to its immunosuppressive activity and the triggering of T-cell dysfunction (Yang et al., 2025).

A dual role of RNases is also observed within the tumor environment (Shlyakhovenko and Samoylenko, 2019). While mostly RNases inhibit cell proliferation, induce apoptosis, and inhibit the growth of tumors, some RNases stimulate vascular growth, proliferation, and tumor development (Shlyakhovenko and Samoylenko, 2019).

Another key actor is the RNase inhibitor (RNH1), a highly abundant ubiquitous protein in the cell cytosol, which not only binds RNases with pico-nanomolar range affinity but also participates directly in inflammasome activation and erythropoiesis (Stillinovic et al., 2024). Indeed, RNH1 genetic deficiencies have severe clinical consequences (Hedberg-Oldfors et al., 2023). An interesting study correlated RNases with hematopoiesis and angiogenesis. Gonzalez and coworkers reported specific differences in RNases associated with adaptation to altitude (González-Buenfil et al., 2024). Differences in RNases gene pattern in Pacific islanders revealed predominance of RNase genes related to angiogenesis at high altitude, in contrast to enrichment of RNase genes associated with immune response against pathogens in lowland population, suggesting a selective pressure by the environment, with interesting significance for human evolution studies.

RNases, secreted upon inflammatory conditions, are considered as useful disease biomarkers, with drug targeting perspectives (Canestrari and Paroo, 2018). A recent case study of cardiac complications following COVID-19 mRNA vaccination revealed an enhanced hyperinflammatory response mediated by RNase2 (Ong et al., 2023). In contrast, angiogenin has been associated with the suppression of the NLRP3 inflammasome activation. The mechanism will be mediated by release of tRFs that will promote stress granule formation and sequestration within of an inflammasome activator protein (Cai et al., 2024). Levels of angiogenin are also associated with the autoimmune disease rheumatoid arthritis, where angiogenesis promotes the disease development, a process dependent on the vascular endothelial growth factor (VEGF) and PI3K/Akt pathway activation (Ao et al., 2022). High levels of RNases were also reported in inflammatory bowel diseases, whereas RNase4 is associated with microbiota protection, as detailed in the next chapter.

In **Chapter 10**, Jinghao Sheng exhaustively reports how RNases can be involved in maintaining intestinal homeostasis, contribute to defining gut microbiota, or cause inflammatory bowel disorders (Yao et al., 2024). On one side, antimicrobial and angiogenic activities of RNase5 can participate in protecting the epithelial barrier integrity, whereas the protein overexpression induced by the presence of Gram-negative bacteria lipopolysaccharides can lead to pathological conditions (Bai et al., 2020; Yu et al., 2017; Yao et al., 2024). On the other hand, eosinophil RNases2 and 3, also called the eosinophil cationic protein (ECP) and the eosinophil-derived neurotoxin

(EDN), are proposed for diagnosis and disease monitoring (Saitoh et al., 1999). Interestingly, some eosinophil RNase variants have been associated with predisposition to inflammatory bowel diseases (Blom et al., 2012).

And this leads us to **Chapter 11**, which focuses on RNase natural variants. Specific human natural variants are associated with disease predisposition. A better knowledge of their frequency will contribute to personalized medicine. One of the best studied examples of RNase variants linked to disease is amyotrophic lateral sclerosis (ALS). Many variants of RNase5, some related to its catalytic activity, and some unrelated, have been associated with the onset of ALS (Bradshaw et al., 2017). We also find some infection predisposition linked to RNase variants. For example, a single-nucleotide substitution in RNase6 significantly reduces the protein antimicrobial activity with no impact on the enzymatic activity (Anguita et al., 2024). Besides, RNase7 variants have been linked to urinary tract infection propensity (Pierce et al., 2021). Last, some single-nucleotide polymorphisms (SNP) of RNase3 associated with distinct cytotoxic and antiparasitic activities show a significantly differentiated geographical distribution, suggesting a key role in either the infection susceptibility or the disease prognosis (Adu et al., 2011; Eriksson et al., 2007). One of the identified RNase3 SNPs implies an Arg to Thr substitution and is related to a distinct glycosylation pattern (Trulson et al., 2007). The R97T SNP is located at motif 3 (see Fig. 1.1) and may have implications related to tyrosine kinase receptor binding (Lu et al., 2021).

Up to this point, we hope to have convinced the readers that RNases, by themselves, have therapeutic potential. **Chapter 12** is taking one more step ahead and explore their applicability for drug design. RNase-based drugs have been proposed as therapeutic molecules, based on either RNases enzymatic properties or on other protein biological activities exerted through mechanisms independent of their ribonucleolytic action.

A first section of Chapter 12 focuses on the antimicrobial properties of RNase3. This protein is the most cationic member of the family, with an isoelectric point higher than 10. RNase3 exposes a high number of cationic residues at the protein surface together with hydrophobic patches (Carreras et al., 2003), which confer the protein amphipathic properties, characteristic of antimicrobial peptides (Boix and Nogués, 2007). There is abundant literature that demonstrates RNase3's ability to bind, destabilize the bacterial wall, and induce cell death (Pulido et al., 2013b; Torrent et al., 2008; Torrent et al., 2012a; Torrent et al., 2009). The principal antimicrobial region of the protein was located at the N-terminus (see motif 1 in Fig. 1.1), and several protein-derived peptides were designed that encompassed the antimicrobial region (Sánchez et al., 2011; Torrent et al., 2013). Thanks to a novel algorithm ideated by Torrent et al. (2012b), who ascribe to each sequence residue an antimicrobial propensity, the protein's putative antimicrobial region was located. This sequence included a hydrophobic patch that participates in the protein membrane binding and destabilization (Pulido et al., 2013a; Torrent et al.,

2010). One of the key features of RNase3 was its high lipopolysaccharide-binding and selective activity against Gram-negative bacteria (Torrent et al., 2008). Some novel peptides were designed with therapeutic potential and proven activity against planktonic, biofilm, and macrophage intracellular infections (Lu et al., 2019). Novel variants of the peptides with promising antimicrobial activity were synthesized by the incorporation of nonstandard amino acids to provide an enhanced half-life (Sandín et al., 2021). Interestingly, a novel synthetic 30 amino acid peptide based on RNase3 sequence (ECPep) had a high lipopolysaccharide-binding affinity and could block the proinflammatory response in a mouse model (Li et al., 2022).

Another approach to designing RNase-based antimicrobials is the engineering of hybrid chimera proteins. Prats-Ejarque et al. designed several chimeras that combined the high catalytic activity of RNase1 together with the bacteriolytic activity of RNase3. Interestingly, the RNase3/1 constructs were able to hinder the emergence of colistin resistance in an in vitro evolution assay (Prats-Ejarque et al., 2019; Fernández-Millán et al., 2022; Prats-Ejarque et al., 2022). The potential use of protein chimera was also studied by Doucet and coworkers, based on the recovery of ancestral RNase sequences and fine-tuning engineering of novel protein variants with selected properties (Tran et al., 2024; Boix and Li, 2024). Resurrected ancestors can serve to define a common core scaffold, where flexible functional regions of interest can be grafted.

Another interesting property of some RNase members is the antiviral action. Among the RNaseA family members, antiviral activity is mainly ascribed to human RNase2 (Li & Boix 2021; Anguita et al., 2025). The protein's antiviral activity depends on the ribonucleolytic activity (Rosenberg, 2008), although a specific binding to virus capsid was also reported (Domachowske et al., 1998). Raines and coworkers ideated an HIV antiviral strategy based on the construction of an RNase-based zymogen (Windsor et al., 2021). The construct included an RNase1 variant that evaded the RNH1 inhibitor, together with a target sequence to be cleaved by an HIV-1 protease and release the cytotoxic RNase within the infected cells. A similar strategy was proposed by the same research group to engineer circular zymogens, where RNase1 is inserted in an inactive form due to active site occlusion by a short sequence extension and gets activated by taking advantage of the HIV-1 protease (Shepard et al., 2019). Equally, an RNase1 circular zymogen is activated by SARS-CoV-2 protease (Wralstad and Raines, 2024). Last, the applicability of RNase antiviral action was also exploited by the development of artificial RNases, as synthetic small molecules that reproduce the RNase catalytic triad conformation (Fedorova et al., 2012; Goncharova et al., 2009; Tamkovich et al., 2016). However, these smaller synthetic RNases could not reach the catalytic efficiency of the parental protein.

Following, a second section of **Chapter 12** explores the therapeutic potential of antitumoral RNases. Mostly, some cancer therapy trials have exploited the antitumor activity of Onconase and bovine seminal RNase (BS-RNase). Onconase is a frog

RNase that was named after its selective cytotoxicity for cancer cells upon its discovery (Ardelt et al., 2009). Onconase is probably the first example of an RNase therapeutic application that entered clinical trials to test its efficacy for patients with mesothelioma (Ardelt et al., 2009). BS-RNase has also been reported to have toxicity for cancer cells (Fiorini et al., 2014). Interestingly, as commented above, BS-RNase is the only natural dimer within the family. For both proteins, Onconase and BS-RNase, their cytotoxicity relies mostly on their highly cationic nature, which promotes high affinity for tumoral cells, with enriched anionic groups at the extracellular side of the cell membranes (Fiorini et al., 2015). Secondarily, their antitumoral activity is due to their ability to evade the cytosolic Ribonuclease Inhibitor RNH1 (Boix et al., 1996; Lomax et al., 2014). Some fusion protein constructs have been designed to enhance the cytotoxicity of these proteins, such as linkage to cell surface recognition domains (Suzuki et al., 1999). The advantage of human RNases is that they can overcome potential immunogenicity. Newton and Ryback proposed an immunotoxin by linking an RNase to a monoclonal antibody that could recognize human B-cells, a strategy that proved effective in a lymphoma mouse model (Rybak and Newton, 1999).

Beyond direct cytotoxicity, some human RNases (e.g., RNase1, 5, and 7) are now recognized as signaling molecules. They can activate oncogenic pathways, mostly through direct binding to tyrosine kinase receptors, such as EGFR for RNase5 (Wang et al., 2018), Ephrin4 for RNase1 (Lee et al., 2021), and ROS1 for RNase7 (Liu et al., 2021), with contradictory reports for the last case (Jones et al., 2024). In this context, the therapeutic strategy is totally the opposite, here the RNases are proposed as novel therapeutic targets to block the activation of kinase-dependent pathways.

Last, we find other promising novel applications, such as the use of RNase1 for vascular disease treatment or angiogenin for neurological disorders. RNase1 is abundantly secreted by endothelial cells, and its degradation of extracellular RNA in blood is proposed as an antiinflammatory therapy and strategy to avoid cardiac complications (Cabrera-Fuentes et al., 2014; Zernecke and Preissner, 2016). On its side, RNase5/angiogenin plays a crucial role in various physiological processes in the brain and is associated with the development of some neurological diseases (Zhang et al., 2024). There are several neuroprotection strategies proposed to palliate neurological disorders linked to RNase5. Curiously, another RNase with reported neurotoxicity, RNase3 (Navarro et al., 2010), has recently been highlighted as a biomarker for schizophrenia and associated with neuroinflammation processes (Obeagu, 2024).

A completely different approach, with potential biotechnological applications, proposes the use of snake RNases that work as defense toxins. Interestingly, studies on the content of venom from the Vietnam cobra *Naja atra* identified an RNase endowed with unusual properties, such as a high catalytic activity at an acidic pH (of about 2.5) and an extremely high conformational stability, thermostability, and protease resistance, together with evasion from the cell cytosolic RNH1 inhibitor (Nguyen and Osipov, 2017).

1.6 RNases as moonlight proteins

I hope this book compendium will convince the readers that RNases are pleiotropic proteins. We can suggest that, as multifunctional molecules, they behave as moonlight proteins. Thus their multifaceted properties would allow them to exert tailored host defense tasks upon request. According to the cell type and stress stimuli, selected RNases will be secreted to target specific pathogens, anomalous cell types, or RNA substrates.

Protein moonlighting is an alternative strategy to multidomain proteins. In protein moonlighting, a single scaffold encompasses diverse motives. In RNaseA family members, functional regions differentiated from the enzyme active site are slowly being discovered. Still, further evolutionary analysis may unravel novel regions in the near future. In Fig. 1.4, we have outlined the shared identified motives in the family member sequences (see Fig. 1.1) and depicted them in the RNaseA protein's 3D architecture. We observe how residues related to enzymatic activity are conserved within the whole family (shown in dark red in Fig. 1.4), whereas the exposed surface regions can exert novel singular properties, as highlighted in blue by the conservation scale coloring.

Supporting the diversity of properties that complement the basic ribonucleolytic activity of RNases, we can recall again the multiple names that RNases were given upon discovery, in many cases even before ascription to the RNaseA superfamily by sequence analysis. Among this family, we find lectins, angiogenins, antimicrobial proteins, alarmins, and tyrosine kinase receptor agonists. Thus we can talk about "moonlight RNases," proteins that can easily adapt to the cellular context and contribute to its immune surveillance. A pleiotropic character is thus frequent among immune system secretory proteins.

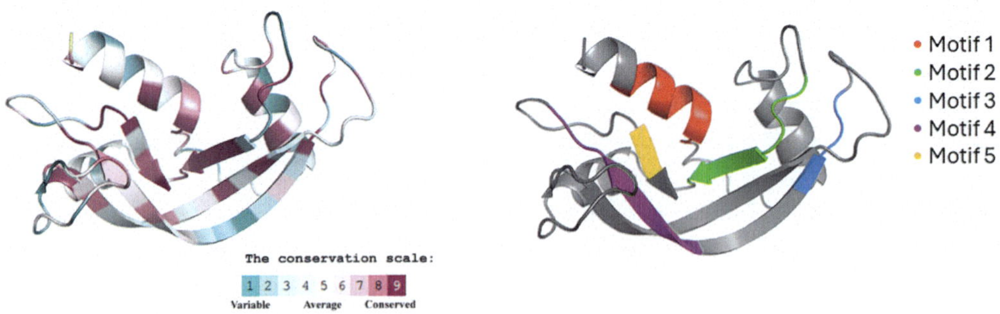

Figure 1.4 Protein motives within the RNaseA scaffold. Depiction of RNaseA 3D structure colored according to conservation score by the use of the consurf server: https://consurf.tau.ac.il/ (Left). Coloring of identified motives at RNaseA family members (Right), based on 3D RNaseA structure (PDB ID 7RSA). For motif identification, see sequence alignment in Fig. 1.1.

Interestingly, the protein ribonucleolytic activity might serve a variety of goals, from random removal of bulk RNA from damaged tissues to selective cleavage and release of signaling molecules. From an evolutionary point of view, there are still many open questions to answer. It is still unclear which was the most ancestral property from the original protein scaffold. Which came first, the antipathogen surveillance or the tissue remodeling activities? It is pertinent here to recall James Lee's thoughts on eosinophil RNases: eosinophils that were originally recruited to sites of viral infection because of the induced tissue damage and cell death may have subsequently acquired anti-viral activities that were selectively advantageous to the host (Lee and Lee, 2005).

1.7 Latest discoveries

As novel methodologies become available, we keep discovering new roles within this family of proteins. One of the most surprising recent reports is the role of angiogenin during translation. Thanks to recent advances in cryo-electron microscopy, Loveland and coworkers revealed angiogenin-specific binding at the ribosomal cavities and cleavage of tRNA, thereby directly intervening in translation regulation (Loveland et al., 2024). This interaction at the ribosomal complex can shift the angiogenin C-terminus, otherwise blocking the protein active site and thereby facilitating the tRNA binding and cleavage at the anticodon loop. Indeed, tRNA halves are now considered as key regulators of selective translation, adapted to cell type and environment. An ingenious methodology has been proposed to test angiogenin's activation by the ribosome using an assay termed as the **Ri**bosome **S**timulated **A**ngiogenin **N**uclease **A**ssay (RiSANA) (Sholi et al., 2025; Li et al., 2025). RiSANA can be used to test the activity of angiogenin, or any other RNase, against different tRNAs and with different ribosome complexes. The protocol may serve to design new diagnostic or therapeutic tools. Specific action of RNases on ncRNAs can also take place at the extracellular level or at the endophagosomal compartment (Ostendorf et al., 2020) and can generate specific signaling molecules that mediate cell-to-cell communication (Pawar et al., 2020; Pawar et al., 2024). Last, a very intriguing role of secretory RNases is their potential action on the recently discovered glycosylated RNAs exposed at the cell surface (Flynn et al., 2021; Chai et al., 2023). GlycoRNAs participate in immune cell recognition, and their cleavage might be key for tissue remodeling, protection against infection, or tumoral processes (Zhang et al., 2023; Montag et al., 2025).

1.8 Conclusions and future perspectives

Members of the vertebrate-specific RNaseA superfamily are multifunctional proteins that participate in the host response to disease conditions, such as infection or cancer. As secretory extracellular players endowed with a ribonucleolytic activity, they might constitute hub control centers to regulate intercellular RNA networks. Their multifaceted properties—including binding to cell receptors, tRNA cleavage, and translation regulation—make them critically important for understanding immune evolution in vertebrates. A fragile balance distinguishes between their role in cell protection against external stress factors and their tissue-damaging action during chronic diseases. Interestingly, their beneficial/detrimental dual role still poses many challenges. We do not know yet how they can combine or switch functions. Other pending questions to solve include the role of posttranslational modifications, such as phosphorylation or glycosylation, as an additional layer of complexity for the fine-tuning regulation of the protein selective action. Overall, RNases are promising candidates for diagnostics and therapeutics. Future research aims to use novel techniques to understand how these small proteins combine properties or switch from one function to another. Further work will assist the design of new drugs, such as targeted antibiotics or tailored enzymes to specifically cleave RNA substrates and generate signaling molecules. As RNases keep writing their evolving story, we trace back their wandering paths as compliant travel companions, trying not to get lost along the journey.

References

Adu, B., Dodoo, D., Adukpo, S., Gyan, B.A., Hedley, P.L., Goka, B., et al., 2011. Polymorphisms in the RNASE3 gene are associated with susceptibility to cerebral malaria in Ghanaian children. PLoS One 6 (12). Available at https://doi.org/10.1371/journal.pone.0029465Denmark, http://www.plosone.org/article/fetchObjectAttachment.action?uri=info%3Adoi%2F10.1371%2Fjournal.pone.0029465&representation=PDF.

Akiyama, Y., Tomioka, Y., Abe, T., Anderson, P., Ivanov, P., 2021. In lysate RNA digestion provides insights into the angiogenin's specificity towards transfer RNAs. RNA Biology 18 (12), 2546–2555. Available at https://doi.org/10.1080/15476286.2021.1930758.

Anguita, R., Ali, Y., Becknell, B., Boix, E., 2025. Ribonucleases of mice and men: unveiling the roles of the rnase a superfamily in host defence. Journal of Immunology Research 2025 (1). Available at https://doi.org/10.1155/jimr/3940139, https://onlinelibrary.wiley.com/journal/1607.

Anguita, R., Prats-Ejarque, G., Moussaoui, M., Becknell, B., Boix, E., 2024. A common polymorphism in RNASE6 impacts its antimicrobial activity toward uropathogenic *Escherichia coli*. International Journal of Molecular Sciences 25 (1), 604. Available at https://doi.org/10.3390/ijms25010604.

Ao, L., Gao, H., Jia, L., Liu, S., Guo, J., Liu, B., et al., 2022. Matrine inhibits synovial angiogenesis in collagen-induced arthritis rats by regulating HIF-VEGF-Ang and inhibiting the PI3K/Akt signaling pathway. Molecular Immunology 141, 13–20. Available at https://doi.org/10.1016/j.molimm.2021.11.002.

Aravind, L., Koonin, E.V., 2001. A natural classification of ribonucleases. Methods in Enzymology 341, 3–28. Available at https://doi.org/10.1016/S0076–6879(01)41142-6.

Ardelt, W., Ardelt, B., Darzynkiewicz, Z., et al., 2009. Ribonucleases as potential modalities in anticancer therapy. European Journal of Pharmacology 625 (1-3), 181–189. Available at https://doi.org/10.1016/j.ejphar.2009.06.067.

Artamonov, A.A., Kondratov, K.A., Bystritsky, E.A., Nikitin, Y.V., Velmiskina, A.A., Mosenko, S.V., et al., 2024. Changes in the repertoire of tRNA-derived fragments in different blood cell populations. Federation Life 14 (10). Available at https://doi.org/10.3390/life14101294, www.mdpi.com/journal/life/.

Bai, R., Sun, D., Chen, M., Shi, X., Luo, L., Yao, Z., et al., 2020. Myeloid cells protect intestinal epithelial barrier integrity through the angiogenin/plexin-B2 axis. The EMBO Journal 39 (13). Available at https://doi.org/10.15252/embj.2019103325.

Blom, K., Rubin, J., Halfvarson, J., Törkvist, L., Rönnblom, A., Sangfelt, P., et al., 2012. Eosinophil associated genes in the inflammatory bowel disease 4 region: correlation to inflammatory bowel disease revealed. World Journal of Gastroenterology 18 (44), 6409–6419. Available at https://doi.org/10.3748/wjg.v18.i44.6409, http://www.wjgnet.com/1007–9327/pdf/v18/i44/6409.pdf.

Boix, E., Blanco, J.A., Nogués, M.V., Moussaoui, M., 2013. Nucleotide binding architecture for secreted cytotoxic endoribonucleases. Biochimie 95 (6), 1087–1097. Available at https://doi.org/10.1016/j.biochi.2012.12.015.

Boix, E., Wu, Y.N., Vasandani, V.M., Saxena, S.K., Ardelt, W., Ladner, J., et al., 1996. Role of the N terminus in RNase a homologues: differences in catalytic activity, ribonuclease inhibitor interaction and cytotoxicity. Journal of Molecular Biology 257 (5), 992–1007. Available at https://doi.org/10.1006/jmbi.1996.0218, https://www.journals.elsevier.com/journal-of-molecular-biology.

Boix, E., Torrent, M., Sánchez, D., Nogués, M.V., 2008. The antipathogen activities of eosinophil cationic protein. Current Pharmaceutical Biotechnology 9 (3), 141–152. Available at https://doi.org/10.2174/138920108784567353Spain, http://www.ingentaconnect.com/content/ben/cpb/2008/00000009/00000003/art00003.

Boix, E., Nogués, M.V., 2007. Mammalian antimicrobial proteins and peptides: overview on the RNase A superfamily members involved in innate host defence. Molecular Biosystems 3 (5), 317–335. Available at https://doi.org/10.1039/b617527a.

Boix, E., Li, J., 2024. Ancestral ribonucleases back in motion for evolutionary-dynamics guided protein design. Trends in Biochemical Sciences 49 (8), 660–662. Available at https://doi.org/10.1016/j.tibs.2024.06.005.

Bradshaw, W.J., Rehman, S., Pham, T.T.K., Thiyagarajan, N., Lee, R.L., Subramanian, V., et al., 2017. Structural insights into human angiogenin variants implicated in Parkinson's disease and amyotrophic lateral sclerosis. Scientific Reports 7. Available at https://doi.org/10.1038/srep41996, www.nature.com/srep/index.html.

Cabrera-Fuentes, H.A., Ruiz-Meana, M., Simsekyilmaz, S., Kostin, S., Inserte, J., Saffarzadeh, M., et al., 2014. RNase1 prevents the damaging interplay between extracellular RNA and tumour necrosis factor-α in cardiac ischaemia/reperfusion injury. Thrombosis and Haemostasis 112 (6), 1110–1119. Available at https://doi.org/10.1160/TH14-08-0703, http://dx.doi.Org/10.1160/TH14-08-0703.

Cai, J., Li, C., Liu, S., Tan, M., Sun, Y., Sun, X., et al., 2024. Angiogenin-mediated tsRNAs control inflammation and metabolic disorder by regulating NLRP3 inflammasome. Cell Death and Differentiation 31 (8), 1057–1069. Available at https://doi.org/10.1038/s41418-024-01311-8, https://www.nature.com/cdd/.

Canestrari, E., Paroo, Z., 2018. Ribonucleases as drug targets. Trends in Pharmacological Sciences 39 (10), 855–866. Available at https://doi.org/10.1016/j.tips.2018.07.005.

Carreras, E., Boix, E., Rosenberg, H.F., Cuchillo, C.M., Nogués, M.V., 2003. Both aromatic and cationic residues contribute to the membrane-lytic and bactericidal activity of eosinophil cationic protein. Biochemistry 42 (22), 6636–6644. Available at https://doi.org/10.1021/bi0273011.

Chai, P., Lebedenko, C.G., Flynn, R.A., 2023. RNA Crossing Membranes: Systems and Mechanisms Contextualizing Extracellular RNA and Cell Surface GlycoRNAs. Annu Rev Genomics Hum Genet 24, 85–107. Available at https://doi.org/10.1146/annurev-genom-101722-101224.

Cheng, G.Z., Li, J.Y., Li, F., Wang, H.Y., Shi, G.X., 2009. Human ribonuclease 9, a member of ribonuclease A superfamily, specifically expressed in epididymis, is a novel sperm-binding protein. Asian Journal of Andrology 11 (2), 240–251. Available at https://doi.org/10.1038/aja.2008.30.

Cho, S., Beintema, J.J., Zhang, J., et al., 2005. The ribonuclease A superfamily of mammals and birds: identifying new members and tracing evolutionary histories. Genomics 85 (2), 208–220. Available at https://doi.org/10.1016/j.ygeno.2004.10.008.

Cuchillo, C.M., Nogués, M.V., Raines, R.T., 2011. Bovine Pancreatic Ribonuclease: Fifty Years of the First Enzymatic Reaction Mechanism. Biochemistry 50 (37), 7835–7841. Available at https://doi.org/10.1021/bi201075b.

Di Fazio, A., Gullerova, M., 2023. An old friend with a new face: tRNA-derived small RNAs with big regulatory potential in cancer biology. British Journal of Cancer 128 (9), 1625–1635. Available at https://doi.org/10.1038/s41416-023-02191-4, https://www.nature.com/bjc/.

Diop, G., Derbois, C., Loucoubar, C., Mbengue, B., Ndao, B.N., Thiam, F., et al., 2018. Genetic variants of RNASE3 (ECP) and susceptibility to severe malaria in Senegalese population. Malaria Journal 17 (1). Available at https://doi.org/10.1186/s12936-018-2205-9, http://www.malariajournal.com/home/.

Domachowske, J.B., Dyer, K.D., Bonville, C.A., Rosenberg, H.F., 1998. Recombinant human eosinophii-derived neurotoxin/RNase 2 functions as an effective antiviral agent against respiratory syncytial virus. Journal of Infectious Diseases 177 (6), 1458–1464. Available at https://doi.org/10.1086/515322, http://jid.oxfordjournals.org/content/current.

Elder, J.J.H., Papadopoulos, R., Hayne, C.K., Stanley, R.E., 2024. The making and breaking of tRNAs by ribonucleases. Trends in Genetics 40 (6), 511–525. Available at https://doi.org/10.1016/j.tig.2024.03.007, https://www.sciencedirect.com/science/journal/01689525.

Ercole, C., Colamarino, R.A., Pizzo, E., Fogolari, F., Spadaccini, R., Picone, D., 2009. Comparison of the structural and functional properties of RNase a and BS-RNase: a stepwise mutagenesis approach. Biopolymers - Peptide Science Section 91 (12), 1009–1017. Available at https://doi.org/10.1002/bip.21176Italy, http://www3.interscience.wiley.com/cgi-bin/fulltext/122240118/PDFSTART.

Eriksson, J., Reimert, C.M., Kabatereine, N.B., Kazibwe, F., Ireri, E., Kadzo, H., et al., 2007. The 434(G>C) polymorphism within the coding sequence of eosinophil cationic protein (ECP) correlates with the natural course of *Schistosoma mansoni* infection. International Journal for Parasitology 37 (12), 1359–1366. Available at https://doi.org/10.1016/j.ijpara.2007.04.001.

Fedorova, A.A., Goncharova, E.P., Kovpak, M.P., Vlassov, V.V., Zenkova, M.A., 2012. Influenza virus inactivated by artificial ribonucleases as a prospective killed virus vaccine. Vaccine 30 (19), 2973–2980. Available at https://doi.org/10.1016/j.vaccine.2012.02.051.

Fernández-Millán, P., Vázquez-Monteagudo, S., Boix, E., Prats-Ejarque, G., 2022. Exploring the RNase a scaffold to combine catalytic and antimicrobial activities. Structural characterization of RNase 3/1 chimeras. Frontiers in Molecular Biosciences 9. Available at https://doi.org/10.3389/fmolb.2022.964717.

Fiorini, C., Gotte, G., Donnarumma, F., Picone, D., Donadelli, M., 2014. Bovine seminal ribonuclease triggers Beclin1-mediated autophagic cell death in pancreatic cancer cells. Biochimica et Biophysica Acta - Molecular Cell Research 1843 (5), 976–984. Available at https://doi.org/10.1016/j.bbamcr.2014.01.025.

Fiorini, C., Cordani, M., Gotte, G., Picone, D., Donadelli, M., 2015. Onconase induces autophagy sensitizing pancreatic cancer cells to gemcitabine and activates Akt/mTOR pathway in a ROS-dependent manner. Biochimica et Biophysica Acta (BBA) - Molecular Cell Research 1853 (3), 549–560. Available at https://doi.org/10.1016/j.bbamcr.2014.12.016.

Flynn, R.A., Pedram, K., Malaker, S.A., Batista, P.J., Smith, B.A.H., Johnson, A.G., et al., 2021. Small RNAs are modified with N-glycans and displayed on the surface of living cells. Cell 184 (12), 3109. Available at https://doi.org/10.1016/j.cell.2021.04.023, https://www.sciencedirect.com/science/journal/00928674.

Fontecilla-Camps, J.C., de Llorens, R., le Du, M.H., Cuchillo, C.M., 1994. Crystal structure of ribonuclease A.d(ApTpApApG) complex. Direct evidence for extended substrate recognition. Journal of Biological Chemistry 269 (34), 21526–21531. Available at https://doi.org/10.1016/s0021-9258(17)31836-7.

Goncharova, E.P., Kovpak, M.P., Ryabchikova, E.I., Konevets, D.A., Sil'nikov, V.N., Zenkova, M.A., et al., 2009. Viral genome cleavage with artificial ribonucleases: a new method to inactivate RNA-containing viruses. Doklady. Biochemistry and Biophysics 427 (1), 221–224. Available at https://doi.org/10.1134/s1607672909040140.

González-Buenfil, R., Vieyra-Sánchez, S., Quinto-Cortés, C.D., Oppenheimer, S.J., Pomat, W., Laman, M., et al., 2024. Genetic signatures of positive selection in human populations adapted to high altitude in papua new Guinea. Genome Biology and Evolution 16 (8). Available at https://doi.org/10.1093/gbe/evae161, http://gbe.oxfordjournals.org/.

Goo, S.M., Cho, S., 2013. The expansion and functional diversification of the mammalian ribonuclease a superfamily epitomizes the efficiency of multigene families at generating biological novelty. Genome Biology and Evolution 5 (11), 2124–2140. Available at https://doi.org/10.1093/gbe/evt161, http://gbe.oxfordjournals.org/.

Gotte, G., Laurents, D.V., Libonati, M., 2006. Three-dimensional domain-swapped oligomers of ribonuclease A: identification of a fifth tetramer, pentamers and hexamers, and detection of trace heptameric, octameric and nonameric species. Biochim Biophys Acta 1764 (1), 44–54. Available at https://doi.org/10.1016/j.bbapap.2005.10.011.

Harder, J, Schroder, JM, 2002. RNase 7, a novel innate immune defense antimicrobial protein of healthy human skin. J Biol Chem 277 (48), 46779–46784. Available at https://doi.org/10.1074/jbc.M207587200.

Hedberg-Oldfors, C., Mitra, S., Molinaro, A., Visuttijai, K., Fogelstrand, L., Oldfors, A., et al., 2023. Ribonuclease inhibitor 1 (RNH1) deficiency cause congenital cataracts and global developmental delay with infection-induced psychomotor regression and anemia. European Journal of Human Genetics 31 (8), 887–894. Available at https://doi.org/10.1038/s41431-023-01327-7, https://www.nature.com/ejhg/.

Hernandez, F.J., 2025. Nucleases: from primitive immune defenders to modern biotechnology tools. Immunology 174 (3), 279–286. Available at https://doi.org/10.1111/imm.13884, http://onlinelibrary.wiley.com/journal/10.1111/(ISSN)1365–2567.

Igic, B, Kohn, JR, 2001. Evolutionary relationships among self-incompatibility RNases. Proc Natl Acad Sci U S A 98 (23), 13167–13171. Available at https://doi.org/10.1073/pnas.231386798.

Jermann, T.M., Opitz, J.G., Stackhouse, J., Benner, S.A., 1995. Reconstructing the evolutionary history of the artiodactyl ribonuclease superfamily. Nature 374 (6517), 57–59. Available at https://doi.org/10.1038/374057a0.

Jones, K., Keddy, C., Jenkins, C., Nicholson, K., Shinde, U., Davare, M.A., 2024. Novel insight into mechanisms of ROS1 catalytic activation via loss of the extracellular domain. Scientific Reports 14 (1). Available at https://doi.org/10.1038/s41598-024-71687-7, https://www.nature.com/srep/.

Jonkhout, N., Tran, J., Smith, M.A., Schonrock, N., Mattick, J.S., Novoa, E.M., 2017. The RNA modification landscape in human disease. RNA (New York, N.Y.) 23 (12), 1754–1769. Available at https://doi.org/10.1261/rna.063503.117, http://rnajournal.cshlp.org/content/23/12/1754.full.pdf+html.

Koczera, P., Martin, L., Marx, G., Schuerholz, T., 2016. The ribonuclease a superfamily in humans: canonical RNases as the buttress of innate immunity. International Journal of Molecular Sciences 17 (8), 1278. Available at https://doi.org/10.3390/ijms17081278.

Lee, H.H., Wang, Y.N., Yang, W.H., Xia, W., Wei, Y., Chan, L.C., et al., 2021. Human ribonuclease 1 serves as a secretory ligand of ephrin A4 receptor and induces breast tumor initiation. Nature Communications 12 (1). Available at https://doi.org/10.1038/s41467-021-23075-2, http://www.nature.com/ncomms/index.html.

Li, J., Boix, E., 2021. Host defence RNases as antiviral agents against enveloped single stranded RNA viruses. Virulence 12 (1), 444–469. Available at https://doi.org/10.1080/21505594.2021.1871823.

Lee, J.J., Lee, N.A., 2005. Eosinophil degranulation: an evolutionary vestige or a universally destructive effector function? Clinical & Experimental Allergy 35 (8), 986–994. Available at https://doi.org/10.1111/j.1365-2222.2005.02302.x.

Li, G., Sun, L., Xin, C., Hao, T., Kharel, P., Manning, A.C., et al., 2025. A hypoxia-responsive tRNA-derived small RNA confers renal protection through RNA autophagy. Science (New York, N.Y.) 389 (6763). Available at https://doi.org/10.1126/science.adp5384, https://www.science.org/toc/science/389/6763.

Li, G., Manning, A.C., Bagi, A., Yang, X., Gokulnath, P., Spanos, M., et al., 2022. Distinct stress-dependent signatures of cellular and extracellular tRNA-derived small RNAs. Advanced Science 9 (17). Available at https://doi.org/10.1002/advs.202200829, http://onlinelibrary.wiley.com/journal/10.1002/(ISSN)2198-3844.

Li, J., Kang, X., Guidi, I., Lu, L., Fernández-Millán, P., Prats-Ejarque, G., et al., 2024. Structural determinants for tRNA selective cleavage by RNase 2/EDN. Structure (London, England 1993) 32

(3), 328. Available at https://doi.org/10.1016/j.str.2023.12.012, https://www.sciencedirect.com/science/journal/09692126.

Liu, C., Zha, Z., Zhou, C., Chen, Y., Xia, W., Wang, Y.N., et al., 2021. Ribonuclease 7-driven activation of ROS1 is a potential therapeutic target in hepatocellular carcinoma. Journal of Hepatology 74 (4), 907–918. Available at https://doi.org/10.1016/j.jhep.2020.09.030, http://www.sciencedirect.com/science/journal/01688278.

Lomax, J.E., Bianchetti, C.M., Chang, A., Phillips, G.N., Fox, B.G., Raines, R.T., 2014. Functional evolution of ribonuclease inhibitor: insights from birds and reptiles. Journal of Molecular Biology 426 (17), 3041–3056. Available at https://doi.org/10.1016/j.jmb.2014.06.007, https://www.journals.elsevier.com/journal-of-molecular-biology.

Loveland, A.B., Koh, C.S., Ganesan, R., Jacobson, A., Korostelev, A.A., 2024. Structural mechanism of angiogenin activation by the ribosome. Nature 630 (8017), 769–776. Available at https://doi.org/10.1038/s41586-024-07508-8, https://www.nature.com/nature/.

Lu, L., Arranz-Trullén, J., Prats-Ejarque, G., Pulido, D., Bhakta, S., Boix, E., 2019. Human antimicrobial RNases inhibit intracellular bacterial growth and induce autophagy in mycobacteria-infected macrophages. Frontiers in Immunology 10. Available at https://doi.org/10.3389/fimmu.2019.01500, https://www.frontiersin.org/journals/immunology#.

Lu, L., Wei, R.L., Prats-Ejarque, G., Goetz, M., Wang, G., Torrent, M., et al., 2021. Human RNase3 immune modulation by catalytic-dependent and independent modes in a macrophage-cell line infection model. Cellular and Molecular Life Sciences 78 (6), 2963–2985. Available at https://doi.org/10.1007/s00018-020-03695-5, http://link.springer.de/link/service/journals/00018/index.htm.

Lu, L., Li, J., Moussaoui, M., Boix, E., 2018. Immune modulation by human secreted RNases at the extracellular space. Frontiers in Immunology 9 (MAY). Available at https://doi.org/10.3389/fimmu.2018.01012, https://www.frontiersin.org/articles/10.3389/fimmu.2018.01012/full.

Lu, L., Li, J., Wei, R., Guidi, I., Cozzuto, L., Ponomarenko, J., et al., 2022. Selective cleavage of ncRNA and antiviral activity by RNase2/EDN in THP1-induced macrophages. Cellular and Molecular Life Sciences 79 (4). Available at https://doi.org/10.1007/s00018-022-04229-x, http://link.springer.de/link/service/journals/00018/index.htm.

Martin, L., Koczera, P., Simons, N., Zechendorf, E., Hoeger, J., Marx, G., et al., 2016. The human host defense ribonucleases 1, 3 and 7 are elevated in patients with sepsis after major surgery—a pilot study. International Journal of Molecular Sciences 17 (3), 294. Available at https://doi.org/10.3390/ijms17030294.

Montag, N., Gousis, P., Wittmann, J., et al., 2025. The emerging role of GlycoRNAs in immune regulation and recognition. Immunology Letters 276, 107048. Available at https://doi.org/10.1016/j.imlet.2025.107048.

Moussaoui, M., Guasch, A., Boix, E., Cuchillo, C.M., Nogués, M.V., 1996. The role of non-catalytic binding subsites in the endonuclease activity of bovine pancreatic ribonuclease A. Journal of Biological Chemistry 271 (9), 4687–4692. Available at https://doi.org/10.1074/jbc.271.9.4687.

Narayan, M., 2011. The case of oxidative folding of ribonuclease A: factors impacting fold maturation of ER-processed proteins. Protein Reviews 14, 23–42. Available at https://doi.org/10.1007/978-1-4419-7273-6_2, http://www.springer.com/series/6876?detailsPage=titles.

Navarro, S., Boix, E., Cuchillo, C.M., Nogués, M.V., 2010. Eosinophil-induced neurotoxicity: the role of eosinophil cationic protein/RNase 3. Journal of Neuroimmunology 227 (1-2), 60–70. Available at https://doi.org/10.1016/j.jneuroim.2010.06.012.

Nguyen, T.V., Osipov, A.V., 2017. A study of ribonuclease activity in venom of vietnam cobra. Journal of Animal Science and Technology 59 (1). Available at https://doi.org/10.1186/s40781-017-0145-5.

Nogués, M.V., Moussaoui, M., Boix, E., Vilanova, M., Ribó, M., Cuchillo, C.M., 1998. The contribution of noncatalytic phosphate-binding subsites to the mechanism of bovine pancreatic ribonuclease A. Cellular and Molecular Life Sciences 54 (8), 766–774. Available at https://doi.org/10.1007/s000180050205.

Noro, I., Bettin, I., Fasoli, S., Smania, M., Lunardi, L., Giannini, M., et al., 2023. Human RNase 1 can extensively oligomerize through 3D domain swapping thanks to the crucial contribution of its C-terminus. International Journal of Biological Macromolecules 249, 126110. Available at https://doi.org/10.1016/j.ijbiomac.2023.126110.

Nunes, I.V., Breitenbach, L., Pawusch, S., Eigenbrod, T., Ananth, S., Schad, P., et al., 2024. Bacterial RNA sensing by TLR8 requires RNase 6 processing and is inhibited by RNA 2'O-methylation. EMBO Reports 25 (11), 4674–4692. Available at https://doi.org/10.1038/s44319-024-00281-9, https://www.embopress.org/page/journal/14693178/about.

Obeagu, E.I., 2024. Exploring the role of eosinophil cationic protein (ECP) in schizophrenia: insights and implications. Medicine (United States) 103 (22), E38380. Available at https://doi.org/10.1097/MD.0000000000038380, https://journals.lww.com/md-journal/pages/default.aspx.

Ong, E.Z., Koh, C.W.T., Tng, D.J.H., Ooi, J.S.G., Yee, J.X., Chew, V.S.Y., et al., 2023. RNase2 is a possible trigger of acute-on-chronic inflammation leading to mRNA vaccine-associated cardiac complication. Medicine 4 (6), 353. Available at https://doi.org/10.1016/j.medj.2023.04.001, www.cell.com/med.

Opitz, J.G., Ciglic, M.I., Haugg, M., Trautwein-Fritz, K., Raillard, S.A., Jermann, T.M., et al., 1998. Origin of the catalytic activity of bovine seminal ribonuclease against double-stranded RNA. Biochemistry 37 (12), 4023–4033. Available at https://doi.org/10.1021/bi9722047.

Ostendorf, T., Zillinger, T., Andryka, K., Schlee-Guimaraes, T.M., Schmitz, S., Marx, S., et al., 2020. Immune sensing of synthetic, bacterial, and protozoan RNA by toll-like receptor 8 requires coordinated processing by RNase T2 and RNase 2. Immunity 52 (4), 591. Available at https://doi.org/10.1016/j.immuni.2020.03.009, www.immunity.com.

Parés, X., Nogués, M.V., de Llorens, R., Cuchillo, C.M., 1991. Structure and function of ribonuclease A binding subsites. Essays in Biochemistry 26, 89–103.

Pawar, K., Shigematsu, M., Sharbati, S., Kirino, Y., Mathieu, J.M., 2020. Bertrand, Infection-induced 5′-half molecules of tRNAHisGUG activate Toll-like receptor 7. PLoS Biology 18 (12), e3000982. Available at https://doi.org/10.1371/journal.pbio.3000982.

Pawar, K., Kawamura, T., Kirino, Y., et al., 2024. The tRNA ^{Val} half: A strong endogenous Toll-like receptor 7 ligand with a 5′-terminal universal sequence signature. Proceedings of the National Academy of Sciences 121 (19). Available at https://doi.org/10.1073/pnas.2319569121.

Pierce, K.R., Eichler, T., Vasquez, C.M., Schwaderer, A.L., Simoni, A., Creacy, S., et al., 2021. Ribonuclease 7 polymorphism rs1263872 reduces antimicrobial activity and associates with pediatric urinary tract infections. Journal of Clinical Investigation 131 (22). Available at https://doi.org/10.1172/JCI149807, https://www.jci.org/articles/view/149807.

Pizzo, E., Varcamonti, M., Maro, A.D., Zanfardino, A., Giancola, C., D'Alessio, G., 2008. Ribonucleases with angiogenic and bactericidal activities from the Atlantic salmon. FEBS Journal 275 (6), 1283–1295. Available at https://doi.org/10.1111/j.1742-4658.2008.06289.x.

Pizzo, E., Merlino, A., Turano, M., Russo Krauss, I., Coscia, F., Zanfardino, A., Varcamonti, M., Furia, A., Giancola, C., Mazzarella, L., Sica, F., D'Alessio, G., et al., 2010. A new RNase sheds light on the RNase/angiogenin subfamily from zebrafish. Biochemical Journal 433 (2), 345–355. Available at https://doi.org/10.1042/bj20100892.

Pizzo, E., D'Alessio, G., 2007. The success of the RNase scaffold in the advance of biosciences and in evolution. Gene 406 (1-2), 8–12. Available at https://doi.org/10.1016/j.gene.2007.05.006.

Prats-Ejarque, G., Lorente, H., Villalba, C., Anguita, R., Lu, L., Vázquez-Monteagudo, S., et al., 2022. Structure-based design of an rnase chimera for antimicrobial therapy. International Journal of Molecular Sciences 23 (1). Available at https://doi.org/10.3390/ijms23010095, https://www.mdpi.com/1422-0067/23/1/95/pdf.

Prats-Ejarque, G., Li, J., Ait-Ichou, F., Lorente, H., Boix, E., 2019. Testing a human antimicrobial RNase chimera against bacterial resistance. Frontiers in Microbiology 10. Available at https://doi.org/10.3389/fmicb.2019.01357.

Pulido, D., Arranz-Trullén, J., Prats-Ejarque, G., Velázquez, D., Torrent, M., Moussaoui, M., et al., 2016. Insights into the antimicrobial mechanism of action of human RNase6: structural determinants for bacterial cell agglutination and membrane permeation. International Journal of Molecular Sciences 17 (4). Available at https://doi.org/10.3390/ijms17040552, http://www.mdpi.com/1422-0067/17/4/552/pdf.

Pulido, D., Moussaoui, M., Nogués, M.V., Torrent, M., Boix, E., 2013a. Towards the rational design of antimicrobial proteins: single point mutations can switch on bactericidal and agglutinating activities on the RNase A superfamily lineage. FEBS Journal 80 (22), 5841–5852. Available at https://doi.org/10.1111/febs.12506.

Pulido, D., Torrent, M., Andreu, D., Nogues, M.V., Boix, E., 2013b. Two human host defense ribonucleases against mycobacteria, the eosinophil cationic protein (RNase 3) and RNase 7. Antimicrobial Agents and Chemotherapy 57 (8), 3797–3805. Available at https://doi.org/10.1128/AAC.00428-13Spain, http://aac.asm.org/content/57/8/3797.full.pdf.

Raines, R.T., 1998. Ribonuclease A. Chemical Reviews 98 (3), 1045–1066. Available at https://doi.org/10.1021/cr960427h.

Rosenberg, H., 2008. Eosinophil-derived neurotoxin / RNase 2: connecting the past, the present and the future. Current Pharmaceutical Biotechnology 9 (3), 135–140. Available at https://doi.org/10.2174/138920108784567236.

Rybak, S.M., & Newton, D.L. (1999). REVIEW Natural and Engineered Cytotoxic Ribonucleases: Therapeutic Potential 1. http://www.idealibrary.com.

Saitoh, O., Kojima, K., Sugi, K., Matsuse, R., Uchida, K., Tabata, K., et al., 1999. Fecal eosinophil granule-derived proteins reflect disease activity in inflammatory bowel disease. American Journal of Gastroenterology 94 (12), 3513–3520. Available at https://doi.org/10.1016/S0002-9270(99)00699-1, http://www.nature.com/ajg/archive/index.html.

Sánchez, D., Moussaoui, M., Carreras, E., Torrent, M., Nogués, V., Boix, E., 2011. Mapping the eosinophil cationic protein antimicrobial activity by chemical and enzymatic cleavage. Biochimie 93 (2), 331–338. Available at https://doi.org/10.1016/j.biochi.2010.10.005.

Sandín, D., Valle, J., Chaves-Arquero, B., Prats-Ejarque, G., Larrosa, M.N., González-López, J.J., et al., 2021. Rationally modified antimicrobial peptides from the N-Terminal domain of human RNase 3 show exceptional serum stability. Journal of Medicinal Chemistry 64 (15), 11472–11482. Available at https://doi.org/10.1021/acs.jmedchem.1c00795, http://pubs.acs.org/jmc.

Saxena, S.K., Sirdeshmukh, R., Ardelt, W., Mikulski, S.M., Shogen, K., Youle, R.J., 2002. Entry into cells and selective degradation of tRNAs by a cytotoxic member of the RNase A family. Journal of Biological Chemistry 277 (17), 15142–15146. Available at https://doi.org/10.1074/jbc.M108115200.

Shaffer, J.F., Gupta, A., Kharkwal, G., Linares, E.E., Holmes, A.D., Swartz, J.R., et al., 2024. Epididymis-specific RNase A family genes regulate fertility and small RNA processing. Journal of Biological Chemistry 300 (12). Available at https://doi.org/10.1016/j.jbc.2024.107933, https://www.sciencedirect.com/science/journal/00219258.

Sharma, U., Conine, C.C., Shea, J.M., Boskovic, A., Derr, A.G., Bing, X.Y., et al., 2016. Biogenesis and function of tRNA fragments during sperm maturation and fertilization in mammals. Science (New York, N.Y.) 351 (6271), 391–396. Available at https://doi.org/10.1126/science.aad6780, http://science.sciencemag.org/content/sci/351/6271/391.full.pdf.

Shepard, S.M., Windsor, I.W., Raines, R.T., Cummins, C.C., 2019. Nucleoside tetra- and pentaphosphates prepared using a tetraphosphorylation reagent are potent inhibitors of ribonuclease A. Journal of the American Chemical Society 141 (46), 18400–18404. Available at https://doi.org/10.1021/jacs.9b09760, http://pubs.acs.org/journal/jacsat.

Shlyakhovenko, V.O., Samoylenko, OA., 2019. The dual role of ribonucleases in tumor-host relationship. Experimental Oncology 41 (3), 195–199. Available at https://doi.org/10.32471/exp-oncology.2312-8852.

Sholi, E., Loveland, A.B., Korostelev, A.A., et al., 2025. Assay for ribosome stimulation of angiogenin nuclease activity. Methods in Enzymology,TRNA-derived RNAs. Elsevier, pp. 381–404. Available at https://doi.org/10.1016/bs.mie.2024.11.007.

Sorrentino, S., 2010. The eight human "canonical" ribonucleases: molecular diversity, catalytic properties, and special biological actions of the enzyme proteins. FEBS Letters 584 (11), 2194–2200. Available at https://doi.org/10.1016/j.febslet.2010.04.018.

Stillinovic, M., Sarangdhar, M.A., Andina, N., Tardivel, A., Greub, F., Bombaci, G., et al., 2024. Ribonuclease inhibitor and angiogenin system regulates cell type–specific global translation. Science Advances 10 (22). Available at https://doi.org/10.1126/sciadv.adl0320.

Su, Z., Kuscu, C., Malik, A., Shibata, E., Dutta, A., 2019. Angiogenin generates specific stress-induced tRNA halves and is not involved in tRF-3–mediated gene silencing. Journal of Biological Chemistry 294 (45), 16930–16941. Available at https://doi.org/10.1074/jbc.ra119.009272.

Suzuki, M., Saxena, S.K., Boix, E., Prill, R.J., Vasandani, V.M., Ladner, J.E., et al., 1999. Engineering receptor-mediated cytotoxicity into human ribonucleases by steric blockade of inhibitor interaction. Nature Biotechnology 17 (3), 265–270. Available at https://doi.org/10.1038/7010.

Tamkovich, N., Koroleva, L., Kovpak, M., Goncharova, E., Silnikov, V., Vlassov, V., et al., 2016. Design, RNA cleavage and antiviral activity of new artificial ribonucleases derived from mono-, di- and tripeptides connected by linkers of different hydrophobicity. Bioorganic & Medicinal Chemistry 24 (6), 1346–1355. Available at https://doi.org/10.1016/j.bmc.2016.02.007.

Torrent, M., Odorizzi, F., Nogués, M.V., Boix, E., 2010. Eosinophil cationic protein aggregation: identification of an N-terminus amyloid prone region. Biomacromolecules 11 (8), 1983–1990. Available at https://doi.org/10.1021/bm100334u.

Torrent, M., Navarro, S., Moussaoui, M., Nogués, M.V., Boix, E., 2008. Eosinophil cationic protein high-affinity binding to bacteria-wall lipopolysaccharides and peptidoglycans. Biochemistry 47 (11), 3544–3555. Available at https://doi.org/10.1021/bi702065b.

Torrent, M, de la Torre, BG, Nogués, VM, Andreu, D, Boix, E, 2009. Bactericidal and membrane disruption activities of the eosinophil cationic protein are largely retained in an N-terminal fragment. Biochem J 421 (3), 425–434. Available at https://doi.org/10.1042/BJ20082330.

Torrent, M., Pulido, D., Nogués, M.V., Boix, E., 2012a. Exploring new biological functions of amyloids: bacteria cell agglutination mediated by host protein aggregation. PLoS Pathogens 8 (11). Available at https://doi.org/10.1371/journal.ppat.1003005Spain, http://www.plospathogens.org/article/fetchObjectAttachment.action?uri=info%3Adoi%2F10.1371%2Fjournal.ppat.1003005&representation=PDF.

Torrent, M., Di Tommaso, P., Pulido, D., Nogués, M.V., Notredame, C., Boix, E., et al., 2012b. AMPA: an automated web server for prediction of protein antimicrobial regions. Bioinformatics (Oxford, England) 28 (1), 130–131. Available at https://doi.org/10.1093/bioinformatics/btr604.

Torrent, M., Pulido, D., Valle, J., Nogués, M.V., Andreu, D., Boix, E., 2013. Ribonucleases as a host-defence family: evidence of evolutionarily conserved antimicrobial activity at the N-terminus. Biochemical Journal 456 (1), 99–108. Available at https://doi.org/10.1042/BJ20130123Spain, http://www.biochemj.org/bj/456/0099/4560099.pdf.

Tran, T.T.Q., Narayanan, C., Loes, A.N., Click, T.H., Pham, N.T.H., Létourneau, M., et al., 2024. Ancestral sequence reconstruction dissects structural and functional differences among eosinophil ribonucleases. Journal of Biological Chemistry 300 (5). Available at https://doi.org/10.1016/j.jbc.2024.107280, https://www.sciencedirect.com/science/journal/00219258.

Trulson, A., Byström, J., Engström, Å., Larsson, R., Venge, P., 2007. The functional heterogeneity of eosinophil cationic protein is determined by a gene polymorphism and post-translational modifications. Clinical & Experimental Allergy 37 (2), 208–218. Available at https://doi.org/10.1111/j.1365-2222.2007.02644.x.

Venge, Byström, Carlson, Håkansson, Karawacjzyk, Peterson, et al., 1999. Eosinophil cationic protein (ECP): molecular and biological properties and the use of ECP as a marker of eosinophil activation in disease. Clinical & Experimental Allergy 29 (9), 1172–1186. Available at https://doi.org/10.1046/j.1365-2222.1999.00542.x.

Wang, Y.-N., Lee, H.-H., Hung, M.-C., et al., 2018. A novel ligand-receptor relationship between families of ribonucleases and receptor tyrosine kinases. Journal of Biomedical Science 25 (1). Available at https://doi.org/10.1186/s12929-018-0484-7.

Windsor, I.W., Dudley, D.M., O'Connor, D.H., Raines, R.T., 2021. Ribonuclease zymogen induces cytotoxicity upon HIV-1 infection. AIDS Research and Therapy 18 (1). Available at https://doi.org/10.1186/s12981-021-00399-z, http://www.aidsrestherapy.com/home/.

Wralstad, E.C., Raines, R.T., 2024. Sensitive detection of SARS-CoV-2 main protease 3CLpro with an engineered ribonuclease zymogen. Protein Science 33 (4). Available at https://doi.org/10.1002/pro.4916, http://onlinelibrary.wiley.com/journal/10.1002/(ISSN)1469-896X.

Yang, W.H., Huang, B.Y., Rao, H.Y., Ye, P., Chen, B., Wang, H.C., et al., 2025. Ribonuclease 1 induces T-cell dysfunction and impairs CD8+ T-cell cytotoxicity to benefit tumor growth through hijacking STAT1. Advanced Science 12 (13). Available at https://doi.org/10.1002/advs.202404961, http://onlinelibrary.wiley.com/journal/10.1002/(ISSN)2198-3844.

Yao, Z., Bai, R., Liu, W., Liu, Y., Zhou, W., Xu, Z., et al., 2024. Activation of angiogenin expression in macrophages by lipopolysaccharide via the TLR4/NF-κB pathway in colitis. Acta Biochimica et Biophysica Sinica 56 (6), 857–865. Available at https://doi.org/10.3724/abbs.2024013, https://www.sciengine.com/abbs.

Yi, C., Zhang, H., Yang, J., Chen, D., Jiang, S., 2023. Elucidating common pathogenic transcriptional networks in infective endocarditis and sepsis: integrated insights from biomarker discovery and single-cell RNA sequencing. Frontiers in Immunology 14. Available at https://doi.org/10.3389/fimmu.2023.1298041, https://www.frontiersin.org/journals/immunology#.

Yu, W., Goncalves, K.A., Li, S., Kishikawa, H., Sun, G., Yang, H., et al., 2017. Plexin-B2 mediates physiologic and pathologic functions of angiogenin. Cell 171 (4), 849. Available at https://doi.org/10.1016/j.cell.2017.10.005, https://www.sciencedirect.com/journal/cell.

Zechendorf, E., O'Riordan, C.E., Stiehler, L., Wischmeyer, N., Chiazza, F., Collotta, D., et al., 2020. Ribonuclease 1 attenuates septic cardiomyopathy and cardiac apoptosis in a murine model of polymicrobial sepsis. JCI Insight 5 (8). Available at https://doi.org/10.1172/jci.insight.131571, https://insight.jci.org/articles/view/131571.

Zegers, I., Maes, D., Dao-Thi, M., Wyns, L., Poortmans, F., Palmer, R., 1994. The structures of rnase a complexed with 3′-CMP and d(CpA): active site conformation and conserved water molecules. Protein Science 3 (12), 2322–2339. Available at https://doi.org/10.1002/pro.5560031217.

Zernecke, A., Preissner, K.T., 2016. Extracellular ribonucleic acids (RNA) enter the stage in cardiovascular disease. Circulation Research 118 (3), 469–479. Available at https://doi.org/10.1161/CIRCRESAHA.115.307961, http://circres.ahajournals.org.

Zhang, N., Tang, W., Torres, L., Zhu, L., Wang, X., Ajaj, Y., et al., 2023. Cell surface RNAs control neutrophil function. Blood 142 (Supplement 1). Available at https://doi.org/10.1182/blood-2023-187570 674–674.

Zhang, Y, Li, Y, Bin, S, Cheng, X, Niu, Q, 2024. A Neglected Gene: The Role of the ANG Gene in the Pathogenesis of Amyotrophic Lateral Sclerosis. Aging Dis 16 (1), 13–32. Available at https://doi.org/10.14336/AD.2024.0107.

CHAPTER 2

Catalytic mechanism of action and substrate-binding sites

Xincheng Kang and Mohammed Moussaoui Keribii
Department of Biochemistry and Molecular Biology, Universitat Autònoma de Barcelona, Cerdanyola del Vallès, Spain

2.1 Introduction

2.1.1 Background

Ribonuclease A (RNaseA) is one of the first enzymes that scientists isolated and characterized. In the early 20th century, researchers found that some substance in a cow's pancreas could break down RNA (Jones, 1920). A huge amount of RNaseA protein was purified and contributed to the development of the main biochemistry methodologies at that time. RNaseA was one of the earliest and most studied enzymes. Following later research, other homologous proteins were found in mammals and other vertebrates, sharing the same 3D fold and catalytic mechanism. All members were classified collectively, taking RNaseA as the family reference (Beintema and Blank, 1988). This is a family of secretory proteins, with a mature protein size between 14 and 16 Da, sharing a minimum of 30% sequence identity and common kidney-shaped fold with an $\alpha+\beta$ type of three-dimensional folding structure (Aravind and Koonin, 2001; Cuchillo and Nogués, 2011). All members include six or eight cysteine residues that form three to four disulfide bonds (Raines, 1998). All members of the family retain two histidines and one lysine that form the catalytic triad, the lysine being located at the conserved family signature CKXXNTF. Some members of the RNaseA family apart to degrade RNA can also play an important role in immune defense, with antibacterial, antiviral, and immunomodulatory functions (Lu and Li, 2018). Understanding the interaction mechanisms of RNases with their physiological substrates is critical for uncovering their roles in health and disease and for developing potential therapeutic applications.

In the 1950s and 1960s bovine pancreatic ribonuclease (RNaseA), thanks to its small size and high thermal stability, provided an ideal model for protein folding and structure–function relationship studies. Christian Anfinsen's experiments with RNaseA proved that a protein's amino acid sequence determines its three-dimensional structure, and was awarded the Nobel Prize in Chemistry in 1972. Bovine pancreatic RNaseA was one of the first enzymes to be purified in large quantities and obtained in crystalline form and was also one of the first small proteins to be analyzed by X-ray diffraction. It

RNases in Health and Diseases
DOI: https://doi.org/10.1016/B978-0-443-27312-4.00008-0

provided an early key model for studying the relationship between the amino acid sequence, 3D structure, and catalytic properties. In addition, RNaseA became an important tool for studying protein folding and thermodynamic stability due to its small molecule size, high stability, and reversible folding properties (Anfinsen, 1973) (see Chapter 4). Further structural studies of RNases provided valuable information on RNA–protein interactions (Leonidas et al., 2003). More recently, RNaseA is being widely used as a tool in molecular biology experiments to remove undesired RNA contaminants (Sun et al., 2022).

2.1.2 Biological properties and significance

RNaseA catalyzes the degradation of single-stranded RNA (ssRNA) by cleaving the 3',5' phosphodiester bond of RNA to form a 2′,3′-cyclic phosphate, which ultimately generates a 3′-phosphate end (Cuchillo and Nogués, 2011). RNaseA was initially attributed a biological role in the pancreas: to digest and remove the large amount of RNA produced by gastrointestinal microorganisms in ruminants. Later on (Raines, 1998), RNaseA homologs were reported to have other biological properties, such as antibacterial and antiviral effects. Besides, RNA recognition and removal can play an important role in host defense mechanisms, especially by protecting the body from foreign RNAs (Lu and Li, 2018). RNases can have immunomodulatory activities, such as degrading exogenous RNA and activating Toll-like receptors (Greulich et al., 2019; Nunes et al., 2024; Ostendorf et al., 2020). They can enhance the immune response in the early stages of infection and promote dendritic cells (DCs) to secrete repair factors in the later stages of infection, which helps tissue remodelling and homeostasis restoration (Yang et al., 2004). In the field of biotechnology, RNaseA is commonly used for RNA degradation, RNA protection assays, and diagnostic techniques (Sun et al., 2022). In drug development, it also shows potential anticancer value (Mironova et al., 2013). Overall, RNaseA not only plays a key role in RNA metabolism but also attracts interest in basic research, biotechnology, and biomedicine.

2.1.3 A model to study RNA catalysis and substrate-binding mode

2.1.3.1 Catalytic mechanism of RNaseA

RNaseA catalyzes the hydrolysis of phosphodiester bonds; the reaction is carried out in two steps. First, RNaseA cleaves the RNA substrate to generate a 2',3'-cyclic phosphate intermediate; subsequently, the intermediate is hydrolyzed to produce a 3'-phosphate end (Cuchillo and Nogués, 2011; Shigematsu et al., 2018). In this process, His12 and His119 in RNaseA act as acid–base catalysts. In the initial stage of the reaction, His12 in its basic form initiates the reaction by deprotonating the 2'-hydroxyl group of RNA. On the other hand, His119 in its acidic form protonates the leaving group during the reaction and facilitates the cleavage of the phosphodiester bond. In addition, the positively charged side chain residue of Lys41 helps to position the phosphate group through electrostatic interactions and stabilizes the reaction intermediates. In addition, catalytic efficiency and substrate specificity rely on other complementary residues that form the enzyme active site groove. RNaseA maintains its 3D structure by four disulfide bonds, and the active site is located in a cavity located between the first α-helix and last β-strand (Fig. 2.1).

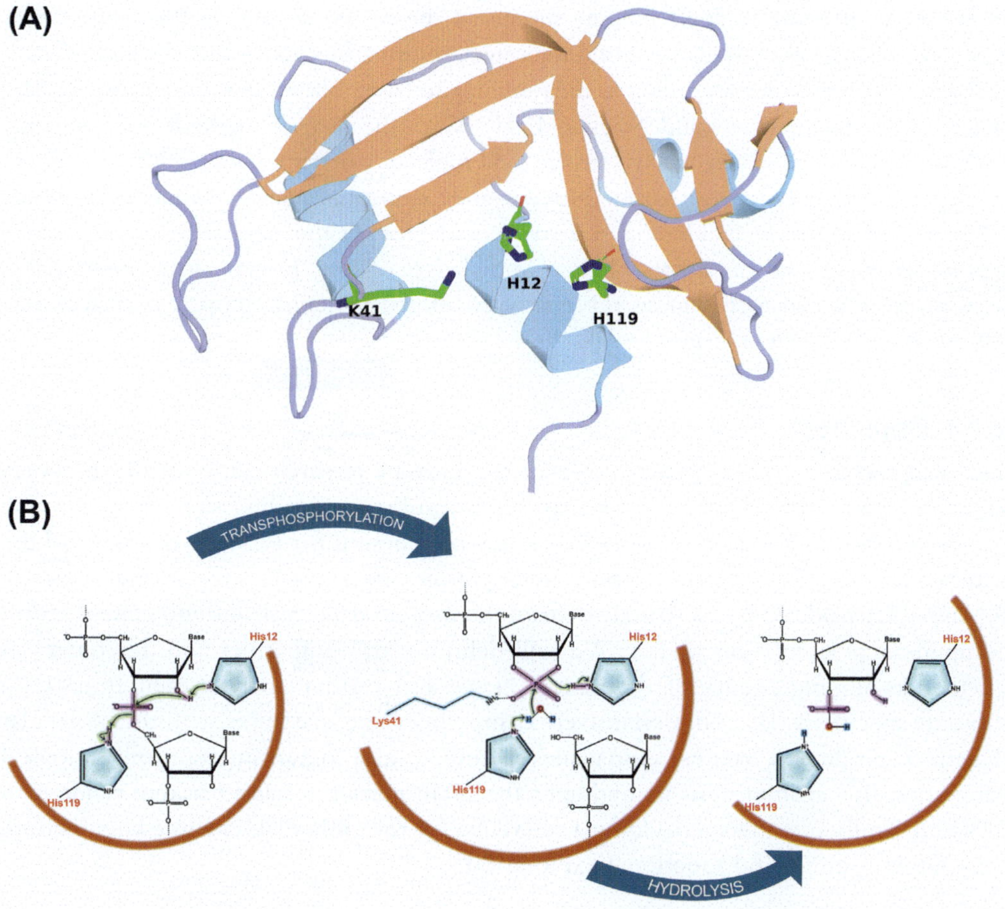

Figure 2.1 (A) Ribbon depiction of RNaseA 3D structure; (B) Diagram of the catalytic mechanism of RNaseA. (A) Highlighting secondary structure elements and catalytic residues. α-Helices are shown in blue, β-strands in orange, and loops/coils in purple. The three residues essential for catalytic activity (stick representation in green) are indicated. (B) The catalytic triad is shown with cyan shading and red labels, and the reacting phosphate group is indicated by purple shading (*RNaseA*, ribonuclease A).

When the RNA substrate is accommodated at the active center, a variety of noncovalent interactions are formed between the enzyme and the nucleotide to ensure that the nucleotide phosphodiester backbone is precisely positioned (Raines, 2004). By studying site-directed mutagenesis variants of RNaseA and other family members, the role of each key amino acid has been explored, and the substrate-binding diversity within the family has been approached (Boix et al., 2013; Nogués and Moussaoui, 1998).

2.1.3.2 Studying RNaseA substrate-binding mode

In RNaseA, there are multiple binding subsites at the 5' and 3' ends of the active center where the cleaved phosphodiester bond is located, ensuring substrate selectivity and efficient cleavage. Sites are named according to their binding to bases, ribose, and phosphates (B_n, R_n, and P_n, respectively) (Parés and Nogués, 1991). With the help of methods such as X-ray crystallography, nuclear magnetic resonance, and molecular dynamics simulation, the subsite architecture has been characterized (Nogués and Moussaoui, 1998). Understanding the RNaseA family members' binding patterns should provide the basis for designing more specific enzymes and inhibitors. Better knowledge of the protein structure–function relationship will lead to designing RNases with tailored biological properties (Fernández-Millán et al., 2022; Prats-Ejarque et al., 2021).

2.1.4 Objectives

This chapter aims to systematically detail the current research status of the RNaseA superfamily in terms of catalytic mechanism and substrate-binding sites. First, we will explore the key amino acid residues that RNaseA relies on for RNA hydrolysis and the structural basis of enzyme–substrate recognition. Next, we will review existing RNase–nucleotide complex structures from the protein data bank and integrate relevant computational simulation studies. We will delineate the similarities and differences in nucleotide-binding modes of different RNases and further analyze their impact on substrate specificity. By comprehensively comparing these structures with the molecular dynamics prediction results, mutagenesis studies, and determination of biological properties, this chapter aims to highlight the main conserved and variable residues in RNaseA family members, providing a reference for researchers in enzyme engineering, drug target mining, and functional exploration.

2.2 Structure of RNaseA

2.2.1 Overview

RNaseA is a single-chain polypeptide of 124 amino acid residues. Its overall 3D kidney-shaped structure is stabilized by four disulfide bonds. RNaseA has a compact structure, and its secondary structure mainly consists of four antiparallel β-sheets and three short α-helices (Fig. 2.1). The active site is located in a central cavity conformed by the catalytic triad (His12, His119, and Lys41) (Chou and Fasman, 1977; Richardson, 1981).

2.2.2 Binding sites of RNaseA

Several structural and kinetic studies demonstrated the presence in RNaseA of noncatalytic substrate-binding centers, in addition to the active site (Raines, 1998; Nogués and Moussaoui, 1998). Several substrate-binding sites that interact with the phosphate and the nitrogenous base of the nucleotides are important for the stabilization of the polymeric RNA substrate at the active site region. These sites include the main base site (B_1) formed by the residues Thr45, Ser123, and Phe120, the secondary base site (B_2) formed by Gln69, Asn71, and Glu111, and other less characterized additional sites, such as B_0 and B_3 (see Fig. 2.2). Among them, the B_1 subsite specifically binds to pyrimidine bases, especially cytosine, with a binding efficiency about 30 times higher than that of uracil. The B_2 and B_3 subsites are more favored to bind to purine bases. Regarding the phosphate-binding subsites, we find Lys-66 in the P_0, Lys-7 and Arg-10 in the P_2 subsites, while the P_1 site at the active site includes the catalytic triad together with complementary residues, such as Gln11 (Boix et al., 1994; Fisher et al., 1998). Next, we will describe the binding site characteristics of the eight canonical human RNase types within the RNaseA family (Sorrentino, 2010) (Fig. 2.3).

2.3 Nucleotide-binding patterns within the RNaseA family

2.3.1 RNase1

Human RNase1 was originally identified in pancreatic fluid and shares 70% sequence similarity with bovine RNaseA (Beintema and Wietzes, 1984; Sorrentino and Libonati, 1997). Later studies indicated that in humans, RNase1 is abundantly expressed in vascular endothelial cells (Landré et al., 2002; Barrabés et al., 2007). The B_1 site of RNase1 has the highest affinity for cytosine (C), and its ordered binding affinity preference is poly(C)>poly(U)>poly(A) (Garnett and Raines, 2022; Landré et al., 2002). This enzyme has three N-glycosylation sites (Asn34, Asn76, and Asn88) with the consensus sequence Asn-Xaa-Thr/Ser, which enhance the enzyme's thermal stability and resistance to protease degradation but reduce its catalytic activity (Beintema and Blank, 1988; Ressler 2019; Yamashita et al., 1986). Tissue-specific differences in the glycosylation pattern were observed. Apart from ssRNA, RNase1 can also degrade double-stranded RNA and RNA–DNA hybrids, which is attributed to the large amount of positive charges distributed on its surface (Beintema and

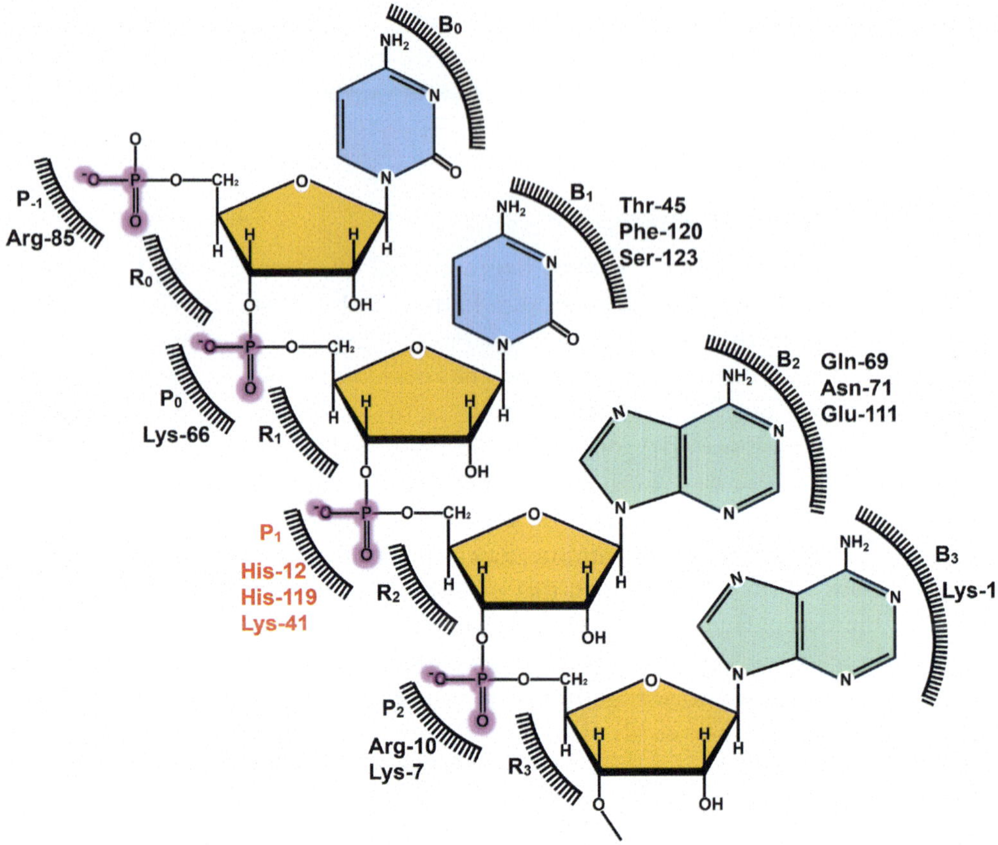

Figure 2.2 Schematic diagram of the RNA-binding site in RNaseA. The P_1 corresponding to the cleaved phosphodiester bond is highlighted with a red label, and the phosphodiester bonds are shown with purple shading. The identified protein residues interacting with bases (Bn) and phosphates (pn) are shown (*RNaseA*, ribonuclease A).

Wietzes, 1984). The B_2 site plays a stabilizing role mainly through Gln69 and Asn71 during the binding process with RNA, and changes in loop4 are also reported to affect B_2's ability to bind to different nucleotides (Boix et al., 2013; Raines, 1998; Gagné and Doucet, 2013; Fernández-Millán et al., 2022). RNase1 activity is inhibited by the proteinaceous RNase inhibitor, and Lys7 at the P_2 site plays an important role in the protein-inhibitor interaction (Fernández-Millán et al., 2022). RNase1 has been reported to play an important role in immune regulation (Beintema and Kleineidam, 1998; Lu and Li, 2018). Computational simulations suggest that RNase1 can act as a ligand for the ephrin4 receptor, regulate cell signaling pathways, and play a role in cancer occurrence, cell proliferation, and apoptosis (Li and Yamaguchi, 2022).

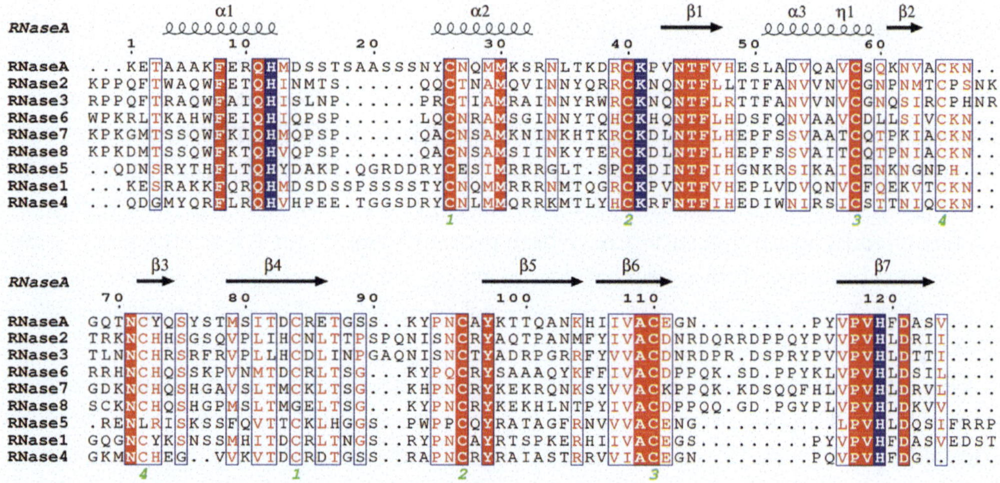

Figure 2.3 Sequence alignment of bovine RNaseA with the eight canonical human RNase homologs. The three residues required for catalytic activity are shown in blue, fully conserved amino acid residues are indicated in red, and the eight cysteine residues forming disulfide bonds are marked with green numbers beneath the sequence (*RNaseA*, ribonuclease A).

2.3.2 RNase2

Human RNase2, also known as the eosinophil-derived neurotoxin (EDN) for its abundance in the secondary eosinophil granules and neurotoxicity upon its discovery (Durack et al., 1981) shares with RNaseA the overall three-dimensional fold and catalytic site groove. The P_1 site of RNase2 includes the equivalent catalytic triad counterparts (His15, His129, and Lys38). In the high-resolution 3D structure of RNaseA, the catalytic residue His119 has been shown to adopt dual conformations in different pH environments (Berisio et al., 2002). Likewise, the protein 3D structure of EDN at atomic resolution reveals that His29 can also adopt two conformations as previously reported for RNaseA (Berisio et al., 2002), which can either contribute to substrate cleavage or secondary base binding (Gagné and Doucet, 2013; Swaminathan et al., 2002).

Regarding the B_1 site, responsible for recognizing the main nucleotide upstream of the cleavage site, Thr42 and Leu130 are identified in RNase2 for specific recognition, where uridine is favored over cytosine (Sikriwal and Seth, 2007; Sorrentino, 1998; Terzyan et al., 1999). An extended RNA chain enhances EDN-binding capacity, and additional phosphate-binding sites were identified (Baker and Holloway, 2006). Similar to RNaseA, whose B_2 and B_3 subsites bind all bases but show a marked preference for purines, especially adenine, at the B_2 (Raines, 1998), the B_2 site of EDN is also more prone to bind purines than pyrimidines, with a marked preference for adenine. Compared with RNaseA, RNase2 shows differences at the P_0 (Lys66 in RNaseA) and

P_2 (Lys7, Arg10 in RNaseA) sites, which may modify its binding mode to RNA. Reduced interactions at P_0 and P_2 are predicted for RNase2–nucleotide complexes. Instead, there is a P_{-1} site in EDN, comprising residues 35–40, that may play a key role in substrate specificity and binding affinity (Leonidas et al., 2003). Recent studies have established that RNase2 selectively generates specific tRNA-derived fragments by cleaving tRNAs at sites with a uracil/cytosine (B_1) and an adenine (B_2). This specificity is determined by its unique structure, where protein loops 3 and 4 form the key P_{-1} and B_2 substrate-binding subsites, respectively, enabling precise recognition and cleavage (Li et al., 2024; Lu et al., 2022). For the role of ribonucleases in regulating noncoding RNA, please refer to Chapter 7.

Glycosylation modification of RNase2 may also intervene in the protein's biological properties, such as its antiviral activity and immune regulation. Trp-7 in RNase2 is modified by C-glycosylation (C-mannosylation), and its Trp-X-X-Trp tag determines the specific covalent attachment of mannose, which may contribute to the stability of RNase2 during secretion and its extracellular function. In addition, the C-glycosylation site is close to the P_2 substrate-binding site of RNase2, which may regulate its nucleic acid-binding pattern (Hofsteenge et al., 1994; Krieg et al., 1998). EDN also has a strong glycosaminoglycan-binding capacity, with key binding sites, including Arg35, Arg36, and Arg38 (Hung et al., 2013). In addition to RNA degradation, RNase2 also has neurotoxic and antiviral activities associated with its catalytic activity, and these properties might be related to unique interactions with specific RNA substrates. For example, RNase2 can target viral RNA and inhibit viral replication, while also exerting immunomodulatory effects (Li and Boix, 2021; Lu et al., 2022; Rosenberg, 2015). On its side, structural determinants for EDN neurotoxicity are still to be unveiled.

2.3.3 RNase3

Human RNase3, also known as the eosinophil cationic protein (ECP), is another of the main eosinophil proteins stored in secondary granules. It is a highly cationic protein with cytotoxicity and antibacterial activities. It has a high sequence identity (67%) with EDN but has clearly distinct biological properties (Gleich and Loegering, 1986; Rosenberg and Dyer, 1995). ECP has 19 arginine residues (Arg), resulting in a pI value of 10.8, much higher than that of EDN (Mallorquí-Fernández et al., 2000). RNase3 has physico-chemical properties characteristic of an antimicrobial peptide (Nguyen and Haney, 2011). It binds to negatively charged bacterial membranes through surface positive charges and a hydrophobic patch located at the N-terminal region, which mediates its antimicrobial effects and promotes bacterial aggregation and membrane destruction. Experimental results showed that ECP mainly acts on bacterial membranes through interactions at the protein N-terminus, rather than relying on catalytic activity to exert its bactericidal effect (Lien et al., 2013; Pulido et al., 2018;

Torrent et al., 2009). The catalytic efficiency of ECP is lower than that of EDN (Boix, 2001). This might be attributed to changes in the position of Asn70 at the B_2 site and the loss of Lys7 at the P_2 site (Boix et al., 1999). The ECP subsite arrangement favors an exonuclease-type activity versus an endonuclease in comparison to RNaseA, releasing shorter-length nucleotide products (Boix et al., 1999). Subsite disposition in ECP was further determined by X-ray crystallography, with solved 3D structures of protein-nucleotide and protein-sulfate complexes (Boix et al., 2012; Mohan et al., 2002). In addition, glycosylation modification of RNase3 may affect its nucleic acid binding ability and cytotoxicity (Gleich et al., 1986; Mallorquí-Fernández et al., 2000; Salazar et al., 2014). In summary, the biological properties of RNase3 are not merely dependent on its nucleic acid-binding and cleavage ability, but a more dynamic mechanism should be envisaged, which is associated partly with its bacterial wall-specific binding, membrane disruption, and cell aggregation to promote its antimicrobial action (Boix et al., 2012).

2.3.4 RNase4

RNase4 is a ribonuclease that specifically cleaves the 3' side of uridine. It consists of 119 amino acid residues, contains four disulfide bonds, and has a pyroglutamyl modification at the N-terminus. Its amino acid sequence has high identity to RNase1 (43%) and RNase5/angiogenin (39%). Evolutionary studies suggest that RNase4, together with both RNases1 and 5, suffered a massive divergence expansion during the early mammal diversification (Goo and Cho, 2013). RNase4 type members display conserved structural and tissue expression profiles within species, suggesting they exert a key cellular role (Futami et al., 1997; Zhou and Strydom 1993). Structural studies confirm the conservation of catalytic residues in RNase4 (His12, Lys40, and His116) and identify key substitutions affecting its specificity (e.g., Lys7 → Arg, Asp14 → His) and a unique two-residue deletion corresponding to positions 77–78 in RNaseA (Liang and Acharya, 2016; Terzyan et al., 1999). RNase4 has a higher selectivity for uridine versus cytosine for B_1 in comparison to RNase1, attributed to Arg101 and specific residue substitutions at the C-terminus (Terzyan et al., 1999). RNase4 exhibits a high catalytic efficiency for extended poly(U) substrates, with an estimated rate of 10× over poly(C) (Boix et al., 2013). Besides, RNase4 shares a B_2 preference for adenine with its close counterparts (Boix et al., 2013). RNase4 has been reported to exhibit cytotoxic effects on certain cancer cells and to promote angiogenesis *in vitro*, as observed in the bovine enzyme (Di Liddo et al., 2010). Specific mutations in RNase4 resulting in reduced enzyme activity are also associated with amyotrophic lateral sclerosis, suggesting that the RNA metabolism of RNase4 may play a key role in the occurrence of diseases and provide potential targets for future therapeutic-oriented research (Padhi et al., 2019).

2.3.5 RNase5

RNase5 is considered the most ancestral type within the RNaseA superfamily (Cho and Beintema, 2005). Although its overall structure is close to RNase4, some differences in key regions allow it to have special functions in angiogenesis, cancer progression, and neuroprotection. RNase5 was called angiogenin (ANG) upon the discovery of its angiogenic properties (Hu et al., 1991). ANG keeps the characteristic catalytic triad of RNaseA (His13, Lys40, and His114), but Gln117 at the C-terminus blocks substrate entry, making its ribonuclease activity lower (Acharya et al., 1994; Russo et al., 1994). The ribosome triggers a conformational rearrangement in angiogenin's C-terminal tail, enabling cleavage of tRNA at the anticodon loop within the ribosomal A-site (Loveland et al., 2024). ANG has a nuclear localization signal (Arg-Arg-Arg) that allows it to actively enter the nucleus and promote rRNA metabolism in the nucleolus, thereby enhancing cell proliferation and angiogenesis (Pizzo et al., 2013). The protein has only three conserved disulfide bonds (Cys26-Cys81, Cys40-Cys93, and Cys58-Cys108) instead of the four disulfide bonds that are characteristic of the RNaseA family. Nonetheless, the protein structure is very stable and remains functional in serum and extracellular environments(Hu et al., 1991; Acharya et al., 1994). Overall, ANG controls angiogenesis and promotes tumor growth through its limited RNase activity, nuclear transport ability, and receptor-binding. It also plays a protective role in neurodegenerative diseases like ALS, making it an important target for cancer and neurodisease treatment (Hoang et al., 2019). Compared with RNase1, ANG can specifically cleave tRNA and rRNA and play an important role in stress response and neuroprotection. RNase5 (ANG) generates tiRNAs by cleaving the tRNA anticodon loop; these tiRNAs are generated under conditions of cellular stress and play a key role in inhibiting protein translation and promoting cell survival (Ivanov et al., 2011). tRNA methylation modification (such as 5-methylcytosine m5C) plays an important role in regulating ANG-mediated tRNA degradation (Lyons et al., 2017).

2.3.6 RNase6

RNase6, also known as RNase k6, as it was first identified in the kidney, shows a highly conserved sequence in mammalian evolution (Rosenberg and Dyer, 1996). The crystal 3D structure of RNase6 shows the characteristic kidney-shaped globular fold of the RNaseA family. His15, Lys38, and His122 form the catalytic triad at the active site, while His36 and His39 create a new secondary catalytic site (Prats-Ejarque et al., 2016; Prats-Ejarque et al., 2019). In addition, molecular dynamics simulations showed that RNase6 can bind to extended oligomeric nucleotide substrates, with a more pronounced endonuclease cleavage pattern for poly(C), in comparison to the typical cleavage pattern of RNaseA (Prats-Ejarque et al., 2019). The secondary catalytic site of

RNase6, formed by His36, His39, and Lys87, may enhance the enzyme's ability to degrade long-chain RNA substrates. Also, studies show that His36 and His39 at the enzyme's secondary active site are crucial for the overall enzyme activity. The His15A mutant still retains some catalytic ability for polynucleotides, but the double mutant (His15A/His36A) completely loses activity. This suggests that the secondary site has a catalytic function in RNase6 (Prats-Ejarque et al., 2019). RNase6 prefers to cleave substrates containing uridine (U) at the B_1 position and adenosine (A) at the B_2 site. Gln40 forms a hydrogen bond with Asp80, which may regulate substrate selectivity at the B_1 site. At the B_2 site, Asn68 and Arg66 together influence the binding pattern selectivity for purine substrates (Hamann et al., 1990). Overall, RNase6 biological properties are associated with antimicrobial activity and host immunity (Deming and Dyer, 1998; Lu et al., 2018; Pulido et al., 2016).

2.3.7 RNase7

Human ribonuclease 7 (RNase7), a representative antimicrobial protein of the RNaseA superfamily, was first discovered in healthy skin and is widely expressed in various epithelial tissues (Harder and Schroder 2002). It also possesses a classic RNase A-type structure, featuring a conserved His15–Lys38–His123 catalytic triad and four pairs of disulfide bonds. RNase7 exhibits a significantly higher RNA-hydrolyzing activity in comparison to other family members (approximately 50-fold higher than ECP) (Zhang and Dyer, 2003). It also exhibits broad-spectrum antimicrobial activity that is independent of its enzymatic activity (Köten et al., 2009). The antimicrobial effect of RNase7 primarily relies on the high positive charge on its molecular surface (isoelectric point ~10.5), enabling it to bind to negatively charged components of the bacterial membrane via its positively charged lysine residues. Although it is a cationic protein with properties similar to ECP, its sequence identity is only 40%. In particular, a flexible lysine cluster on the surface (K1, K3, K111, and K112) plays a key role in the membrane-binding and permeabilization process, rapidly leading to bacterial death (Huang et al., 2007). As part of the epithelial innate immune system, RNase7 expression can be induced by bacterial infection or inflammatory factors, and it has significant bactericidal activity against a variety of pathogens, including vancomycin-resistant enterococci (Harder and Schroder 2002). By binding to lipopolysaccharide and peptidoglycan on the bacterial surface, it causes local membrane damage and leakage of cellular contents; but unlike RNase3, it does not cause bacterial agglutination (Torrent et al., 2010). Furthermore, a main fragment responsible for the antimicrobial activity of RNase7 is located at the N-terminus (Torrent et al., 2013). RNase7 is considered one of the most potent human antimicrobial RNases and possesses potential therapeutic value against drug-resistant pathogens (Becknell and Spencer, 2016).

2.3.8 Other RNase families

Besides the RNaseA superfamily, we briefly introduce here other unrelated RNase families with distinct 3D structure, nucleotide-binding patterns, and substrate specificity, such as RNaseT1, RNaseT2, RNaseH, RNaseP, and RNaseIII. Among a large family of microbial ribonucleases found in both prokaryotes and eukaryotes, Ribonuclease T1 (RNaseT1) from the fungus *Aspergillus oryzae* is the most well-known and studied member (Takahashi and Moore, 1982; Takahashi et al., 1970). RNaseT1 has high selectivity for guanine (G) nucleotide residues at the B_1 site. In contrast to the RNaseA family, its active site contains a metal ion-binding site that stabilizes the reaction transition state and interacts electrostatically with phosphate groups (Höschler et al., 1999; Steyaert 1997).

The RNaseT2 family is a class of acidic RNases that are widely distributed in plants, animals, and microorganisms, with a unique member in humans. Its binding pattern relies on conserved catalytic residues (2 His and 1 Glu) and a unique RNA-binding loop, which allows it to recognize and cleave ssRNA. It also plays a role in immune regulation and the stress response (Luhtala and Parker, 2010; Thorn et al., 2012).

On its side, the RNaseH family specifically recognizes and degrades the RNA portion of RNA–DNA hybrids. Its binding pattern depends on a highly conserved active site, including Asp, Glu, and His residues, and it stabilizes the substrate through interactions with the DNA strand. RNaseH plays a key role in DNA replication, repair, and retroviral replication (Pallan and Egli, 2008).

RNaseP is a ribonucleoprotein enzyme responsible for processing the 5' end of tRNA precursors. Its binding pattern involves the coordinated action of protein and RNA components. It achieves substrate specificity by recognizing specific tRNA domains, such as the T-loop and acceptor stem (Kazantsev and Pace, 2006). Coevolution of the RNA ribozyme and several protein cofactor subunits has taken place through bacteria, archaea, and eukaryotes, where bacterial enzymes show a large RNA component and more evolved species have a simpler RNA and more complex protein subunits (Zhou et al., 2024).

Last, the RNaseIII family mainly recognizes and cleaves double-stranded RNA. Its binding pattern relies on the double-stranded RNA-binding domain and the catalytic domain. It functions by recognizing specific sequences and structures of double-stranded RNA (Blaszczyk et al., 2001).

Unlike the RNaseA or RNaseT2 families, which uses an acid–base catalysis mediated by His, some RNases, such as RNaseH, rely on metal ions, or specific RNA recognition, as RNaseP for phosphodiester bond hydrolysis. This gives them unique selectivity in substrate recognition and regulation of catalytic efficiency. Studying the nucleotide-binding patterns of these non-RNaseA enzyme families contributes to understanding RNase diversity and illustrates the distinct structural features of RNase enzymes to achieve RNA catalysis.

2.4 Current perspectives

RNases have become a hot topic of research due to their key roles in RNA metabolism, immune defense, and disease regulation. In recent years, significant progress has been made in protein engineering and high-resolution structural analysis, but biotechnological challenges still exist in improving specificity and delivery efficiency. Researchers are working to optimize RNases for applications in cancer, antiviral, and immune regulation, such as RNase5 and RNase7 in cancer diagnosis, prognosis assessment, and treatment (Liu et al., 2021; Siraj 2022; Wang et al., 2018). Protein engineering improves the stability and specificity of RNases, making them RNA-targeted therapeutic tools and driving the development of biosensors, such as RNA virus detection and RNA metabolism monitoring. High-resolution techniques such as cryo-EM and X-ray crystallography, combined with computational simulations, accelerate RNase-related drug development. Although important breakthroughs have been made in the study of RNases, their interactions with complex RNA structures still need to be further explored. In the future, multidisciplinary integration will further reveal the roles of RNases in RNA regulation and intercellular communication and promote their application in disease diagnosis and therapy.

AI disclosure

During the preparation of this work, the author(s) used ChatGPT in order to polish the English language and improve the logical flow and readability of the manuscript. After using this tool/service, the author(s) reviewed and edited the content as needed and take (s) full responsibility for the content of the publication.

References

Acharya, K R, Shapiro, R, Allen, S C, Riordan, J F, Vallee, B L, 1994. Crystal structure of human angiogenin reveals the structural basis for its functional divergence from ribonuclease. Proceedings of the National Academy of Sciences 91 (8), 2915–2919. Available at https://doi.org/10.1073/pnas.91.8.2915.

Anfinsen, CB, 1973. Principles that govern the folding of protein chains. Science 181 (4096), 223–230. Available at https://doi.org/10.1126/science.181.4096.223.

Aravind, L, Koonin, EV, 2001. A natural classification of ribonucleases. Methods Enzymol 341, 3–28. Available at https://doi.org/10.1016/s0076-6879(01)41142-6.

Baker, MD, Holloway, DE, Swaminathan, GJ, Acharya, KR, 2006. Crystal structures of eosinophil-derived neurotoxin (EDN) in complex with the inhibitors 5'-ATP, Ap3A, Ap4A, and Ap5A. Biochemistry 45 (2), 416–426. Available at https://doi.org/10.1021/bi0518592.

Barrabés, S, Pagès-Pons, L, Radcliffe, CM, Tabarés, G, Fort, E, Peracaula, R, et al., 2007. Glycosylation of serum ribonuclease 1 indicates a major endothelial origin and reveals an increase in core fucosylation in pancreatic cancer. Glycobiology 17 (4), 388–400. Available at https://doi.org/10.1093/glycob/cwm002.

Beintema, JJ, Blank, A, Schieven, GL, Dekker, CA, orrentino S, S, Libonati, M, 1988. Differences in glycosylation pattern of human secretory ribonucleases. Biochem J 255 (2), 501–505.

Beintema, JJ, Kleineidam, RG, 1998. The ribonuclease A superfamily: general discussion. Cell Mol Life Sci 54 (8), 825–832. Available at https://doi.org/10.1007/s000180050211.

Beintema, JJ, Wietzes, P, Weickmann, JL, Glitz, DG, 1984. The amino acid sequence of human pancreatic ribonuclease. Anal Biochem 136 (1), 48–64. Available at https://doi.org/10.1016/0003-2697(84)90306-3.

Berisio, R., Sica, F., Lamzin, V.S., Wilson, K.S., Zagari, A., Mazzarella, L., et al., 2002. Atomic resolution structures of ribonuclease A at six pH values. Acta Crystallographica Section D Biological Crystallography 58 (3), 441–450. Available at https://doi.org/10.1107/s0907444901021758.

Blaszczyk, J, Tropea, JE, Bubunenko, M, Routzahn, KM, Waugh, DS, Court, DL, Ji, X, et al., 2001. Crystallographic and modeling studies of RNase III suggest a mechanism for double-stranded RNA cleavage. Structure 9 (12), 1225–1236. Available at https://doi.org/10.1016/s0969-2126(01)00685-2.

Boix, E, 2001. Eosinophil cationic protein. Methods Enzymol 341, 287–305. Available at https://doi.org/10.1016/s0076-6879(01)41159-1.

Boix, E, Blanco, JA, Nogués, MV, oussaoui M, M, 2013. Nucleotide binding architecture for secreted cytotoxic endoribonucleases. Biochimie 95 (6), 1087–1097. Available at https://doi.org/10.1016/j.biochi.2012.12.015.

Boix, E, Nogués, MV, Schein, CH, Benner, SA, Cuchillo, CM, 1994. Reverse transphosphorylation by ribonuclease A needs an intact p2-binding site. Point mutations at Lys-7 and Arg-10 alter the catalytic properties of the enzyme. J Biol Chem 269 (4), 2529–2534.

Boix, E, Leonidas, DD, Nikolovski, Z, Nogués, MV, Cuchillo, CM, Acharya, KR, et al., 1999. Crystal structure of eosinophil cationic protein at 2.4 Å resolution'. Biochemistry 38 (51), 16794–16801. Available at https://doi.org/10.1021/bi9919145.

Boix, E, Nikolovski, Z, Moiseyev, GP, Rosenberg, HF, Cuchillo, CM, Nogués, MV, et al., 1999. Kinetic and product distribution analysis of human eosinophil cationic protein indicates a subsite arrangement that favors exonuclease-type activity. Journal of Biological Chemistry 274 (22), 15605–15614. Available at https://doi.org/10.1074/jbc.274.22.15605.

Boix, E, Salazar, VA, Torrent, M, Pulido, D, Nogués, MV, oussaoui M, M, 2012. Structural determinants of the eosinophil cationic protein antimicrobial activity. Biol Chem 393 (8), 801–815. Available at https://doi.org/10.1515/hsz-2012-0160.

Boix, E, Pulido, D, Moussaoui, M, Victòria Nogués, M., Russi, S, et al., 2012. The sulfate-binding site structure of the human eosinophil cationic protein as revealed by a new crystal form. Journal of Structural Biology 179 (1), 1–9. Available at https://doi.org/10.1016/j.jsb.2012.04.023.

Cho, S, Beintema, JJ, Zhang, J, 2005. The ribonuclease A superfamily of mammals and birds: identifying new members and tracing evolutionary histories. Genomics 85 (2), 208–220. Available at https://doi.org/10.1016/j.ygeno.2004.10.008.

Chou, PY, Fasman, GD, 1977. Beta-turns in proteins. J Mol Biol 115 (2), 135–175. Available at https://doi.org/10.1016/0022-2836(77)90094-8.

Cuchillo, CM, Nogués, MV, Raines, RT, 2011. Bovine pancreatic ribonuclease: fifty years of the first enzymatic reaction mechanism. Biochemistry 50 (37), 7835–7841. Available at https://doi.org/10.1021/bi201075b.

Deming, MS, Dyer, KD, Bankier, AT, Piper, MB, Dear, PH, Rosenberg, HF, 1998. Ribonuclease k6: chromosomal mapping and divergent rates of evolution within the RNase A gene superfamily. Genome Res 8 (6), 599–607. Available at https://doi.org/10.1101/gr.8.6.599.

Di Liddo, R, alzoppo D, D, Baiguera, S, Conconi, MT, Dettin, M, Parnigotto, PP, Grandi, C, et al., 2010. In vitro biological activity of bovine milk ribonuclease-4. Mol Med Rep 3 (1), 127–132. Available at https://doi.org/10.3892/mmr_00000229.

Durack, DT, Ackerman, SJ, Loegering, DA, Gleich, GJ, 1981. Purification of human eosinophil-derived neurotoxin. Proc Natl Acad Sci U S A 78 (8), 5165–5169. Available at https://doi.org/10.1073/pnas.78.8.5165.

ecknell B, B, Spencer, JD, 2016. A Review of Ribonuclease 7's Structure, Regulation, and Contributions to Host Defense. Int J Mol Sci 17 (3), 423. Available at https://doi.org/10.3390/ijms17030423.

Fernández-Millán, P, Vázquez-Monteagudo, S, Boix, E, Prats-Ejarque, G, 2022. Exploring the RNase A scaffold to combine catalytic and antimicrobial activities. Structural characterization of RNase 3/1 chimeras. Front Mol Biosci 9, 964717. Available at https://doi.org/10.3389/fmolb.2022.964717.

Fisher, BM, Schultz, LW, Raines, RT, 1998. Coulombic effects of remote subsites on the active site of ribonuclease A. Biochemistry 37 (50), 17386–17401. Available at https://doi.org/10.1021/bi981369s.

Futami, J, Tsushima, Y, Murato, Y, Tada, H, Sasaki, J, Seno, M, Yamada, H, et al., 1997. Tissue-specific expression of pancreatic-type RNases and RNase inhibitor in humans. DNA Cell Biol 16 (4), 413–419. Available at https://doi.org/10.1089/dna.1997.16.413.

Gagné, D, Doucet, N, 2013. Structural and functional importance of local and global conformational fluctuations in the RNase A superfamily. FEBS J 280 (22), 5596–5607. Available at https://doi.org/10.1111/febs.12371.

Garnett, ER, Raines, RT, 2022. Emerging biological functions of ribonuclease 1 and angiogenin. Crit Rev Biochem Mol Biol 57 (3), 244–260. Available at https://doi.org/10.1080/10409238.2021.2004577.

Gleich, GJ, Loegering, DA, Bell, MP, Checkel, JL, Ackerman, SJ, McKean, DJ, 1986. Biochemical and functional similarities between human eosinophil-derived neurotoxin and eosinophil cationic protein: homology with ribonuclease. Proc Natl Acad Sci U S A 83 (10), 3146–3150. Available at https://doi.org/10.1073/pnas.83.10.3146.

Goo, SM, Cho, S, 2013. The expansion and functional diversification of the mammalian ribonuclease a superfamily epitomizes the efficiency of multigene families at generating biological novelty. Genome Biol Evol 5 (11), 2124–2140. Available at https://doi.org/10.1093/gbe/evt161.

Greulich, W, Wagner, M, Gaidt, MM, Stafford, C, Cheng, Y, Linder, A, Carell, T, Hornung, V, et al., 2019. TLR8 Is a Sensor of RNase T2 Degradation Products. Cell 179 (6), 1264–1275.e13. Available at https://doi.org/10.1016/j.cell.2019.11.001.

Höschler, K, oier H, H, Hubner, B, Saenger, W, Orth, P, Hahn, U, 1999. Structural analysis of an RNase T1 variant with an altered guanine binding segment. J Mol Biol 294 (5), 1231–1238. Available at https://doi.org/10.1006/jmbi.1999.3324.

Hamann, KJ, Ten, RM, Loegering, DA, Jenkins, RB, Heise, MT, Schad, CR, Pease, LR, Gleich, GJ, Barker, RL, et al., 1990. Structure and chromosome localization of the human eosinophil-derived neurotoxin and eosinophil cationic protein genes: evidence for intronless coding sequences in the ribonuclease gene superfamily. Genomics 7 (4), 535–546. Available at https://doi.org/10.1016/0888-7543(90)90197-3.

Harder, J, Schröder, J-M, 2002. RNase 7, a novel innate immune defense antimicrobial protein of healthy human skin. Journal of Biological Chemistry 277 (48), 46779–46784. Available at https://doi.org/10.1074/jbc.m207587200.

Hoang, TT, Johnson, DA, Raines, RT, Johnson, JA, 2019. Angiogenin activates the astrocytic Nrf2/antioxidant-response element pathway and thereby protects murine neurons from oxidative stress. J Biol Chem 294 (41), 15095–15103. Available at https://doi.org/10.1074/jbc.RA119.008491.

Hofsteenge, J, Mueller, DR, de Beer, T, Loeffler, A, Richter, WJ, Vliegenthart, JFG, et al., 1994. New type of linkage between a carbohydrate and a protein: C-glycosylation of a specific tryptophan residue in human RNase Us. Biochemistry 33 (46), 13524–13530. Available at https://doi.org/10.1021/bi00250a003.

Hu, G F, Chang, S I, Riordan, J F, Vallee, B L, et al., 1991. An angiogenin-binding protein from endothelial cells. Proceedings of the National Academy of Sciences 88 (6), 2227–2231. Available at https://doi.org/10.1073/pnas.88.6.2227.

Huang, YC, Lin, YM, Chang, TW, Wu, SJ, Lee, YS, Chang, MD, hen C, C, Wu, SH, Liao, YD, et al., 2007. The flexible and clustered lysine residues of human ribonuclease 7 are critical for membrane permeability and antimicrobial activity. J Biol Chem 282 (7), 4626–4633. Available at https://doi.org/10.1074/jbc.M607321200.

Hung, TJ, Chang, WT, Tomiya, N, Lee, YC, Chang, HT, Chen, CJ, Kuo, PH, Fan, TC, Chang, MD, et al., 2013. Basic amino acid residues of human eosinophil derived neurotoxin essential for glycosaminoglycan binding. Int J Mol Sci 14 (9), 19067–19085. Available at https://doi.org/10.3390/ijms140919067.

Ivanov, P, Emara, MM, Villen, J, Gygi, SP, Anderson, P, 2011. Angiogenin-induced tRNA fragments inhibit translation initiation. Mol Cell 43 (4), 613–623. Available at https://doi.org/10.1016/j.molcel.2011.06.022.

Jones, W, 1920. The action of boiled pancreas extract on yeast nucleic acid. American Journal of Physiology-Legacy Content 52 (1), 203–207. Available at https://doi.org/10.1152/ajplegacy.1920.52.1.203.

Köten, B, Simanski, M, Gläser, R, Podschun, R, Schröder, J-M, Harder, J, et al., 2009. RNase 7 contributes to the cutaneous defense against Enterococcus faecium. PLoS ONE 4 (7), e6424. Available at https://doi.org/10.1371/journal.pone.0006424.

Kazantsev, AV, Pace, NR, 2006. Bacterial RNase P: a new view of an ancient enzyme. Nat Rev Microbiol 4 (10), 729–740. Available at https://doi.org/10.1038/nrmicro1491.

Krieg, J, Hartmann, S, Vicentini, A, Gläsner, W, Hess, D, Hofsteenge, J, et al., 1998. Recognition signal for C-mannosylation of Trp-7 in RNase 2 consists of sequence Trp-x-x-Trp. Molecular Biology of the Cell 9 (2), 301–309. Available at https://doi.org/10.1091/mbc.9.2.301.

Landré, JB, Hewett, PW, Olivot, JM, Friedl, P, Ko, Y, Sachinidis, A, oenner M, M, et al., 2002. Human endothelial cells selectively express large amounts of pancreatic-type ribonuclease (RNase 1). J Cell Biochem 86 (3), 540–542. Available at https://doi.org/10.1002/jcb.10234.

Leonidas, DD, Chavali, GB, Oikonomakos, NG, Chrysina, ED, Kosmopoulou, MN, Vlassi, M, Frankling, C, Acharya, KR, et al., 2003. High-resolution crystal structures of ribonuclease A complexed with adenylic and uridylic nucleotide inhibitors. Implications for structure-based design of ribonucleolytic inhibitors. Protein Sci 12 (11), 2559–2574. Available at https://doi.org/10.1110/ps.03196603.

Li, J, Boix, E, 2021. Host Defence RNases as Antiviral Agents against Enveloped Single Stranded RNA Viruses. Virulence 12 (1), 444–469. Available at https://doi.org/10.1080/21505594.2021.1871823.

Li, YC, Yamaguchi, H, Liu, YY, Hsu, KC, Sun, TH, Sun, PC, Hung, MC, 2022. Structural insights into EphA4 unconventional activation from prediction of the EphA4 and its complex with ribonuclease 1. Am J Cancer Res 12 (10), 4865–4878.

Li, J, Kang, X, Guidi, I, u L, L, Fernández-Millán, P, Prats-Ejarque, G, Boix, E, et al., 2024. Structural determinants for tRNA selective cleavage by RNase 2/EDN. Structure 32 (3), 328–341.e4. Available at https://doi.org/10.1016/j.str.2023.12.012.

Liang, S, Acharya, KR, 2016. Structural basis of substrate specificity in porcine RNase 4. FEBS J 283 (5), 912–928. Available at https://doi.org/10.1111/febs.13646.

Lien, PC, Kuo, PH, Chen, CJ, Chang, HH, Fang, SL, Wu, WS, Lai, YK, Pai, TW, Chang, MD, et al., 2013. In silico prediction and in vitro characterization of multifunctional human RNase3. Biomed Res Int 2013, 170398. Available at https://doi.org/10.1155/2013/170398.

Liu, C, Zha, Z, Zhou, C, Wang, S-C, Ren, N, Hung, M-C, et al., 2021. Ribonuclease 7-driven activation of ROS1 is a potential therapeutic target in hepatocellular carcinoma. Journal of Hepatology 74 (4), 907–918. Available at https://doi.org/10.1016/j.jhep.2020.09.030.

Loveland, AB, Koh, CS, Ganesan, R, Jacobson, A, Korostelev, AA, 2024. Structural mechanism of angiogenin activation by the ribosome. Nature 630 (8017), 769–776. Available at https://doi.org/10.1038/s41586-024-07508-8.

Lu, L, Li, J, Moussaoui, M, Boix, E, 2018. Immune Modulation by Human Secreted RNases at the Extracellular Space. Front Immunol 9, 1012. Available at https://doi.org/10.3389/fimmu.2018.01012.

u L, L, Li, J, Wei, R, Guidi, I, Cozzuto, L, Ponomarenko, J, Prats-Ejarque, G, Boix, E, et al., 2022. Selective cleavage of ncRNA and antiviral activity by RNase2/EDN in THP1-induced macrophages. Cell Mol Life Sci 79 (4), 209. Available at https://doi.org/10.1007/s00018-022-04229-x.

Luhtala, N, Parker, R, 2010. T2 Family ribonucleases: ancient enzymes with diverse roles. Trends Biochem Sci 35 (5), 253–259. Available at https://doi.org/10.1016/j.tibs.2010.02.002.

Lyons, SM, Fay, MM, Akiyama, Y, Anderson, PJ, Ivanov, P, 2017. RNA biology of angiogenin: Current state and perspectives. RNA Biol 14 (2), 171–178. Available at https://doi.org/10.1080/15476286.2016.1272746.

Mallorquí-Fernández, G, Pous, J, Peracaula, R, Aymamí, J, Maeda, T, Tada, H, et al., 2000. Three-dimensional crystal structure of human eosinophil cationic protein (RNase 3) at 1.75 A resolution. J Mol Biol 300 (5), 1297–1307. Available at https://doi.org/10.1006/jmbi.2000.3939.

Mironova, N, Patutina, O, Brenner, E, Kurilshikov, A, Vlassov, V, Zenkova, M, et al., 2013. MicroRNA drop in the bloodstream and microRNA boost in the tumour caused by treatment with ribonuclease A leads to an attenuation of tumour malignancy. PLoS One 8 (12). Available at https://doi.org/10.1371/journal.pone.0083482.

Mohan, CG, Boix, E, Evans, HR, Nikolovski, Z, Nogués, MV, Cuchillo, CM, Acharya, KR, et al., 2002. The crystal structure of eosinophil cationic protein in complex with 2',5'-ADP at 2.0 Å resolution reveals the details of the ribonucleolytic active site. Biochemistry 41 (40), 12100–12106. Available at https://doi.org/10.1021/bi0264521.

Nguyen, LT, Haney, EF, Vogel, HJ, 2011. The expanding scope of antimicrobial peptide structures and their modes of action. Trends Biotechnol 29 (9), 464–472. Available at https://doi.org/10.1016/j.tibtech.2011.05.001.

Nogués, MV, Moussaoui, M, Boix, E, Vilanova, M, Ribó, M, Cuchillo, CM, 1998. The contribution of noncatalytic phosphate-binding subsites to the mechanism of bovine pancreatic ribonuclease A. Cell Mol Life Sci 54 (8), 766–774. Available at https://doi.org/10.1007/s000180050205.

Nunes, IV, Breitenbach, L, Pawusch, S, Eigenbrod, T, Ananth, S, Schad, P, Fackler, OT, Butter, F, Dalpke, AH, Chen, LS, et al., 2024. Bacterial RNA sensing by TLR8 requires RNase 6 processing and is inhibited by RNA 2'O-methylation. EMBO Rep 25 (11), 4674–4692. Available at https://doi.org/10.1038/s44319-024-00281-9.

Nunes, IV, Breitenbach, L, Pawusch, S, Eigenbrod, T, Ananth, S, Schad, P, Fackler, OT, Butter, F, Dalpke, AH, Chen, L-S, et al., 2024. Bacterial RNA sensing by TLR8 requires RNase 6 processing and is inhibited by RNA 2'O-methylation. EMBO Reports 25 (11), 4674–4692. Available at https://doi.org/10.1038/s44319-024-00281-9.

Ostendorf, Thomas, Zillinger, Thomas, Andryka, Katarzyna, Schlee-Guimaraes, Thais Marina, Schmitz, Saskia, Marx, Samira, Bayrak, Kübra, Linke, Rebecca, Salgert, Sarah, Wegner, Julia, Grasser, Tatjana, Bauersachs, Sonja, Soltesz, Leon, Hübner, Marc P., Nastaly, Maximilian, Coch, Christoph, Kettwig, Matthias, Roehl, Ingo, Henneke, Marco, Hoerauf, Achim, Barchet, Winfried, Gärtner, Jutta, Schlee, Martin, Hartmann, Gunther, Bartok, Eva, et al., 2020. Immune Sensing of Synthetic, Bacterial, and Protozoan RNA by Toll-like Receptor 8 Requires Coordinated Processing by RNase T2 and RNase 2. Immunity 52 (4), 591–605.e6. Available at https://doi.org/10.1016/j.immuni.2020.03.009.

Padhi, AK, Narain, P, Gomes, J, 2019. Rare Angiogenin and Ribonuclease 4 variants associated with amyotrophic lateral sclerosis exhibit loss-of-function: a comprehensive in silico study. Metab Brain Dis 34 (6), 1661–1677. Available at https://doi.org/10.1007/s11011-019-00473-6.

Pallan, PS, Egli, M, 2008. Insights into RNA/DNA hybrid recognition and processing by RNase H from the crystal structure of a non-specific enzyme-dsDNA complex. Cell Cycle 7 (16), 2562–2569. Available at https://doi.org/10.4161/cc.7.16.6461.

Parés, X, Nogués, MV, de Llorens, R, Cuchillo, CM, 1991. Structure and function of ribonuclease A binding subsites. Essays Biochem 26, 89–103.

Pizzo, E, Sarcinelli, C, Sheng, J, Fusco, S, ormiggini F, F, Netti, P, Yu, W, D'Alessio, G, Hu, GF, et al., 2013. Ribonuclease/angiogenin inhibitor 1 regulates stress-induced subcellular localization of angiogenin to control growth and survival. J Cell Sci 126 (Pt 18), 4308–4319. Available at https://doi.org/10.1242/jcs.134551.

Prats-Ejarque, G, Blanco, JA, Salazar, VA, Nogués, VM, oussaoui M, M, Boix, E, 2019. Characterization of an RNase with two catalytic centers. Human RNase6 catalytic and phosphate-binding site arrangement favors the endonuclease cleavage of polymeric substrates. Biochim Biophys Acta Gen Subj 1863 (1), 105–117. Available at https://doi.org/10.1016/j.bbagen.2018.09.021.

Prats-Ejarque, G, Arranz-Trullén, J, Blanco, JA, Pulido, D, Nogués, MV, oussaoui M, M, Boix, E, et al., 2016. The first crystal structure of human RNase 6 reveals a novel substrate-binding and cleavage site arrangement. Biochem J 473 (11), 1523–1536. Available at https://doi.org/10.1042/BCJ20160245.

Prats-Ejarque, G, Lorente, H, Villalba, C, Anguita, R, u L, L, Vázquez-Monteagudo, S, Fernández-Millán, P, Boix, E, et al., 2021. Structure-Based Design of an RNase Chimera for Antimicrobial Therapy. Int J Mol Sci 23 (1). Available at https://doi.org/10.3390/ijms23010095.

Pulido, D, Arranz-Trullén, J, Prats-Ejarque, G, Velázquez, D, Torrent, M, oussaoui M, M, Boix, E, et al., 2016. Insights into the Antimicrobial Mechanism of Action of Human RNase6: Structural Determinants for Bacterial Cell Agglutination and Membrane Permeation. Int J Mol Sci 17 (4), 552. Available at https://doi.org/10.3390/ijms17040552.

Pulido, D, Prats-Ejarque, G, Villalba, C, Albacar, M, Moussaoui, M, Andreu, D, et al., 2018. Positional scanning library applied to the human eosinophil cationic protein/RNase3 N-terminus reveals novel and potent anti-biofilm peptides. Eur J Med Chem 152, 590–599. Available at https://doi.org/10.1016/j.ejmech.2018.05.012.

Raines, RT, 1998. Ribonuclease A. Chem Rev 98 (3), 1045–1066. Available at https://doi.org/10.1021/cr960427h.

Raines, R.T., 2004. Artificial Nucleases. Active Site of Ribonuclease A. Springer Berlin Heidelberg, Berlin, Heidelberg, pp. 19–32. Available at https://doi.org/10.1007/978-3-642-18510-6_3.

Ressler, VT., Raines, RT., et al., 2019. Consequences of the endogenous *N*-glycosylation of human ribonuclease 1. Biochemistry 58 (7), 987–996. Available at https://doi.org/10.1021/acs.biochem.8b01246.

Richardson, JS, 1981. The anatomy and taxonomy of protein structure. Adv Protein Chem 34, 167–339. Available at https://doi.org/10.1016/s0065-3233(08)60520-3.

Rosenberg, HF, 2015. Eosinophil-Derived Neurotoxin (EDN/RNase 2) and the Mouse Eosinophil-Associated RNases (mEars): Expanding Roles in Promoting Host Defense. Int J Mol Sci 16 (7), 15442–15455. Available at https://doi.org/10.3390/ijms160715442.

Rosenberg, HF, Dyer, KD, 1996. Molecular cloning and characterization of a novel human ribonuclease (RNase k6): increasing diversity in the enlarging ribonuclease gene family. Nucleic Acids Res 24 (18), 3507–3513. Available at https://doi.org/10.1093/nar/24.18.3507.

Rosenberg, HF, Dyer, KD, 1995. Eosinophil cationic protein and eosinophil-derived neurotoxin. Evolution of novel function in a primate ribonuclease gene family. J Biol Chem 270 (37), 21539–21544. Available at https://doi.org/10.1074/jbc.270.37.21539.

Russo, N, Shapiro, R, Acharya, KR, Riordan, JF, Vallee, BL, 1994. Role of glutamine-117 in the ribonucleolytic activity of human angiogenin. Proc Natl Acad Sci U S A 91 (8), 2920–2924. Available at https://doi.org/10.1073/pnas.91.8.2920.

Russo, N, Shapiro, R, Acharya, K R, Riordan, J F, Vallee, B L, 1994. Role of glutamine-117 in the ribonucleolytic activity of human angiogenin. Proceedings of the National Academy of Sciences 91 (8), 2920–2924. Available at https://doi.org/10.1073/pnas.91.8.2920.

Salazar, VA, Rubin, J, oussaoui M, M, Pulido, D, Nogués, MV, Venge, P, Boix, E, et al., 2014. Protein post-translational modification in host defense: the antimicrobial mechanism of action of human eosinophil cationic protein native forms. FEBS J 281 (24), 5432–5446. Available at https://doi.org/10.1111/febs.13082.

Shigematsu, M, Kawamura, T, Kirino, Y, et al., 2018. Generation of 2′,3′-cyclic phosphate-containing RNAs as a hidden layer of the transcriptome. Frontiers in Genetics 9. Available at https://doi.org/10.3389/fgene.2018.00562.

Sikriwal, D, Seth, D, Dey, P, Batra, JK, 2007. Human eosinophil-derived neurotoxin: involvement of a putative non-catalytic phosphate-binding subsite in its catalysis. Mol Cell Biochem 303 (1-2), 175–181. Available at https://doi.org/10.1007/s11010-007-9471-0.

Siraj, YA, 2022. Promises of eukaryotic ribonucleases for cancer treatment: a systematic review. Translational Medicine Communications 7 (1). Available at https://doi.org/10.1186/s41231-022-00113-9.

Sorrentino, S, Libonati, M, 1997. Structure-function relationships in human ribonucleases: main distinctive features of the major RNase types. FEBS Lett 404 (1), 1–5. Available at https://doi.org/10.1016/s0014-5793(97)00086-0.

Sorrentino, S, 1998. Human extracellular ribonucleases: multiplicity, molecular diversity and catalytic properties of the major RNase types. Cell Mol Life Sci 54 (8), 785–794. Available at https://doi.org/10.1007/s000180050207.

Sorrentino, S, 2010. The eight human "canonical" ribonucleases: molecular diversity, catalytic properties, and special biological actions of the enzyme proteins. FEBS Lett 584 (11), 2194–2200. Available at https://doi.org/10.1016/j.febslet.2010.04.018.

Steyaert, J, 1997. A decade of protein engineering on Ribonuclease T_1— atomic dissection of the enzyme-substrate interactions. European Journal of Biochemistry 247 (1), 1–11. Available at https://doi.org/10.1111/j.1432-1033.1997.t01-1-00001.x.

Sun, D, Han, C, Sheng, J, 2022. The role of human ribonuclease A family in health and diseases: A systematic review. iScience 25 (11), 105284. Available at https://doi.org/10.1016/j.isci.2022.105284.

Sun, D, Han, C, Sheng, J, et al., 2022. The role of human ribonuclease A family in health and diseases: A systematic review. iScience 25 (11), 105284. Available at https://doi.org/10.1016/j.isci.2022.105284.

Swaminathan, GJ, Holloway, DE, Veluraja, K, Acharya, KR, et al., 2002. Atomic resolution (0.98 Å) structure of eosinophil-derived neurotoxin'. Biochemistry 41 (10), 3341–3352. Available at https://doi.org/10.1021/bi015911f.

Takahashi, K, Moore, S, 1982. 13 Ribonuclease T1. The Enzymes,Nucleic Acids Part B. Elsevier, pp. 435–468. Available at https://doi.org/10.1016/s1874-6047(08)60285-1.

Takahashi, K, Uchida, T, Egami, F, 1970. Ribonuclease T1, structure and function. Adv Biophys 1, 53–98.

Terzyan, SS, Peracaula, R, de Llorens, R, Tsushima, Y, Yamada, H, Seno, M, Gomis-Rüth, FX, Coll, M, et al., 1999. The three-dimensional structure of human RNase 4, unliganded and complexed with d (Up), reveals the basis for its uridine selectivity. J Mol Biol 285 (1), 205–214. Available at https://doi.org/10.1006/jmbi.1998.2288.

Thorn, A, Steinfeld, R, Ziegenbein, M, Grapp, M, Krätzner, R, et al., 2012. Structure and activity of the only human RNase T2. Nucleic Acids Research 40 (17), 8733–8742. Available at https://doi.org/10.1093/nar/gks614.

Torrent, M, Pulido, D, Valle, J, Nogués, MV, Andreu, D, Boix, E, 2013. Ribonucleases as a host-defence family: evidence of evolutionarily conserved antimicrobial activity at the N-terminus. Biochem J 456 (1), 99–108. Available at https://doi.org/10.1042/BJ20130123.

Torrent, M, de la Torre, BG, Nogués, VM, Andreu, D, Boix, E, et al., 2009. Bactericidal and membrane disruption activities of the eosinophil cationic protein are largely retained in an N-terminal fragment. Biochemical Journal 421 (3), 425–434. Available at https://doi.org/10.1042/bj20082330.

Torrent, M, Badia, M, Moussaoui, M, Sanchez, D, Nogués, MV, Boix, E, et al., 2010. Comparison of human RNase 3 and RNase 7 bactericidal action at the Gram-negative and Gram-positive bacterial cell wall. The FEBS Journal 277 (7), 1713–1725. Available at https://doi.org/10.1111/j.1742-4658.2010.07595.x.

Wang, Y-N, Lee, H-H, Chou, C-K, Yang, W-H, Maitra, A, Hung, M.-C, et al., 2018. Angiogenin/Ribonuclease 5 is an EGFR ligand and a serum biomarker for Erlotinib sensitivity in pancreatic cancer. Cancer Cell 33 (4), 752–769.e8. Available at https://doi.org/10.1016/j.ccell.2018.02.012.

Yamashita, K., Hitoi, A., Irie, M., Kobata, A., et al., 1986. Fractionation by lectin affinity chromatography indicates that the glycosylation of most ribonucleases in human viscera and body fluids is organ specific. Archives of Biochemistry and Biophysics 250 (1), 263–266. Available at https://doi.org/10.1016/0003-9861(86)90725-3.

Yang, D, Chen, Q, Rosenberg, HF, Rybak, SM, Newton, DL, Wang, ZY, Fu, Q, Tchernev, VT, Wang, M, Schweitzer, B, Kingsmore, SF, Patel, DD, Oppenheim, JJ, Howard, OM, et al., 2004. Human ribonuclease A superfamily members, eosinophil-derived neurotoxin and pancreatic ribonuclease, induce dendritic cell maturation and activation. J Immunol 173 (10), 6134–6142. Available at https://doi.org/10.4049/jimmunol.173.10.6134.

Zhang, J, Dyer, KD, Rosenberg, HF, 2003. Human RNase 7: a new cationic ribonuclease of the RNase A superfamily. Nucleic Acids Res 31 (2), 602–607. Available at https://doi.org/10.1093/nar/gkg157.

Zhou, HM, Strydom, DJ, 1993. The amino acid sequence of human ribonuclease 4, a highly conserved ribonuclease that cleaves specifically on the 3' side of uridine. Eur J Biochem 217 (1), 401–410. Available at https://doi.org/10.1111/j.1432-1033.1993.tb18259.x.

Zhou, B, Wan, F, Lei, KX, Lan, P, Wu, J, Lei, M, 2024. Coevolution of RNA and protein subunits in RNase P and RNase MRP, two RNA processing enzymes. J Biol Chem 300 (3), 105729. Available at https://doi.org/10.1016/j.jbc.2024.105729.

CHAPTER 3

Experimental tools for studying ribonucleases activities

Mohammed Moussaoui Keribii
Department of Biochemistry and Molecular Biology, Universitat Autònoma de Barcelona, Cerdanyola del Vallès, Spain

3.1 Introduction

RNases catalyze the hydrolysis of the 3',5'-phosphodiester bonds of RNA single strands by a two-stage reaction (Richards and Wyckoff, 1971; Eftink and Biltonen, 1987). The first stage is a transphosphorylation reaction from the 5′ of a nucleotide to the 2′ position of the adjacent nucleotide, forming a cyclic 2',3'-phosphate end and a free 5′-OH end. When no susceptible 3′-5′ phosphodiester bonds are left, the second stage, consisting of hydrolysis of the cyclic phosphodiester bond, takes place, forming a 3′-terminal phosphate group (Cuchillo et al., 1993). Consequently, both reactions take place separately, the catalytic efficiency of RNase for the transphosphorylation reaction being much higher than that for the hydrolysis reaction.

RNA and its oligomeric and polymeric analogs are used as substrates in the study of RNase activity. Polymeric substrates are mainly used in the methods of detection of RNase activity, while monomeric and oligomeric analogs are used in kinetic and substrate specificity studies. Here, we describe the main methods used in the study of RNase activity.

All the methods and conditions of the reactions described below correspond to those applied to study bovine pancreatic ribonuclease (RNaseA), the most-studied RNase and considered as a model enzyme for the study of other ribonucleases. These conditions may vary slightly, such as pH or the enzyme concentration, depending on the RNase being studied.

3.2 Detection of RNase activity

3.2.1 Activity staining (Zymogram)

A versatile negative-staining ribonuclease zymogram is described. The method has several advantages, as it combines, by means of different staining procedures, high resolving power, sensitivity, and specificity with rapid, reproducible, and simultaneous analysis of the activity and purity of ribonuclease samples on the same polyacrylamide gel. Activity bands can be visualized at any time during the incubation process before staining the gel. This allows the choice of different staining procedures after incubation. By the use of polycytidylic acid

RNases in Health and Diseases
DOI: https://doi.org/10.1016/B978-0-443-27312-4.00004-3

[poly(C)] as a substrate, less than 1 pg of bovine pancreatic ribonuclease A was detected in less than 2 h after the electrophoretic run. An additional advantage with respect to other methods is that no refrigeration is needed during electrophoresis.

The method of sodium dodecyl sulfate (SDS)–polyacrylamide gel separation and analysis of RNase activity, described by Blank et al. (1982), consists of the electrophoretic separation of proteins by means of a gel containing poly(C) or polyuridylic acid [poly(U)] as an analog substrate of RNA. Polyacrylamide gels are prepared following the method described by Laemli but adding the poly(C) or poly(U) polymer substrate during the preparation of the resolving part of the gel, so that the substrate is immobilized in the gel once polymerized. During the preparation of the resolving layer, the substrate is dissolved in the required volume of water for gel preparation at a concentration of 0.6 mg/mL. To dissolve the polymeric substrate in water, the mixture must be incubated at 55°C until the substrate is completely dissolved. The dissolved substrate is then mixed with a 30% polyacrylamide solution, and the buffer Tris/HCl 1.5 M pH 8.8 containing 0.4% (W/V) SDS, and then tetramethylethylenediamine and ammonium persulfate solution are added to allow polymerization of the gel. The stacking gel contains no substrate and is prepared in the same way described by Laemmli 1970.

The amount of protein to be loaded depends on the level of activity of the RNase being studied and whether the substrate used is poly(C) or poly(U), since RNases have different affinities for both substrates.

The RNase sample to be loaded is mixed with a loading buffer that does not contain a reducing agent, such as β-mercaptoethanol, to prevent the breakdown of the disulfide bridges and allow protein renaturation and its subsequent RNase activity after electrophoresis.

Once the electrophoresis is finished, the gel is removed and prepared for protein renaturation to allow RNase activity. Next, the gel is stained to reveal the bands of activity. This process is performed at room temperature under gentle and constant agitation:

a) Removal of the SDS: The gel is kept for 10 min in a cold solution of 10 mM Tris/HCl, pH 7.5, containing 10% (v/v) 2-propanol. This step is performed twice.

b) Protein renaturation: The gel is washed twice for 15 min each with buffer Tris/HCl 10 mM, pH 7.5, to remove isopropanol and allow protein renaturation.

c) Gel incubation to allow RNase activity: The gel is kept in Tris/HCl buffer 100 mM, pH 7.5, at room temperature. The incubation time extends from 15 to 90 min, depending on the activity and the amount of protein loaded in the gel.

d) Removal of the incubation buffer: The gel is washed for 10 min with buffer Tris/HCl 10 mM, pH 7.5.

e) Gel staining: The gel is kept for 5–10 min under gentle shaking in a solution of Tris/HCl 10 mM, pH 7.5, containing 0.2% (p/v) toluidine blue dye, which stains the undigested substrate because it interacts electrostatically with nucleic acids.

f) Washing: The gel is then washed several times with distilled water until the bands appear.

If the analysis of the purity of ribonuclease samples on the same polyacrylamide gel is needed, the dye can be completely removed with 7% acetic acid, and then the gel is stained with Coomassie blue or silver nitrate to reveal the protein bands. Fig. 3.1 shows an example of a zymogram showing the activity of RNaseA and RNase8.

3.2.2 Measurement of RNA hydrolysate

The method of measurement of RNA hydrolysate as a product of RNase activity (Isobe and Uchiyama, 1986) is useful when there are many samples to be analyzed, such as chromatographic fractions, to identify those containing RNase. This method consists of the determination of the amount of degradation products of a high-molecular-weight RNA by the RNase present in the sample to be analyzed. After digestion, the reaction is immediately stopped by precipitation of the nonhydrolyzed substrate, and the amount of degradation products is determined by the spectrophotometric method.

3.2.2.1 Components of the assay

The samples to be analyzed: (1) Reaction buffer containing RNA as substrate: MOPS 50 mM, pH 7.5; 3 mM EDTA; 0.075 mg/mL BSA solution containing RNA at a concentration of 1.75 mg/mL; (2) Reaction stop reagent: It is made up of two solutions: (1) Phosphotungstic acid 0.75% (w/v) in perchloric acid (2) BSA 0.6% (w/v) in distilled water. Both solutions are mixed 50% each and kept on ice. This mixture causes the precipitation of nonhydrolyzed RNA and consequently the stop of the reaction.

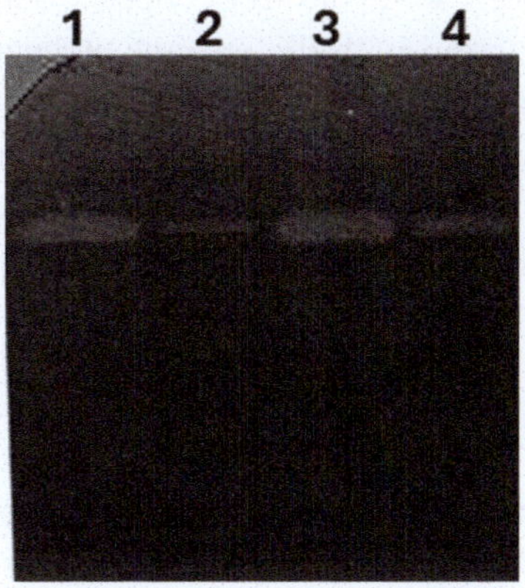

Figure 3.1 Comparison between RNase8 and RNaseA activity by zymogram on a gel containing poly (C) as substrate. Lane 1, 50 pg of RNaseA; lane 2, 40 ng of RNase8; lane 3, 150 pg of RNaseA; and lane 4, 150 ng of RNase8.

3.2.2.2 Procedure

1) Add 0.3 mL of the sample to 0.4 mL reaction buffer containing RNA and incubate the mixture at 37°C. Notice: an additional mixture must be prepared from 0.3 mL of the sample and 0.4 mL of reaction buffer that does not contain RNA. This sample will be used as a blank in the spectrophotometric measurement.
2) After 15 min of the reaction, add 0.3 mL of reaction stop reagent to each sample, mix vigorously, and keep the mixtures on ice for 15 min to allow the precipitation of nonhydrolyzed RNA. After 20 min of centrifugation at 5000*g*, dilute 0.25 mL of each supernatant to 0.75 mL with distilled water, then measure the absorbance at 260 nm.

3.3 Steady-state kinetic studies

Kinetic studies are mainly carried out by means of spectrophotometric methods. Although the two steps of the reaction catalyzed by an RNase can occur in the same reaction mixture, the transphosphorylation and hydrolysis steps can be analyzed separately.

3.3.1 Transphosphorylation reaction

During the cleavage of polymeric substrates by an RNase, each fragment released in the reaction is, in turn, a substrate for the next reaction. Consequently, substrates of different lengths can be used to study the transphosphorylation step of the reaction by means of the spectrophotometric method.

Due to difficulties in kinetic analysis derived from the complex structural features of the RNA molecule, homooligonucleotides such as oligocytidylic acids $(Cp)_nC>p$ or homopolynucleotides such as poly(C) are used as substrates. These kinds of substrates are mainly used in kinetic studies to avoid the effect of the differences between the nitrogenous bases in the interaction with the enzyme, except for those assays that require short oligos, labeled and/or having a specific base sequence, to study the preference of the binding subsites of the enzyme for the nitrogenous bases for the interaction.

Kinetics carried out with homooligonucleotides of different lengths gives information about the binding subsites involved in the interaction with the substrate and then the length of the interaction region.

3.3.1.1 Method

The reaction of transphosphorylation using oligocytidylic or polycytidylic acids as a substrate is carried out in 0.2 M sodium acetate, pH 5.5, at 25°C. The substrate concentration range is 10–500 μM for oligos and up to 3 mg/mL for poly(C). The final enzyme concentration in the reaction mixture varies depending on the native and modified RNase being studied. A linear decrease in absorbance at 286 nm is monitored using either 1-cm or 2-mm path length cells. The first linear 2 min of the progress curves are used for the calculation of initial velocities needed for the determination of the kinetic parameters (Cuchillo et al., 2002).

For each oligonucleotide substrate, the molar absorbance change at the selected wavelength ($\Delta\varepsilon_{286}$) is determined from the maximum absorbance difference between the starting reaction mixture and the complete degradation to C>p.

3.3.2 Hydrolysis reaction

The reaction of hydrolysis takes place as a second step of the reaction catalyzed by RNase. Mononucleotides 2', 3'-cyclic phosphate are used as substrates to analyze the hydrolysis reaction that leads to the formation of the nucleotide 3′-phosphate as the final product.

3.3.2.1 Method

The reaction of hydrolysis is measured by the Crook method (Crook et al., 1960). The reaction is carried out in 0.2 M sodium acetate, pH 5.5, at 25°C. For the substrate cytidine 2', 3'-cyclic phosphate, an increase in absorbance is monitored at 296 nm ($\Delta\varepsilon_{296}$ = 516.4 $M^{-1}\cdot cm^{-1}$). The substrate concentration range is 0.1–3 mM, and the final concentration of enzyme depends on the native or modified RNase being studied. The first linear 3 min of the progress curves are used for the calculation of initial velocities needed for the determination of kinetic parameters (Moussaoui et al., 1998).

3.3.3 Fluorogenic substrates for hypersensitive assays of ribonucleolytic activity

The fluorogenic substrates were designed and synthesized as RNA/DNA chimeras. The substrate is based on the fluorescence quenching of fluorescein held in proximity to rhodamine by a single ribonucleotide embedded within a series of deoxynucleotides. When the substrate is cleaved, the fluorescence of fluorescein is manifested and provides the basis for fast, facile, and extremely sensitive assays for ribonucleolytic activity. The optimal substrate 6-FAM-dArUdAdA-6-TAMRA is a tetranucleotide with a 5',6-carboxyfluorescein label (6-FAM) and a 3',6-carboxy-tetramethylrhodamine (6-TAMRA) (Park et al., 1999).

3.3.3.1 Method

Fluorescence measurement is carried out on a photon-counting fluorometer. For RNaseA, assays are carried out with stirring in 2 mL of 0.1 M MES/NaOH buffer, pH 6.0, containing 0.10 M NaCl, 0.50–100 nM substrate, and 1–200 pM RNaseA. An increase in fluorescence emission at 515 nm, upon excitation at 490 nm, indicates the progress of the reaction. Kinetic parameters and fluorescent properties of the substrate are determined using Eqs. (3.1) and (3.2):

$$I = I_f - (I_f - I_0)\, e^{-kcat/Km)\,[E]\,t} \tag{3.1}$$

$$I = I_0 - (I_f - I_0)\,(kcat/Km)\,(E)\,t \tag{3.2}$$

The fluorescence intensity (*I*), measured at a given time during the reaction, is recorded in units of photon counts per second. The intensity of the product (I_f) is determined by nonlinear least squares analysis (Eq. (3.1)) of data collected. The intensity of substrate (I_0) is determined from data collected prior to the addition of the enzyme. Values of *kcat/Km* are determined either by nonlinear least squares regression analysis or by all data using Eq. (3.1) or by linear least squares regression analysis of initial velocity data using the second equation (Park et al., 1999).

3.4 Analysis of the digestion products by high performance liquid chromatography

The spectrophotometric measurement of the reaction of digestion of an oligomeric or polymeric substrate by RNase can only give an average measure of all the species produced during the reaction, which is summarized in a variation of absorbance at a given wavelength without giving any information about either the size distribution of the products or the characteristics of the broken bond at each moment of the reaction. The use of the high performance liquid chromatography (HPLC) technique has circumvented some of the problems concerning the analysis of the products of the reaction because it allows us to know at each moment of the reaction the size as well as the amount of each product released during the cleavage process, and consequently, the preference of the enzyme for the phosphodiester bonds to be broken.

3.4.1 Protocol summary

The method of digestion by RNaseA of poly(C) substrate and separation by HPLC of the degradation products described below is carried out according to the method of McFarland and Borer (McFarland and Borer, 1979) applied in the separation of the degradation products of polynucleotide acids poly(C) and poly(U) by alkaline hydrolysis at 60°C. The separation of the products of this chemical degradation was carried out by means of reverse-phase chromatography, in which the products are eluted with an organic solvent in ammonium acetate. Under these conditions, the separation takes place by the number of hydrophobic constituents (nitrogenous bases) of the oligonucleotides, so that the elution takes place in an order from the shortest to the longest length of the oligonucleotides. This method allows us to separate oligonucleotides of up to 10 residues. This order of separation is lost for oligonucleotides of more than 10 nucleotides due, probably, to the formation of structures by large oligonucleotides or to the resolution limit of the column. The identification by matrix-assisted laser desorption/ionization - time of flight (MALDI-TOF) mass spectrometry (Cuchillo et al., 2002) of the first eight eluted oligonucleotides confirmed that in such a method of separation, products from mononucleotide to oligonucleotide of eight residues elute in an order from smallest to largest oligo.

3.4.2 Steps of the method

3.4.2.1 Reaction of poly(C) digestion by RNaseA

Prepare a solution of 5 mg/mL of poly(C) in 10 mM Hepes–KOH, pH 7.5. For better dissolution of poly(C), the mixture is kept for up to 5 min at 50°C with reverse shaking of the tube every minute. 50 μL of the poly(C) solution is digested with 10 μL of RNase solution at 25°C. The enzyme concentration depends on the native or modified RNase being used. In general, the final enzyme concentration is low (in the nanomolar range), and thus, to avoid denaturation, a more concentrated enzyme solution is prepared, and adequate dilution is made immediately before the assay. Given the time taken by each assay (~2 h) and to avoid degradation owing either to contamination or to denaturing effects associated with continuous freezing and thawing, aliquots of the substrate stock solution were kept frozen in Eppendorf tubes and only thawed immediately before use (Richards and Wyckoff, 1971).

3.4.2.2 Separation of products of digestion

At different time intervals, between 0 and 45 min, the products of the reaction were analyzed by injecting 20 μL of the reaction mixture onto a reversed-phase HPLC column (Nova-Pak C18) equilibrated with solvent A (10% ammonium acetate (w/v) and 1% acetonitrile (v/v) in water). The elution is carried out at a flow rate of 1 mL/min with an initial 10-min wash and a 50-min linear gradient from 100% solvent A to 10% solvent A plus 90% solvent B (10% ammonium acetate (w/v) and 11% acetonitrile (v/v) in water). After each run, the system was washed for 10 min with 100% acetonitrile and equilibrated with solvent A. Product elution was monitored and quantified from the absorbance at 260 nm (Moussaoui et al., 1996).

3.4.3 Analysis

3.4.3.1 Analysis of the pattern of digestion of a homopolymeric substrate

The analysis of the pattern of digestion at different times of the reaction allows us to determine the preference of RNase for the position in the chain of the phosphodiester bond to be broken. Separation of products released by the digestion of a homopolymeric substrate, such as poly(C), gives overall information about the relative position of the first phosphodiester bond broken in the polymeric chain and then the

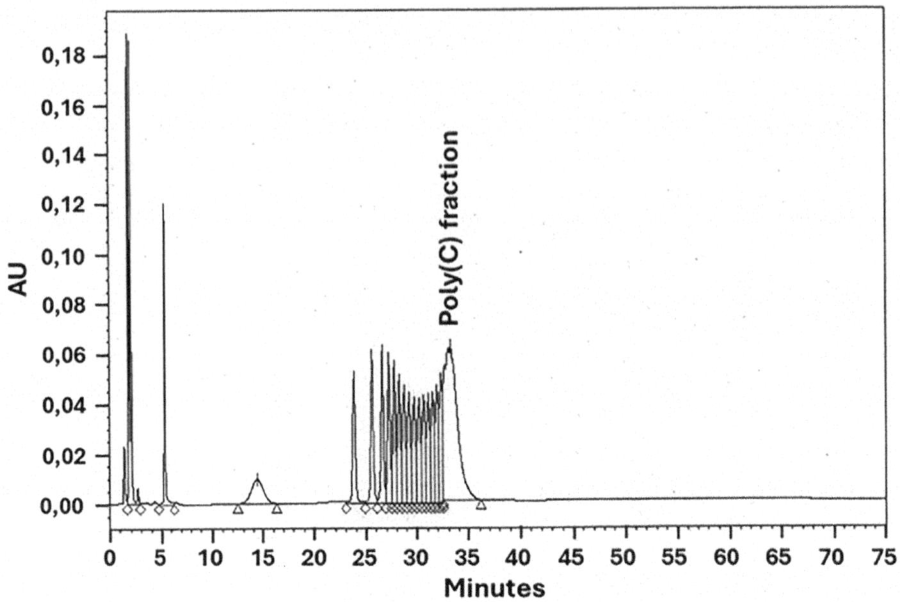

Figure 3.2 Separation by reversed-phase HPLC (Nova-Pak C18 column) of the poly(C) digestion products by RNase8. 150 μL of 5 mg/mL poly(C) solution in 10 mM HEPES–KOH (pH 7.5) was digested with 20 μL of enzyme solution (30 μM) at 25°C for 45 min. 50 μL of the assay mixture was injected into the column.

relative size of the first oligonucleotide released. Fig. 3.2 shows an example of the separation of the products released after the digestion of the poly(C) substrate by RNase8.

3.4.3.2 Analysis of the pattern of digestion of oligonucleotides as substrate

Tetra-, penta-, or hexacytidylic acids can be cleaved by native or modified RNases, and the rate of formation of products at different digestion times can be analyzed by means of a reversed-phase HPLC column. The general reaction conditions described for poly (C) digestion, as well as the separation of the reaction products, can be used. Fig. 3.3 shows an example of the separation of the products released after the digestion of the oligonucleotide $(Cp)_4C>p$ substrate by RNase8.

Digestion pattern analysis consists of determining the amount of each oligonucleotide released during the digestion of the initial substrate. The amount of each oligonucleotide product is calculated by first integrating the peak areas and dividing those values by the corresponding extinction coefficient at 260 nm. For oligocytidylic acids (ε_{260}): 7845 $M^{-1}cm^{-1}$ for C>p, 15175 $M^{-1}cm^{-1}$ for CpC>p, 20745 $M^{-1}cm^{-1}$ for $(Cp)_2C>p$, 24282 $M^{-1}cm^{-1}$ for $(Cp)_3C>p$, 28683 $M^{-1}cm^{-1}$ for $(Cp)_4C>p$, 37711 $M^{-1}cm^{-1}$ for $(Cp)_5C>p$, and 42428 $M^{-1}cm^{-1}$ for $(Cp)_6C>p$.

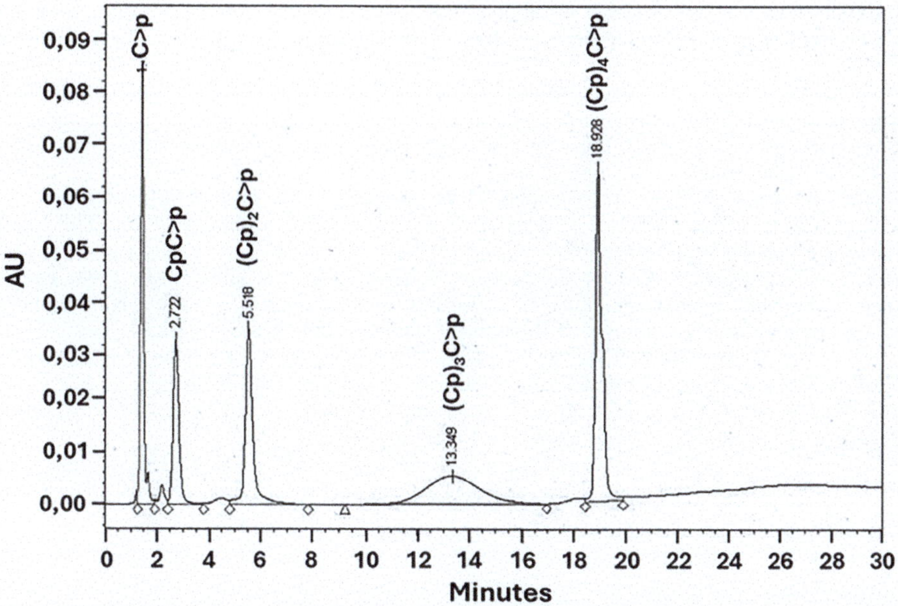

Figure 3.3 Separation by reversed-phase HPLC (Nova-Pak C18 column) of the digestion of a pentanucleotide $(Cp)_4C$>p by RNase8. 100 µL of 100 µM substate solution in 10 mM HEPES–KOH (pH 7.5) was digested with 20 µL of 30 µM enzyme at 25°C for 45 min. 50 µL of the assay mixture was injected into the column.

Quantification of the products released by RNase activity allows us to determine, from the majority of species released, the position in the substrate molecule of the phosphodiester bond preferred by the enzyme to be broken and therefore deduce the endo- or exonucleolytic preference of the RNase being studied.

References

Blank, A., Sugiyama, R.H., Dekker, C.A., 1982. Activity staining of nucleolytic enzymes after sodium dodecyl sulphate- polyacrylamide gel electrophoresis: use of aqueous isopropanol to remove detergent from gel. Analytical Biochemistry 120, 267–275.

Crook, E.M., Mathias, A.P., Rabin, B.R., 1960. Spectrophotometric assay of bovine pancreatic ribonuclease by the use of cytidine 2',3'-phosphate. Biochemistry Journal 74, 234–238.

Cuchillo, C.M., Moussaoui, M., Barman, T., Travers, F., Nogués, M.V., 2002. The exo- or endonucleolytic preference of bovine pancreatic ribonuclease a depends on its subsites structure and the substrate size. Protein Science a Publication of the Protein Society 11 (1), 117–128. Available at https://doi.org/10.1110/ps.13702.

Cuchillo, C.M., Parés, X., Guasch, A., Barman, T.E., Travers, F., Nogués, M.V., 1993. The role of 2',3'-cyclic phosphodiesters in the bovine pancreàtic ribonuclease catalysed cleavage of RNA: intermediates or productes? FEBS Letters 333, 207–210.

Eftink, M.R., Biltonen, R.L., 1987. Pancreatic ribonuclease A: the most studied endoribonuclease. Hydrolytic Enzymes (Neuberger A. and Brocklehurst K. eds.) 7, 333-375. B.V. Amsterdam.

Isobe, K., Uchiyama, S., 1986. An assay for ribonuclease activity, based on ultraviolet absorption of RNA hydrolysate, using phosphotungstic acid. Journal of Biochemical and Biophysical Methods 12, 299–303.

Laemmli, U.K., 1970. Cleavage of structural proteins during the assembly of the head of bacteriophage T4. Nature 227, 680–685.

McFarland, G.D., Borer, P., 1979. Separation of oligo-RNA by reverse-phase HPLC. Nucleic Acids Research 7, 1067–1079.

Moussaoui, M., Guasch, A., Boix, E., Cuchillo, C.M., Nogués, M.V., 1996. The role of non-catalytic binding subsites in the endonuclease activity of bovine pancreatic ribonuclease a. The Journal of Biological Chemistry 271, 4687–4692.

Moussaoui, M., Nogues, M.V., Guasch, A., Barman, T., Travers, F., Cuchillo, C.M., 1998. The subsite structure of bovine pancreatic ribonuclease a accounts for the abnormal kinetic behavior with cytidine 2',3'-cyclic phosphate. The Journal of Biological Chemistry 273, 25565–25572.

Park, C., Kelemen, B.R., Klink, T.A., Sweeney, R.Y., Behlke, M.A., Eubanks, S.R., et al., 1999. Hypersensitive substrate for ribonuclease. Nucleic Acids Research 27, 3696–3701.

Richards, F.M., Wyckoff, H.W., 1971. Bovine pancreatic ribonuclease. In: In: Boyer, P.D. (Ed.), The Enzymes 4. Academic Press, New York, pp. 647–806.

CHAPTER 4

The credentials for Ribonuclease A to be catalytically active: The prerequisite of oxidative protein folding

Mahesh Narayan
Department of Chemistry and Biochemistry, University of Texas at El Paso, El Paso, TX, United States

Abbreviations

ER	endoplamic reticulum
RNaseA	bovine pancreatic ribonuclease A
DTT^{ox}	oxididized dithiothreitol
DTT^{red}	reduced dithiothreitol
GSH	reduced glutathione
GSSG	oxidized glutathione
Des species	a folding intermediate lacking one disulfide bond

4.1 Introduction

RNaseA (Fig. 4.1) is a relatively small, single-chain protein with approximately 124 amino acid residues (Leonidas et al., 2003). The protein's structure features a long four-stranded antiparallel β-sheet and three short α-helices. Functionally, RNaseA is a digestive enzyme that specifically cleaves single-stranded RNA by hydrolyzing the phosphodiester bonds between adjacent nucleotides, primarily on the 3' side of pyrimidine residues (cytosine and uracil) (Raines, 1998; Cuchillo et al., 2011). It is an endoribonuclease with functions in RNA metabolism and in the regulation of gene expression (Raines, 1998). RNaseA has been found to play important roles in diseases such as autoimmune diseases, renal insufficiencies, and pancreas disorders. Some RNaseA family members have cytotoxic and bactericidal properties, potentially playing a role in host defense (Sun et al., 2022).

RNaseA's catalytic activity is dependent upon two key histidine residues, His12 and His119, which act as a general base and a general acid, respectively (Fig. 4.1) (Fedor and Williamson, 2005). Additionally, Lys41 plays a crucial role in substrate binding and catalysis. The active site, where RNA cleavage occurs, is located in a cleft or groove formed by the protein's structure. The enzyme catalyzes RNA cleavage through a two-step mechanism involving transesterification and subsequent hydrolysis (Raines, 1998; Cuchillo et al., 2011).

RNases in Health and Diseases
DOI: https://doi.org/10.1016/B978-0-443-27312-4.00003-1

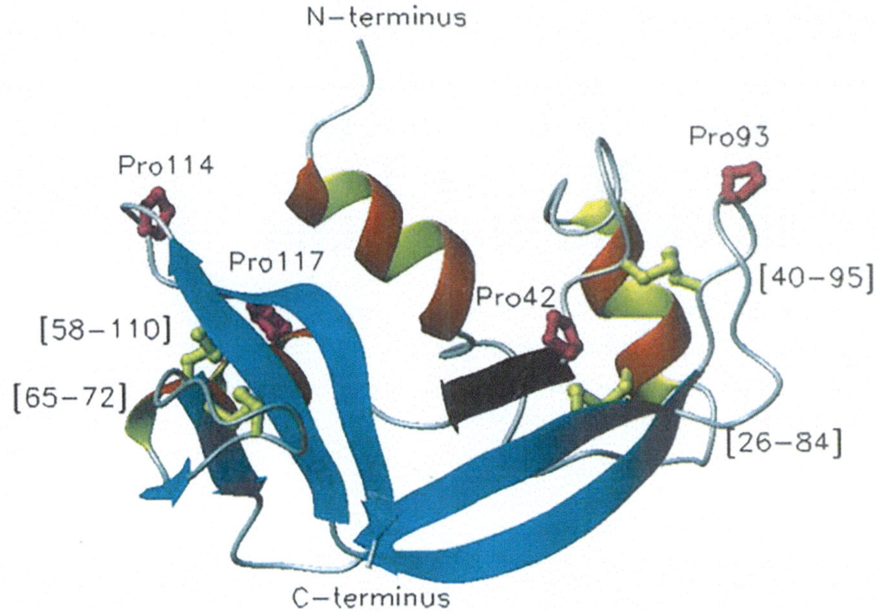

Figure 4.1 Structure of bovine pancreatic RNaseA. Disulfide bonds (yellow ball and stick) and proline residues (red ball and stick) are displayed. *From Scheraga, H.A., Wedemeyer, W.J., Welker, E. Methods in Enzymology. 2001;341:189.*

RNaseA is known for its remarkable stability, even in harsh conditions, due to its disulfide bonds and its overall structure. The relatively small kidney-shaped enzyme is comprised of a single 124-amino acid polypeptide chain, which folds into an intricate structure containing four intrachain disulfide bonds, stabilizing the protein's tertiary structure. The enzyme also features long four-stranded, antiparallel β-sheets and three short α-helices (Leonidas et al., 2003; Wlodawer et al., 1982).

The four disulfide bonds (Fig. 4.1), involving all eight cysteine residues, contribute significantly to the protein's stability and maintain its folded structure. The disulfide bonds in the native protein occur as follows: [26–84], [40–95], [65–72], and [58–110] (Wlodawer et al., 1982).

In addition, there are four proline residues at positions 42, 93, 114, and 117 (Fig. 4.1). Other than Pro 42, the X-Pro peptide bonds of RNaseA play a critical role in the acquisition of its native structure and, therefore, its catalytic activity (Wlodawer et al., 1982).

4.2 The folding of RNaseA to form a biologically active enzyme

The extracellular secretion of RNaseA from the pancreas to the digestive tract and other organs constitutes functioning in an environment that is more harsh than the intracellular confines of the cytosol. In RNaseA and many secreted proteins, in addition to intramolecular Van der Waals interactions the presence of covalent disulfide bonds provides further stability to the tertiary structure of the protein. In human cells, it is estimated that 30%–40% of all cellular proteins (ca. 7000) are destined for the secretory pathway, and of these, approximately 80% contain disulfide bonds (Robinson and Bulleid, 1994).

The presence of disulfide bonds, while advantageous for structural stability and hence functional integrity, complicates the acquisition of the native fold (Pace et al., 1988; Arolas et al., 2006; Thornton, 1981; Narayan et al., 2000; Wedemeyer et al., 2000). This is due to the presence of both numerous on-pathway oxidative intermediates and off-pathway chemical isomers (which result from the formation of nonnative disulfide bonds). The on-pathway intermediates and off-pathway nonnative isomers must converge to the sole native structure with the exclusive complement of native disulfide linkages to permit the protein to become functionally active. The complex process toward the acquisition of the catalytic triad is discussed in detail.

4.2.1 The oxidative folding pathway of RNaseA

RNaseA is synthesized by the ribosome in its fully reduced form, termed R (Narayan et al., 2000). R does not possess disulfide bonds and also lacks structure. As a result, it is not catalytically active. Structure requires the native set of disulfide bonds to form, and for this, R must be threaded into the endoplasmic reticulum (ER), where the oxidizing environment favors the formation of disulfide bonds from cysteines. The redox couple oxidized glutathione (GSSG)/reduced glutathione (GSH) is responsible for the formation of disulfide bonds in the ER, with a ratio of GSH to GSSG ranging from 1:1 to 3:1, which is typically lower than in the cytosol, making the ER environment more oxidizing. By contrast, in the cytosol, the concentration of GSSG is typically in the low micromolar range, while the concentration of GSH is significantly higher, resulting in a GSH:GSSG ratio of around 100:1 or even higher. The reducing environment does not permit the formation of disulfide bonds in the cytosol.

In order to simplify the oxidative folding landscape of RNaseA, DTT^{ox}/DTT^{Red} has been adopted as the redox couple in lieu of GSSG/GSH (Rothwarf and Scheraga, 1993a, b). The latter form mixed disulfide bonds with the protein and complicates the analytical procedures required to separate the on-pathway intermediates and the off-pathway chemical isomers necessary for characterizing the oxidative folding trajectory. By contrast, DTT^{ox}/DDT^{red} do not form mixed disulfides with the protein and thereby facilitate the separation of the aforementioned species. Furthermore, oxidative folding is carried out in the laboratory at a pH of ~8 so that thiol-disulfide exchange is facilitated in light of the fact that the pKa of a cysteine is 8.3. To stop the oxidative folding process, the mixture is subjected to acid quenching or the addition of a thiol-blocking agent such as 2-aminoethyl methanethiosulfonate (AEMTS). Separation of the intermediates, required for a detailed analysis of the folding pathway, its kinetics, rate-limiting step, and mechanism by which the native structure is acquired, can be performed using high-performance liquid chromatography (HPLC). Examination of the separated intermediates also makes possible the identification of kinetically trapped species. Lastly, structural studies of the intermediates, analysis of their stability, and examination of any catalytic activity are also made possible through their fractionation and isolation, also by HPLC, followed by their detailed characterization using X-ray or nuclear magnetic resonance (NMR) spectroscopy.

4.2.2 Formation of a quasiequilibrium ensemble of intermediates

The oxidation of R results in the formation of one-disulfide-bond-containing intermediates termed 1S species (Fig. 4.2) (Rothwarf et al., 1998a, b). The 1S species are an ensemble of isomers representing 28 possible combinations by which eight cysteines can form single disulfide bonds. Of these, four combinations are native, that is,

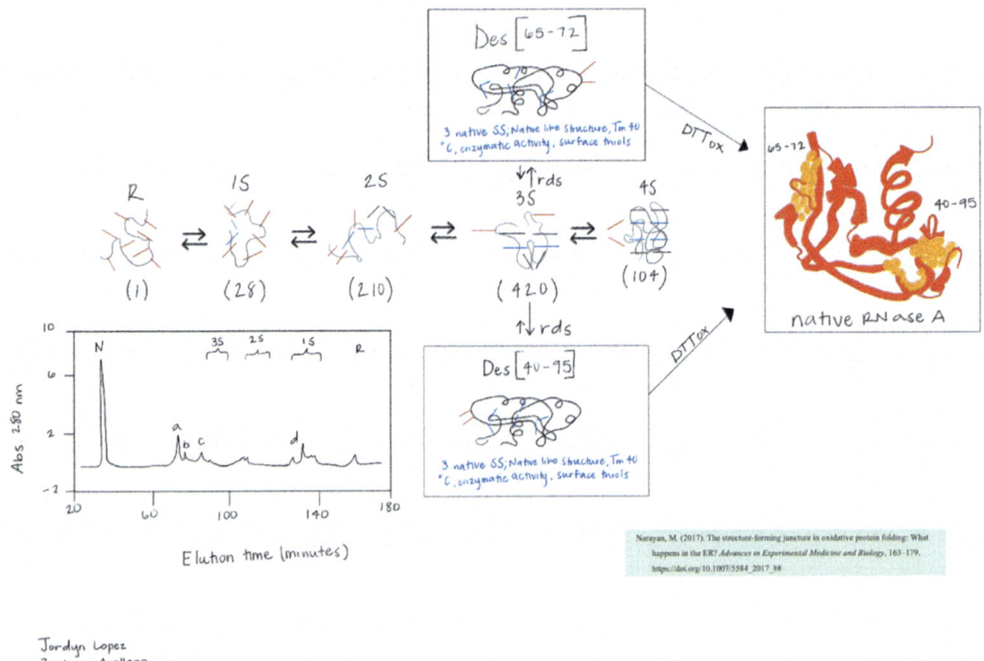

Figure 4.2 The oxidative folding pathway of RNaseA (25°C, pH 8, DTT^{ox}/red) showing the fully-reduced protein (R) and the disulfide bond-containing intermediates (1S–4S). Two structured intermediates, viz. des [65–72] and des [40–97], along with the native protein (N) constitute the remainder of the landscape. Also shown is a typical ion-exchange HPLC chromatogram showing the separation of the aforementioned intermediates using the thiol-blocking agent AEMTS.

they are the same that exist in the native protein: [26–84], [40–95], [58–110], and [65–72].

The formation of 1S from R is rapid because of the lack of structure in R and the favorable oxidizing milieu (Welker et al., 2001a, b). However, because no isomer in the 1S ensemble is structured, 1S is also amenable to reduction. The 28 isomers within the 1S ensemble are also likely to be in rapid interconversion with one another for the same reason. The shuffling between isomeric states takes place via intramolecular thiol-disulfide exchange reactions. Lastly, the 1S ensemble displays no enzymatic activity.

The oxidation of the 1S ensemble results in the formation of two-disulfide-bond-containing intermediates (Rothwarf and Scheraga, 1993a). There exist 210 possible chemical isomers in the 2S ensemble (Fig. 4.2). Of these, only six possess the native complement of disulfide bonds. Similar to the aforementioned scenario, there is no structure in the 2S ensemble. As a result, it can easily be reduced back to the 1S while also undergoing intramolecular thiol-disulfide exchange reactions amongst themselves.

Lacking structure, the 2S ensemble is easily oxidized to the 3S ensemble, which in turn is also readily oxidized to form the 4S ensemble. There are 420 possible 3S isomers, which can be reduced back to 2S, oxidized to 4S, or reshuffle amongst themselves (Fig. 4.2). There are 105 4S isomers (one of them being N) that can (with the exception of N) be easily reduced back to 3S (Fig. 4.2). Since the 4S species are terminally oxidized, there is no further oxidation possible. Furthermore, the lack of free cysteines in the 4S species implies that they cannot intramolecularly reshuffle.

Across the oxidative folding landscape, R-4S, the formation of nonnative covalent bonds or the lack of a sufficient number of native disulfide bonds precludes the formation of the catalytically active biological folds, including the native (N) fold. N itself is an exception to this ensemble. However, the statistical probability of forming disulfides pertaining to N is very small (as discussed later), and hence, the 4S ensemble essentially displays no catalytic activity. Thus in RNaseA, no structure is developed across these ensembles even though the protein is sequentially oxidized to its terminally oxidized state.

4.2.3 The structure-forming step in RNase folding

From the discussion above, it becomes evident that a quasiequilibrium distribution of intermediates pertaining to specific states of oxidation rapidly populates the oxidative folding pathway of RNaseA (Rothwarf et al., 1998a, b). The distribution is described as a quasiequilibrium state because there is a slow “leak” from this population, which continuously drives the equilibrium to the right, resulting in the eventual consumption of the unstructured intermediates.

Eventually, the first inkling of structure appears from within the 3S ensemble (Fig. 4.3) (Narayan, 2011, 2017). When oxidative folding is conducted at 25°C, two

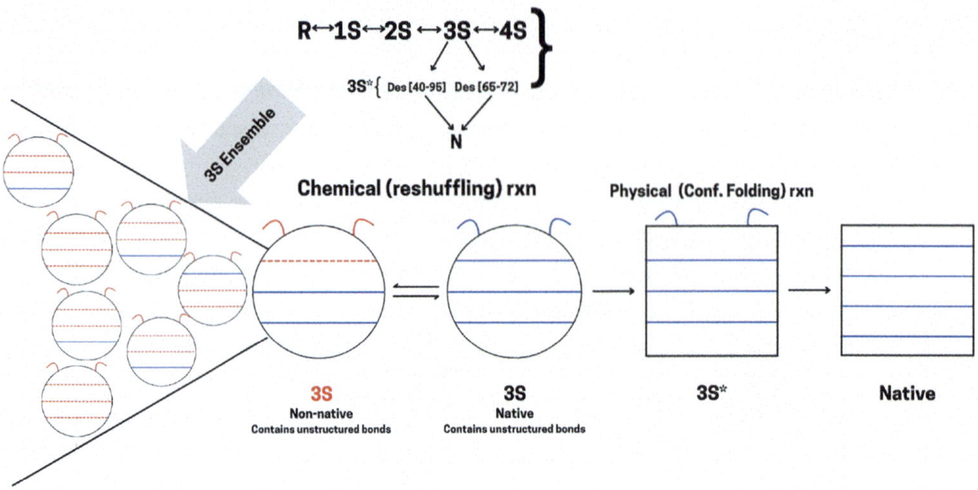

Figure 4.3 The structure-forming step during the oxidative folding of RNaseA. Disulfide bonds are depicted as lines (solid blue, native; dotted-red, nonnative) within the circles (no structure) or squares (structure) which reflect the species participating in this step.

structured 3S species are populated in the folding landscape. These two intermediates possess disulfide combinations of [26–84], [58–110], and [65–72] and [26–84], [40–95], and [58–110] and are termed des [40–95] and des [65–72], respectively. The des [X–Y] terminology denotes the lack of the disulfide bond indicated within the bracket.

Des [40–95] and des [65–72] primarily arise as a consequence of intramolecular thiol-disulfide exchange reactions among the 3S ensemble. A small fraction can also arise from the oxidation of native disulfide-containing combinations of 2S species (Xu and Scheraga, 1998; Iwaoka et al., 1998).

The structure-forming step in RNaseA oxidative folding involves the conformational folding of initially unstructured des [40–95] and des [65–72] to form structured, native-like topologies. In this embodiment, the des species are referred to as 3S* species. The folding of these two species protects their formed disulfides from thiol-disulfide exchange (intramolecular reshuffling) reactions. Thus the formation of 3S* from the 3S species is considered essentially irreversible when oxidative folding is conducted at 25°C, considering that the mid-points of their thermal stability are ~40°C.

The formation of 3S* from 3S, although slow and rate-limiting, slowly consumes the pool of R-4S species, and hence the R-4S distribution qualifies as a quasiensemble. The native protein is formed from each 3S* species by the oxidation of their remaining cysteines. The process is rapid because cysteines 40 and 95 in des [40–95] and cysteines 65 and 72 in des [65–72] are already oriented for facile oxidation in the native-like 3S* species. Furthermore, the oxidation of 3S* to N is considered irreversible when considering the midpoint of the thermal transition of the native protein (65°C).

If oxidative folding is carried out at 15°C, two other structured native-like intermediates are also formed, in addition to des [40–95] and des [65–72]. Des [26–84] and des [58–110] are the remaining two of the four possible 3S species with native disulfide bonds (Welker et al., 1999). They have melting temperatures of ~18°C and are not stable at 25°C. Therefore they simply belong to the pool of 3S species that lack structure.

However, and interestingly, once populated under favorable conditions of temperature, des [26–84] and des [58–110] are unable to become oxidized to N. They persist in the oxidative folding landscape for hours to days and are considered kinetically trapped. This is because when des [26–84] and des [58–110] conformationally fold, the remaining pair of cysteines in each species remains buried within the tertiary structure. As a result, they are inaccessible, and both des species linger in the oxidative folding trajectory. The mechanism by which the kinetically trapped species eventually disappear from the quasiequilibrated landscape is by converting to their unstructured isomers (the 3S ensemble) through thiol-disulfide exchange reactions. It is likely that des [26–84] dismutates by either of its cysteines being attacked by cysteines 40 or 95. Des [26–84] is likely to be reshuffled by an attack from either cysteines 65 or cysteine 72. The 3S ensemble can only form N via the formation of des [40–95] and des [65–72]. Note that a minuscule fraction can directly oxidize the N and then conformationally fold. Statistically, this is likely to be a very small fraction.

4.3 Challenges to the formation of catalytic triad arising in the rate-limiting step

The statistical probability of forming the native complement of disulfide bonds in RNaseA is very low (Xu et al., 1996; Volles et al., 1999). This low probability is also compounded by factors that slow the conformational folding of RNaseA and thereby prevent the preservation of native disulfide bonds from susceptibility to thiol-disulfide exchange reactions. Together, these factors pose challenges to the formation of the catalytic triad of RNaseA. Here, we will discuss in detail the uphill task of constructing a catalytically viable triad that can undertake RNA hydrolysis.

4.3.1 The statistical probability of forming the native set of disulfide bonds in RNaseA

Consider the trajectory of oxidative folding wherein only naïve disulfide bonds are populated. The number of species that populate the oxidative folding landscape (and precede N) includes R, four native disulfide bond-containing 1S species, six 2S species, four 3S species, and N. Therefore the fraction of N that is statistically possible is 1/16 = 6%.

However, it is clear that nonnative disulfide bond-containing species populate the oxidative folding landscape. Therefore under such circumstances, the statistical fraction of a species with four native disulfide bonds is 1/(1 + 28 + 210 + 420 + 105) = 1/764 = 0.13%.

Thus the likelihood of forming the native structure is highly improbable. Even if only native disulfide bonds were to be populated, the presence of a large number of intermediates prior to the formation of N diminishes its statistical yield in oxidative protein folding in comparison with conformational folding. The true statistical yield is reduced even further by a factor of 6/0.13, 46-fold. When mixed disulfides are present, such as with the physiological redox couple GSH/GSSG, the formation of N in the absence of any favorable enthalpic interactions is highly unlikely.

4.3.2 Conformational hindrances to form N

RNaseA possesses four proline residues, viz., Pro 41, Pro 93, Pro 114, and Pro 117. In the unfolded state, X-Pro peptide bonds, where X is any amino acid preceding the Pro residues, display 70%:30% trans:cis characteristics even though the two states are almost isoenergetic. Nevertheless, isomerization across the X-Pro peptide bond is a slow process due to the relatively high energy barrier for the reaction, which is around 14–24 kcal/mol. Furthermore, folding of the protein into the native state locks in the x-Pro peptide bond in either a cis or a trans conformation.

In RNaseA conformational folding, the locking in of three of the four X-Pro peptide bonds (Pro 93, Pro 114, and Pro 117) from a 70:30 trans:cis disposition to a 100% locked-in conformation is rate-limiting (Houry et al., 1994; Houry and Scheraga, 1996). In the native state, the X-Pro peptide bonds around these prolines (93, 114, and 117) are locked in the cis (c), cis (c), and trans (t) state. Each of these three prolines in one of the two possible states when the protein is unfolded gives rise to eight possible dispositions that the protein must refold from to the single native, viz., the cct state.

Thus the statistical chance of the protein conformationally being in the native proline state is only 1/8. This is a simple estimate of the "conformational readiness" of the protein without taking into account other factors that may cause the biopolymer to populate nonnative states. It also does not take into account the "weights" for dispositions of the X-Pro peptide bonds in trans and cis states. We can account for this via the box model of RNaseA folding, which is next discussed (Fig. 4.4).

In order for RNaseA to conformationally fold, prolines 93, 114, and 117 must assume the c, t, t conformations prior to conformational folding. The latter event locks in these conformations via intermolecular interactions, which drive the formation of the native state and the acquisition of biological activity (Fig. 4.2).

In the unfolded protein, the fractional distribution of the three prolines across the eight states is nothing but the product of their probabilities in each chosen state.

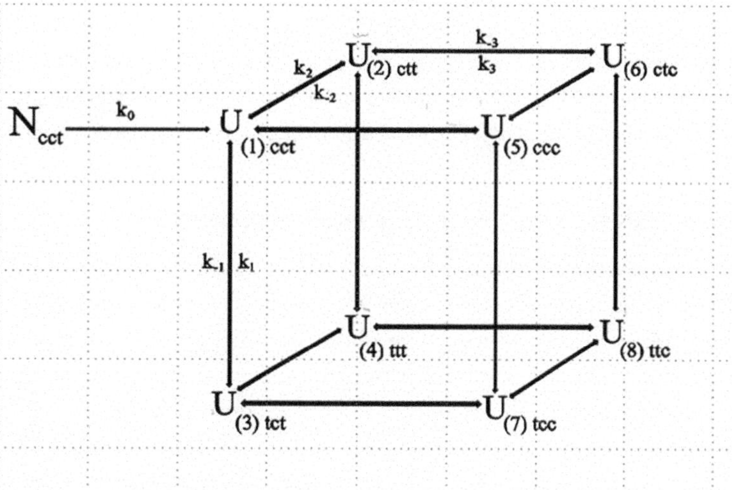

Figure 4.4 The proline isomerization dependent conformational folding of RNaseA. The folding time depends upon the disposition of three essential X-Pro peptide bonds. The refolding time is fastest for the unfolded protein whose prolines happen to be in the c,c,t conformation which is that in N. Thus the unfolded intermediates U_{cct} has shortest (milli-second) refolding time. *From Wedemeyer, W.J., Welker, E., Scheraga, H.A., Biochemistry. 2002;41(50):14637–44.*

For example, the fractional distribution of the three prolines that are all in the trans conformation is 0.7 × 0.7 × 0.7 = 0.343, or 34.3%. By contrast, the fractional distribution of the three prolines that are all in the cis conformation is 0.3 × 0.3 × 0.3 = 0.027, or 2.7%. Considering that the only disposition that can conformationally fold is c,c,t, the fraction of unfolded protein in that conformation is 0.3 × 0.3 × 0.7 = 0.063. Thus those molecules in the unfolded state that possess a c,c,t disposition for prolines 93, 114, and 117 can very rapidly conformationally fold to form N. By contrast, molecules with the prolines in the t,t,c state of isomerization would likely fold the slowest. States in between these extremes would have folding times that are in between.

To further illustrate this aspect, consider the box model shown below. In the box model of proline isomerization, the vertices of the cube represent states of isomerization of the prolines. Against each vertex is the fractional distribution of the combination. The sum of the fractions totals 100%. This is because it is a closed system, and in such a scenario, the sum of all possible events is 1 (or 100%).

There are several folding phases that can be deconvoluted during the conformational folding of RNaseA. The box model permits the reader to easily relate the eight unfolded species types with five refolding phases that have been deconvoluted (Fig. 4.2). For example, N folds rapidly to form a very fast-refolding species termed U_{vf}, which possesses the native X-Pro peptide bond conformations (c,c,t). Other folding phases, viz., U_f, U_m, U_S^{II}, and U_S^{I}, that can be deconvoluted correspond to the fast-, medium-, major slow-, and minor slow-refolding phases.

Thus a t,t,c disposition of prolines 93, 114, and 117 is most removed from their native conformation (c,c,t), and thus species in the t,t,c conformation are likely to fold the slowest. Species such as c,t,c and t,c,c are only one isomerization step away from the c,c,t state. Their position is therefore indicated closest to the c,c,t vertex.

4.3.3 The catalytic triad forming step in RNaseA

We can now consider in-depth the coupling of oxidative folding and conformational folding, both of which are essential for the catalytic activity of RNaseA (Fig. 4.2).

Consider the superimposition of the box model (Fig. 4.5) with an oxidative folding step, which will facilitate underscoring the interplay between redox reactions and conformational folding in the structure-forming step of RNaseA.

Only those 3S species that possess native disulfide bonds, four in all, can conformationally fold. However, to conformationally fold, their prolines must be in the c,c,t disposition. The fraction of species that fit these criteria using probability are 4/420 (=0.95%) 3S species and 6.3% of that population, suggesting a vanishingly small fraction of the folding population that can form 3S*, which is the beginning of the assembly of the catalytic triad. Once the 3S* species is formed, both the native

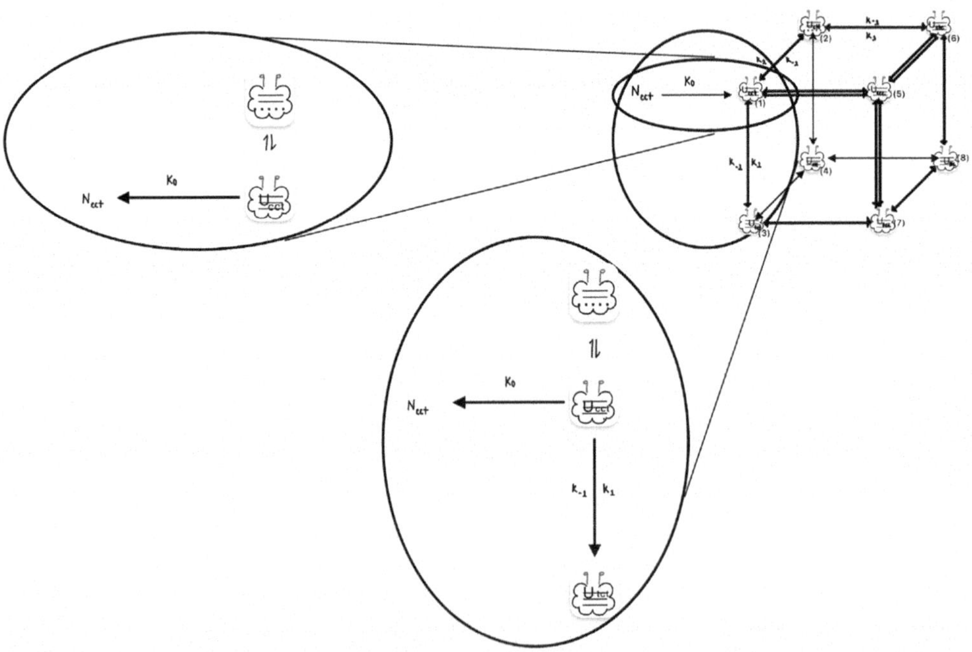

Figure 4.5 The coupling of conformational folding and oxidative folding in RNaseA. Only those 3S species with native disulfide bonds and prolines in the c,c,t can fold in milli-seconds. The remainder of the 3S species with prolines differing from the c,c,t disposition become reshuffled.

complement of three disulfide bonds and the native complement of prolines are locked in within the protective structure. The simple oxidation of the fourth and final disulfide bonds results in the formation of the fully catalytically active N.

Any other species, be it a 3S species containing one or more nonnative disulfide bonds but with prolines in a native disposition, or a 3S species whose disulfide bonds are native but prolines are in the nonnative state, or a species that contains both one or more nonnative disulfide bonds and one or more nonnative proline conformations, cannot fold to acquire catalytic activity. They must channel through those only species that possess both the native complement of three disulfide bonds and the native disposition of prolines.

4.3.4 The competition between thiol-disulfide exchange reactions and proline isomerization-limited conformational folding

We have discussed in detail the factors that are required for RNaseA to fold and become catalytically active. The fraction of species that possess the requirements to fold is vanishingly small. We now consider the structure-forming step of RNaseA by focusing on two competing factors. Consider a 3S species that has a native complement of disulfide bonds and prolines that are c,c,t. This species is almost guaranteed to fold, considering that it can protect both its formed disulfide bonds and its native proline disposition by acquiring structure within milliseconds.

By contrast, consider any other combination of prolines in a 3S species containing all native disulfide bonds. Let us consider 3S species with native disulfides and prolines in a t,c,t state. In this state, two scenarios are at play. The protein can conformationally fold by first acquiring the c,c,t state, or it can reshuffle to a 3S isomer containing at least one nonnative bond. To unveil the competition between these two events, we note that the pKa of a thiol group is ~8.3. Thus if oxidative folding is carried out at pH 8, which is also a suitable pH for conformational folding, thiol-disulfide exchange reaction, being relatively rapid, likely causes reshuffling of a fraction of the 3S species with native disulfide bonds to an isomer containing a nonnative disulfide bond before it can conformationally fold to form 3S*. The reshuffling reaction is likely to proceed further into isomers that contain two or more nonnative disulfide bonds. Due to these reshuffling events, even if the prolines do not isomerize, the probability of forming the original (starting) 3S species with native disulfide bonds is vanishingly small.

Let's consider a milieu at pH 8.5. While conformational folding rates are likely to not be impacted, thiol-disulfide exchange reactions are faster at this pH than at 8 by a factor of 1.4 (antilog of difference in pH). Therefore the likelihood of the native disulfide bond-containing 3S species reshuffling to a nonnative isomer is 1.4× higher, likely lowering the yield of 3S*.

When the experimental conditions are at pH 7, thiol-disulfide exchange is 10× slower than at pH 8. Under such circumstances, a native disulfide bond-containing species with prolines in a c,c,t disposition is highly likely to conformationally fold to form the 3S*. This is not because conformational folding at pH 7 is much faster than at pH 8 but because thiol-disulfide exchange is much slower at pH 8.

4.4 The role of chaperones in facilitating the constitution of the catalytic site

Although thermodynamically downhill, the oxidative folding of RNaseA is clearly an uphill event, as evident from the arguments presented above. Within the cell, the maturation of the protein needs to be rapid to ensure catalytic readiness. The chaperones protein disulfide isomerase (PDI) and peptidyl prolyl isomerase can help facilitate the folding of RNaseA *in vitro*.

4.4.1 Protein disulfide isomerase

PDI is a multifunctional enzyme and chaperone that plays a crucial role in the ER by catalyzing the formation, reduction, and rearrangement of disulfide bonds during protein folding. PDI also acts as a chaperone, helping to prevent misfolded proteins from aggregating and promoting their proper folding.

PDI is a thioredoxin-like protein, meaning it contains a domain with a similar fold to thioredoxin, a protein involved in redox reactions (Wilkinson and Gilbert, 2004; Benham, 2012). The active site of PDI contains a thiol group with a low pKa value; specifically, the N-terminal cysteine residue of its CXXC motif has a pKa of around 6.7. This implies that at a cellular pH of ~7, where thiol-disulfide exchange is slow, PDI can still facilitate thiol-disulfide exchange reactions via its deprotonated thiol group.

Although the presence of PDI catalytic activity does not discriminate between native and nonnative bonds, which can result in the reshuffling of a 3S intermediate containing nonnative bonds to a 3S intermediate containing nonnative bonds, we argue that its catalytic effect is still beneficial for the overall folding of RNaseA.

An immediate consequence of the indiscriminatory nature of disulfide reshuffling activity of PDI is the potential loss in yield of 3S*. This can happen because a 3S species that has native disulfide bonds and prolines in c,c,t position can be rapidly reduced to a 2S species or to a 3S isomer with nonnative disulfide bonds via the catalytic activity of PDI. Thus in such an extreme case, the "catalytic" activity of PDI could be detrimental to the acquisition of the native structure and the gain of catalytic activity.

However, the unique possibility above is overwhelmingly offset by the following consideration. In the absence of structure, the rate of thiol-disulfide exchange is pH-dependent. Within the cellular environment, a pH of 7, coupled with the low statistical probability of forming a 3S intermediate with native disulfide bonds, is likely to make the formation of the native fold and acquisition of catalytic activity highly improbable.

To improve the chance of populating the formation of 3S species with all native bonds, the rate of thiol-disulfide exchange must be rapid. However, at pH 7, the slow rate of reshuffling limits the reshuffling step required to populate the formation of 3S species with native disulfide bonds from their 3S isomers. Note that as previously discussed, the conformational folding step, assuming prolines are c,c,t, is rapid (milliseconds). Even other folding phases, slower than the fastest folding phase, will outcompete thiol-disulfide exchange reactions at pH 7. Thus it is advantageous for the oxidative folding process to have a large fraction of the 3S species with native disulfide bonds, which will conformationally fold to 3S* because their back-reshuffling to nonnative disulfide bond-containing isomers is slow at pH 7. However, as exhaustively discussed, a majority of the 3S species contain nonnative disulfide bonds, and, at pH 7, their ability to reshuffle to form native disulfide-bond-containing isomers is very slow. This is where the presence of PDI with an acidic cysteine is likely catalytically very effective. Its thiolate is able to attack disulfides in the 3S species and reshuffle them. Considering that the pool of 3S species with nonnative disulfide bonds is much larger than those with only native disulfide bonds, the presence of a redox-active catalyst, viz., PDI, becomes extremely beneficial. By "non-discriminate" reshuffling, the "per chance" acquisition of the native diuslfide bonds is simply a sufficient criterion for the conformational folding (which is very rapid when the prolines are c,t,t and even with one conformation removed (say c,t,c) in the context of slow reshuffling). In summary, the catalytic activity of PDI is likely to accelerate severalfold the acquisition of the catalytic activity of RNaseA, particularly by the mechanism discussed here.

4.4.2 Peptidyl-prolyl cis-trans isomerases

Peptidyl-prolyl cis-trans isomerases (PPIases) are enzymes that accelerate the isomerization of peptide bonds preceding proline residues in proteins (Göthel and Marahiel, 1999; Lang et al., 1987). This is often a rate-limiting step in conformational folding. PPIases are categorized into three main families: Cyclophilins, FK506 binding proteins (FKBPs), and Parvulins.

The exact catalytic mechanism of PPIases is still being investigated, but it is thought that they may utilize a conformational twist mechanism, potentially assisted by an enzyme-bound hydrogen bond donor. Some evidence suggests that the active site of PPIases creates a hydrophobic environment that favors the isomerization of the peptide bond. Another mechanism could involve converting the double bond into a single bond, which permits rotation and thus interconversion of the geometric isomers.

As is the case with PDI, PPIase activity can lead to the scrambling of a c,c,t disposition. Thus the burden of protecting the native proline conformation rests on the protein itself, which does so by rapidly folding and protecting the X-Pro peptide bonds from isomerizing. As with PDI, we can envisage a scenario where PPIases can "scramble" c,c,t prolines to nonnative conformations in a 3S intermediate containing native disulfide bonds. However, this is unlikely since the conformational folding of such an intermediate is on the millisecond time scale, effectively providing a means to evade PPIase. Again, we need to consider that in RNaseA, a very small fraction of prolines are in the c,c,t conformation and that slow proline isomerization limits conformational folding of the 3S intermediates with native disulfide bonds and makes them prone to reshuffling back to isomers with nonnative disulfides. In this context, a majority of 3S species, particularly those with native disulfide bonds, will "benefit" from the catalytic activity of PPIase. The perchance formation of c,c,t is an opportunity to conformationally fold.

In both scenarios, the conformational folding of the substrate, that is, RNaseA, to form a stable native structure that buries its native disulfide bonds and locks in the native proline disposition is the only mechanism by which it can "escape" unfolding (reshuffling or reduction of disulfides and isomerization of prolines).

4.5 Conclusion

It is clear that the mechanism by which RNaseA acquires catalytic activity is extremely complex and involves a number of adjuvants (Fig. 4.6). It is a well-concerted process

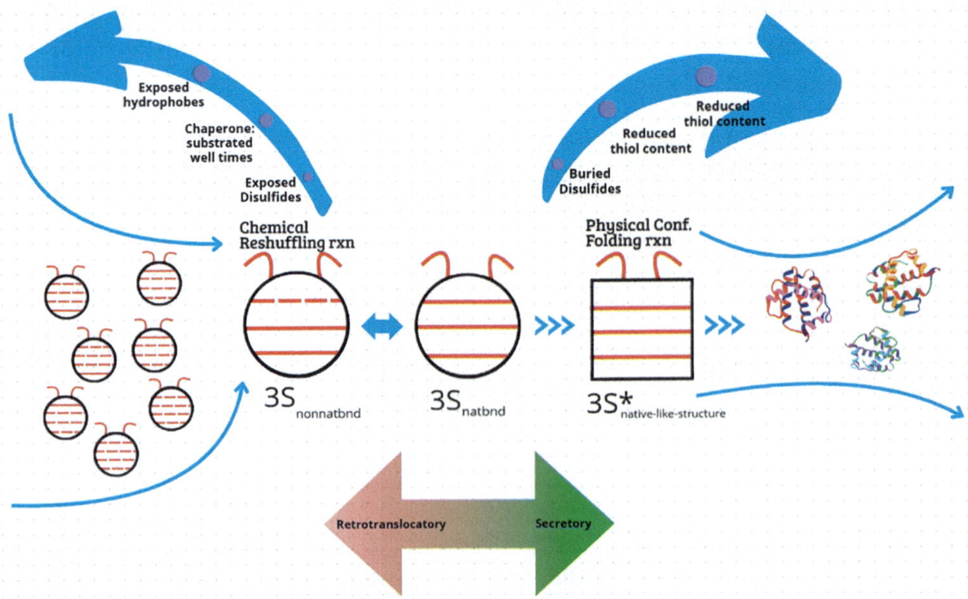

Figure 4.6 Maturation of catalytically active RNaseA process. The maturation of catalytically active protein requires a number of factors and phyisico-chemical characteristics to enable the process.

that has been finely tuned to rapidly acquire the native structure and, correspondingly, the catalytic triad. The risks to retrotranslocation are high (Fig. 4.2). The probability of secretion is low. Yet, as discussed, several factors come into play, enabling the catalytically active protein to finally exit from the ER and hydrolyze RNA.

The author is grateful to Arellano, Zachary; Ariza, Alexia K. Bai, David; Carreon, Arely B.; Chavez, Roxana A.; Garcia, Farica S.; Gutierrez, Natalie A.; Hamer, Emily H.; Hernandez, Luisa F.; Hinojosa, Dante C.; Johnson, Dylan H.; Leinberger, Margaret C.; Lopez, Jordyn B.; Lopez, Samantha G.; Macedo, Julia; Martinez Marquez, Luz A.; Medrano Franco, Juan P.; Morales, Luis R.; Rejino, Victoria J.; Vasquez, Demian E.; Wilson, Milagros L.; and Zaragoza, Anel for scintillating discussions during the preparation of this chapter and for rendering some of the included figures.

References

Arolas, J.L., Aviles, F.X., Chang, J.Y., Ventura, S., 2006. Folding of small disulfide-rich proteins: clarifying the puzzle. Trends in Biochemical Sciences 31 (5), 292–301. Available at https://doi.org/10.1016/j.tibs.2006.03.005.

Benham, A.M., 2012. The protein disulfide isomerase family: key players in health and disease. Antioxidants and Redox Signaling 16 (8), 781–789. Available at https://doi.org/10.1089/ars.2011.4439.

Cuchillo, C.M., Nogués, M.V., Raines, R.T., 2011. Bovine pancreatic ribonuclease: fifty years of the first enzymatic reaction mechanism. Biochemistry 50 (37), 7835–7841. Available at https://doi.org/10.1021/bi201075b.

Fedor, M.J., Williamson, J.R., 2005. The catalytic diversity of RNAs. Nature Reviews. Molecular Cell Biology 6 (5), 399–412. Available at https://doi.org/10.1038/nrm1647.

Göthel, S.F., Marahiel, M.A., 1999. Peptidyl-prolyl cis-trams isomerases, a superfamily of ubiquitous folding catalysts. Cellular and Molecular Life Sciences 55 (3), 423–436. Available at https://doi.org/10.1007/s000180050299.

Houry, W.A., Scheraga, H.A., 1996. Nature of the unfolded state of ribonuclease a: effect of Cis-trans X-pro peptide bond isomerization. Biochemistry 35 (36), 11719–11733. Available at https://doi.org/10.1021/bi960745a.

Houry, W.A., Rothwarf, D.M., Scheraga, H.A., 1994. A very fast phase in the refolding of disulfide-intact ribonuclease a: implications for the refolding and unfolding pathways. Biochemistry 33 (9), 2516–2530. Available at https://doi.org/10.1021/bi00175a022.

Iwaoka, M., Juminaga, D., Scheraga, H.A., 1998. Regeneration of three-disulfide mutants of bovine pancreatic ribonuclease a missing the 65-72 disulfide bond: characterization of a minor folding pathway of ribonuclease a and kinetic roles of Cys65 and Cys72. Biochemistry 37 (13), 4490–4501. Available at https://doi.org/10.1021/bi9725327.

Lang, K., Schmid, F.X., Fischer, G., 1987. Catalysis of protein folding by prolyl isomerase. Nature 329 (6136), 268–270. Available at https://doi.org/10.1038/329268a0.

Leonidas, D.D., Chavali, G.B., Oikonomakos, N.G., Chrysina, E.D., Kosmopoulou, M.N., Vlassi, M., et al., 2003. K.R. Acahrya, high-resolution crystal structures of ribonuclease a complexed with adenylic and uridylic nucleotide inhibitors. Implications for structure-based design of ribonucleolytic inhibitors. Protein Science 12 (11), 2559–2574. Available at https://doi.org/10.1110/ps.03196603.

Narayan, M., 2011. The case of oxidative folding of ribonuclease A: factors impacting fold maturation of ER-processed proteins. Protein Reviews 14, 23–42. Available from. Available at https://doi.org/10.1007/978-1-4419-7273-6_2, http://www.springer.com/series/6876?detailsPage=titles.

Narayan, M., 2017. The structure-forming juncture in oxidative protein folding: what happens in the ER? Advances in Experimental Medicine and Biology 966, 163–179. Available from. Available at https://doi.org/10.1007/5584_2017_88, http://www.springer.com/series/5584.

Narayan, M., Welker, E., Wedemeyer, W.J., Scheraga, H.A., 2000. Oxidative folding of proteins. Accounts of Chemical Research 33 (11), 805–812. Available at https://doi.org/10.1021/ar000063m.

Pace, C.N., Grimsley, G.R., Thomson, J.A., Barnett, B.J., 1988. Conformational stability and activity of ribonuclease T1 with zero, one, and two intact disulfide bonds. Journal of Biological Chemistry 263 (24), 11820–11825. Available at https://doi.org/10.1016/s0021-9258(18)37859-1.

Raines, R.T., 1998. Ribonuclease A. Chemical Reviews 98 (3), 1045–1066. Available at https://doi.org/10.1021/cr960427h.

Robinson, P.J., Bulleid, N.J., 1994. Mechanisms of disulfide bond formation in nascent polypeptides entering the secretory pathway. Cells 9.

Rothwarf, D.M., Li, Y.J., Scheraga, H.A., 1998a. Regeneration of bovine pancreatic ribonuclease A: detailed kinetic analysis of two independent folding pathways. Biochemistry 37 (11), 3767–3776. Available at https://doi.org/10.1021/bi972823f.

Rothwarf, D.M., Li, Y.J., Scheraga, H.A., 1998b. Regeneration of bovine pancreatic ribonuclease A: identification of two nativelike three-disulfide intermediates involved in separate pathways. Biochemistry 37 (11), 3760–3766. Available at https://doi.org/10.1021/bi972822n.

Rothwarf, D.M., Scheraga, H.A., 1993a. Regeneration of bovine pancreatic ribonuclease A. 1. Steady-state distribution. Biochemistry 32 (10), 2671–2679. Available at https://doi.org/10.1021/bi00061a027.

Rothwarf, D.M., Scheraga, H.A., 1993b. Regeneration of bovine pancreatic ribonuclease A. 2. Kinetics of regeneration. Biochemistry 32 (10), 2680–2689. Available at https://doi.org/10.1021/bi00061a028.

Sun, D., Han, C., Sheng, J., 2022. The role of human ribonuclease a family in health and diseases: a systematic review. iScience 25 (11), 105284. Available at https://doi.org/10.1016/j.isci.2022.105284.

Thornton, J.M., 1981. Disulphide bridges in globular proteins. Journal of Molecular Biology 151 (2), 261–287. Available at https://doi.org/10.1016/0022-2836(81)90515-5.

Volles, M.J., Xu, X., Scheraga, H.A., 1999. Distribution of disulfide bonds in the two-disulfide intermediates in the regeneration of bovine pancreatic ribonuclease A: further insights into the folding process. Biochemistry 38 (22), 7284–7293. Available at https://doi.org/10.1021/bi990570f.

Wedemeyer, W.J., Welker, E., Narayan, M., Scheraga, H.A., 2000. Disulfide bonds and protein folding. Biochemistry 39 (15), 4207–4216. Available at https://doi.org/10.1021/bi992922o.

Welker, E., Narayan, M., Volles, M.J., Scheraga, H.A., 1999. Two new structured intermediates in the oxidative folding of RNase A. FEBS Letters 460 (3), 477–479. Available at https://doi.org/10.1016/S0014-5793(99)01391-5.

Welker, E., Narayan, M., Wedemeyer, W.J., Scheraga, H.A., 2001a. Structural determinants of oxidative folding in proteins. Proceedings of the National Academy of Sciences of the United States of America 98 (5), 2312–2316. Available at https://doi.org/10.1073/pnas.041615798.

Welker, E., Wedemeyer, W.J., Narayan, M., Scheraga, H.A., 2001b. Coupling of conformational folding and disulfide-bond reactions in oxidative folding of proteins. Biochemistry 40 (31), 9059–9064. Available at https://doi.org/10.1021/bi010409g.

Wilkinson, B., Gilbert, H.F., 2004. Protein disulfide isomerase. Biochimica et Biophysica Acta - Proteins and Proteomics 1699 (1-2), 35–44. Available from. Available at https://doi.org/10.1016/S1570-9639(04)00063-9, http://www.elsevier.com/locate/bbapap.

Wlodawer, A., Bott, R., Sjölin, L., 1982. The refined crystal structure of ribonuclease A at 2.0 A resolution. Journal of Biological Chemistry 257 (3), 1325–1332. Available at https://doi.org/10.1016/s0021-9258(19)68195-0.

Xu, X., Rothwarf, D.M., Scheraga, H.A., 1996. Nonrandom distribution of the one-disulfide intermediates in the regeneration of ribonuclease A. Biochemistry 35 (20), 6406–6417. Available at https://doi.org/10.1021/bi960090d.

Xu, X., Scheraga, H.A., 1998. Kinetic folding pathway of a three-disulfide mutant of bovine pancreatic ribonuclease A missing the [40-95] disulfide bond. Biochemistry 37 (20), 7561–7571. Available at https://doi.org/10.1021/bi980086x.

CHAPTER 5

Ribonucleases in evolution

Elio Pizzo and Rosanna Culurciello
Department of Biology, University of Naples Federico II, Naples, Italy

5.1 Introduction

Ribonuclease (RNases) were originally described between 1911 and 1920 by Levene and Jones, who detected that a boiled pancreas extract could digest yeast RNA (Jones, 1920). Later, working from a variety of animal sources, Dubos and Thompson (Dubos and Thompson (1938)) were the first to purify a bovine pancreatic ribonuclease (RNaseA), which three years later was successfully crystallized by Kunitz (1940). From that moment on, there was intense attention on the properties of this enzyme, also supported by huge industrial efforts of the Armour company that, starting from the procedure developed by Kunitz at the Rockefeller Institute, prepared 1 kg of the crystalline enzyme from slaughterhouse material. Samples obtained from that purification were then given out free to any laboratory with a research project attributable to properties of RNases. Through the 1960s and 1970s, RNaseA was a favorite to study thanks primarily to its remarkably thermostable and obtainable at high concentration in a particularly accessible source, such as the bovine pancreas. These studies led to the elucidation of its crystal structure (Anfinsen, 1959; Groves, 1966; Scheraga, 1967), determination of its amino acid sequence (Smyth et al., 1963), identification of its catalytic mechanism (Findlay et al., 1961), and clarification of folding pathways (Hantgan et al., 1974). RNaseA was the first enzyme and third protein for which a correct amino acid sequence was determined (Raines, 1998). Four Nobel Prizes have been awarded for work associated with studies of RNaseA (Anfinsen, Moore, Stein, and Merrifield). The vast literature and numerous studies have made RNaseA the most extensively studied enzyme of the 20th century and written a significant chapter in the history of modern protein chemistry (Raines, 1998). However, starting from the 1970s, several other homologous to RNaseA were isolated and characterized, first exclusively in mammals and mostly of pancreatic origin. Then from the late 1980s onwards, there was an intense acceleration in the identification of new members, also in extrapancreatic districts, both in mammalian and nonmammalian vertebrates (Beintema et al., 1977; Beintema, 1998; Cho et al., 2005; Beintema et al., 1997). Finally, above all, thanks to the advent of whole genome sequencing procedures and in general to the enormous progress of informatics tools, it has made it possible to identify further RNaseA homologs in nontetrapod vertebrates (Pizzo et al., 2006; Cho et al.,

RNases in Health and Diseases
DOI: https://doi.org/10.1016/B978-0-443-27312-4.00002-X

2007; Pizzo et al., 2007) but not in basal chordates, such as tunicates and amphioxus, as well as invertebrate metazoans. Based on this, the RNaseA superfamily currently consists of numerous members, however, identified exclusively in vertebrate tetrapods and teleosts. This alone would be enough to make this superfamily particularly intriguing; however, beyond that, there are many additional diverse properties from various members of different classes. All this falls within the objective of this chapter, which is precisely to update the knowledge of RNaseA homologs, and to achieve this, the text will take an unusual path that will concern the most ancient members of the superfamily and then go up across the Vertebrata subphylum to humans.

5.1.1 Ancestral members of the RNaseA superfamily: fish RNases

Fish are the most diverse group among vertebrates, with over 35,000 different species, comprising over half of all known vertebrate species. Furthermore, fish are excellent experimental models for genetic and evolutionary studies. This is especially relevant for studying the RNaseA superfamily, that is, as mentioned before, a vertebrate-specific group of proteins exclusively identified in tetrapods and teleosts (bony fish). Fish RNaseA homologs were first identified relatively recently (Pizzo et al., 2006; Cho et al., 2007), although their initial functional and structural characterization has provided valuable insights into the evolutionary trajectories of these proteins. Known fish RNaseA homologs share two main distinctive features: a low ribonucleolytic activity frequently coupled with the ability to induce endothelial cell differentiation, similarly to the angiogenic activity of human RNase5 (see below), and a compact fold typically stabilized by three disulfide bonds. These features have been experimentally confirmed in RNases from *Danio rerio* and *Salmo salar* (Pizzo et al., 2008; Monti et al., 2009; Zanfardino et al., 2010; Pizzo et al., 2011; Sica et al., 2021; Ferguson, 2019) and are likely conserved across the many other fish proteins identified through bioinformatic approaches but not yet experimentally validated (511 unique sequences deposited in UniProtKB, associated with the ribonuclease A-like domain superfamily, IPR036816, and pancreatic ribonuclease family CKxxNTF signature, IPR023411). Interestingly, some putative fish RNases exhibit a fourth disulfide bond, a distinctive feature found in proteins from *Perca flavescens* (A0A484D7A5), *Oncorhynchus masou formosanus* (B0FFJ3), *Oncorhynchus mykiss* (A0A060X0G1), and *Perca fluviatilis* (A0A6A5FFI6). Notably, this additional disulfide bond is not located in the same positions observed in the amphibian onconase (ONC) or human RNases stabilized by four disulfide bonds, suggesting an independent evolutionary development (Fig. 5.1).

Evolutionary analysis of fish angiogenins reveals specific conservation patterns within key consensus sequences. Among the three recognized conserved regions, two not strictly constrained display a notable degree of variability. In the first consensus region, "FxxQH" emerges as the predominant motif (59 occurrences), followed by less

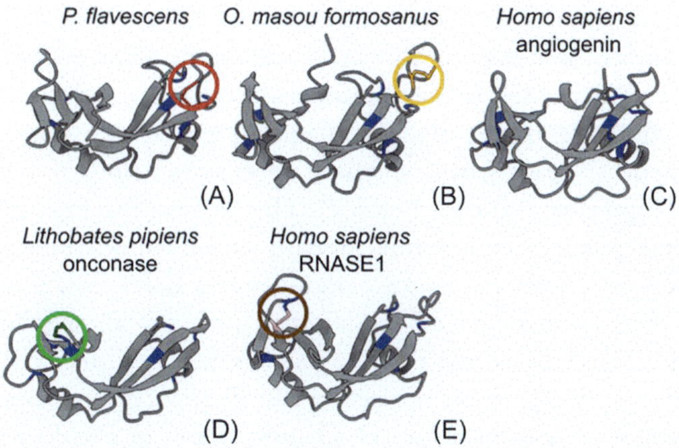

Figure 5.1 Representative RNase fold structures. Representative RNase fold structures from two classes of four-disulfide putative angiogenins in fishes ((A) A0A484D7A5, (B) B0FFJ3), compared with human angiogenin ((c) P03950), northern leopard frog onconase ((D) P22069), and human RNASE1 ((E) P07998). Conserved disulfide-bond-forming cysteines shared with human angiogenin are shown in blue. Colored residues and matching circles highlight class-specific disulfide bonds unique to each structure (*RNase*, ribonuclease).

frequent but tolerated substitutions like "FxxKH" and "FxxRH," suggesting selective flexibility favoring basic or polar residues at these positions. This confirms what we already know from the study of tetrapod RNases. The conservation of a histidine at the final position further supports its likely role in catalytic activity, while the substitutions (Q, K, R) all maintain polar or basic side chains, indicating functional pressure to preserve local charge characteristics. The second variable consensus region exhibits greater variability, predominantly following the pattern "PVHYX." The "PVHYD" sequence, notably the most common variant (15 occurrences), along with multiple less frequent variants, indicates considerable evolutionary flexibility, potentially reflecting species-specific adaptations or ecological specializations. Substitutions at the fifth position include acidic (D, E), polar (Q), small (A, G), and hydrophobic residues, suggesting that this site tolerates a wide range of biochemical environments. This may point to functional diversification in substrate interaction or extracellular signaling. The presence of a fourth disulfide bond in some sequences, though not homologous to those in ONC or human RNases, suggests structural innovation that may enhance protein stability under specific environmental conditions,perhaps relevant to osmotic or temperature stress in aquatic habitats. Whether this feature arose independently in salmonids and perches or reflects an ancestral trait retained only in these lineages remains to be resolved. Additionally, the OMA (Orthologous MAtrix) database, which organizes genes into orthologous groups across species, currently lists 43 unique fingerprints associated with fish RNase sequences. In this context, an OMA fingerprint refers to a set of conserved sequence features or motifs that help define orthologous relationships, aiding in the detection of functionally equivalent genes across divergent taxa. This further underscores the evolutionary expansion and potential functional diversification of the RNaseA superfamily in aquatic vertebrates. Focusing on a robust subset of 26 sequences (those with experimental evidence and/or RefSeq entries), a simple phylogeny is possible (NJ tree, Fig. 5.2). This analysis also highlights the presence of multiple copies of putative angiogenins in many of the considered species and reveals that the black rockcod (*Nototenia coriiceps)* angiogenin is divergent, possibly because of cold adaptation. The comparison of model structures for all sequences against the human angiogenin (Fig. 5.2b) shows that the *Danio rerio* insertion is a distinctive feature of fish angiogenins; zooming in on the active site (Fig. 5.2c) reveals a conserved fold around the catalytic triad, except for an H/Q substitution in *Poecilia formosa*. The web logo of the catalytic triad made on the 26 sequences (Fig. 5.2c, right) confirms the general trend described above.

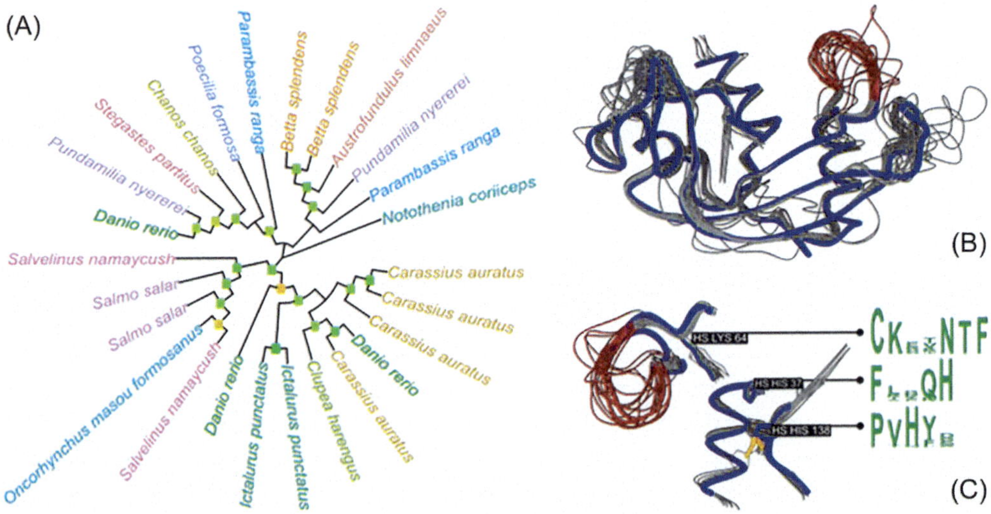

Figure 5.2 Phylogeny tree and overlapped 3D structures. (A) Unrooted NJ tree for a selected subset of 26 proteins (with experimental evidence and/or RefSeq entries). Branch support is color-coded on nodes where the value is ≥0.59, with a scale from orange (0.59) to green (1) to indicate increasing support value. Leaf nodes are colored according to species name. (B) Overlap of the model structures for the 26 proteins (represented in thin gray) plus human angiogenin (represented in thick blue); fish sequence insertion in dark red. (C) Zoom on catalytic residues of human ANG (P03950), color-coded in the same manner as subplot (B), with the addition of orange for a single deviant residue (H/Q in *Poecilia formosa* putative angiogenin). Sequence logos in green for the three consensus sequences, based on the alignment of the 26 structures.

5.2 RNaseA homologs from tetrapods

5.2.1 RNaseA homologs from nonmammalians

Amphibian RNaseA homologs have been isolated mainly from frogs. Early work on *Rana fusca* and *Rana pipiens* eggs focused on the ribonucleolytic activity of the cell extracts (Brachet, 1947; Finamore, 1955). The first amphibian RNaseA homolog was isolated from the liver of bullfrog (*Rana catesbeiana*) in 1976 (Nagano et al., 1976), followed only three years after by a new RNase from *Rana esculenta* liver in 1979 (Malicka-Błaszkiewicz and Kubicz, 1979). The most famous amphibian RNaseA homolog is ONC, isolated from *R. pipiens* oocytes in the late 1980s and named Ranpirnase (for *R. pipiens* RNase), but above all, to hallow its anticancer potential, ONC (Darzynkiewicz et al., 1988; Ardelt et al., 1991; Wu et al., 1993). ONC is the smallest member of the RNaseA superfamily (104 amino acid residues) and is stabilized by four disulfide bonds. Its primary sequence shares 30% identity with that of RNaseA, but its structure is typical of RNaseA homologs (Mosimann et al., 1994). The main differences are in the loop regions, which are shorter than those of other RNase homologs, and at the C-terminus, where a disulfide bridge connects Cys104 to Cys87, a bond absent in mammalian RNases. Another peculiar feature of the ONC structure is the N-terminal pyro-Glu residue, which has been shown to be essential for optimal enzymatic and cytotoxic activity (Newton et al., 1998). ONC is very stable and resistant to proteolysis, with a high T_m (90°C) (Hacke et al., 2013; Leland et al., 1998). This accentuated stability certainly supports its potent cytotoxicity. RNaseA and other members of the superfamily have much higher RNA-degrading activity than ONC but much lower cytotoxicity. Moreover, RNaseA and many other members of the superfamily can be inactivated by RNase inhibitor (RNH1), an abundant protein inherently present in mammalian cells; interestingly, ONC is resistant to the inhibition by RNH1 (Haigis et al., 2003). However, it should be noted that ONC binds to the inhibitor, although with a much lower affinity than that measured for RNaseA or angiogenin, with which RNH1 even shares a femtomolar Kd (Lee et al., 1989). The Kd value for inhibition of ONC by RNH1, determined at low salt concentrations, has been measured to be 0.15 μM (Turcotte et al., 2008), about seven orders of magnitude weaker than that measured for the RNH1/RNaseA or RNH1/angiogenin complexes. This may explain the increased cytotoxicity of ONC. ONC prefers to degrade tRNAs (as angiogenin described below), and this could be the main reason for the toxic properties of this amphibian enzyme. Indeed, it is widely established that in addition to attenuating protein synthesis, tRNA degradation can also trigger apoptosis in cancer cells (Saxena et al., 2002). Moreover, gene expression analysis highlighted that ONC treatment also resulted in downregulation of tumor-associated microRNAs (miRNAs) (Goparaju et al., 2011). Subsequent studies clarified that ONC preferentially degrades precursors of miRNAs, but not miRNAs (Qiao et al., 2012). Downregulation of miRNAs may play an important role in ONC cytotoxicity. ONC is currently in its third phase of clinical trials to determine its effective treatment method for mesothelioma in comparison to chemotherapy drugs (Mikulski et al., 2002) (see also the webpage https://www.mesotheliomahope.com/treatment/chemotherapy/drugs/onconase/). Besides ONC, *R. pipiens* oocytes contain at least three ONC variants, and this has also been verified in other frogs, such as *R. catesbeiana, Rana japonica, Rana dybowskii, and Xenopus leavis* (Singh et al., 2007; Ardelt et al., 2007; Arnold, 2014), in which further homologs of superfamily members have been identified. Interestingly, all amphibian RNAse A homologs share the same arrangement of disulfide bridges, including the fourth bridge located in the C-terminal region and absent in RNases identified in mammals. This distinctive feature could represent a significant evolutionary trace and mark a structural and functional boundary that remains unexplored but is extremely interesting. It should be noted that the fish RNases described above are totally devoid of the fourth disulfide bridge, while in higher tetrapods (starting with reptiles, to be clear), the fourth bridge involves two additional internal cysteine residues.

Only a few RNases from reptiles have been deeply studied. Structurally and phylogenetically, they are very close to mammalian RNases, except for angiogenin, which, in its structure, presents a disulfide bond between residues 65 and 72 (following RNaseA numbering) that is absent in other RNases. Among the most peculiar reptile RNases are the enzymes identified in *Iguana iguana*, in snake venom, and in turtles (Nitto et al., 2005; Katekaew et al., 2010). All show the typical α/β fold of vertebrate RNases, an expression source mainly located in the pancreas, and weak bactericidal or cytotoxic activities. RNaseA homolog from the venom of the Vietnamese cobra *Naja atra* has some interesting properties, such as high catalytic activity in acidic pH range (pH_{opt} 2.58 $\pm$ 0.35) and high conformational stability established through thermostability and protease susceptibility studies (Van Nguyen et al., 2017). In addition, this RNaseA homolog has at least four isoforms, two of which are particularly abundant but of reduced molecular weight (8.95 and 5.93 kDa). To date, the functional properties of these isoforms are unknown.

Also, RNases from birds are little characterized. Those identified in chickens are certainly the most studied. After initial studies (Hayano et al., 1993; Klenova et al., 1992; Levy et al., 1980; Nakano et al., 1992), three RNases from chicken have been identified and well-studied (Cho et al., 2005; Nitto et al., 2006). The cationic RNase A-2, with a higher RNase activity and a very high pH of 11.0, possesses angiogenic and bactericidal activity, the latter not associated with catalytic properties. Surprisingly, the human RNase inhibitor (RNH1) was effective in blocking both RNases A-1 and A-2, despite the phylogenetic distance between Homo and Gallus, and the reported lack of inhibition by RNH1 on amphibian and reptile RNases (Nitto et al., 2005; Wu et al., 1993). Studies on the structural determinants required for the bactericidal activity of

RNase A-2 (Nitto et al., 2006) identified two structural domains, defined by the amino acid sequences 71–76 (domain II) and 89–104 (domain III), as responsible for the activity. The activity was still present when the domains were isolated from the protein. Interestingly, domain II was found to be active even in its scrambled form, and the reason is probably related to the high content of basic and hydrophobic residues in the short 6-residue peptide Arg-Arg-His-Phe-Arg-Ile (Nitto et al., 2006).

5.2.2 Mammalian RNases

As mentioned before, the RNaseA superfamily constitutes a vertebrate-specific gene lineage and certainly underwent substantial expansion during early mammalian evolution, the latter essentially justified by the markedly higher number of RNase genes in extant mammals compared to nonmammalian vertebrates. This is consequently associated with the broad functional repertoire of mammalian RNases, which are implicated in a wide range of physiological processes, including digestion, cytotoxicity, angiogenesis, male reproductive function, and innate immunity. Nevertheless, the precise evolutionary timeframe of this gene superfamily's expansion, as well as the molecular mechanisms driving its functional diversification, remains insufficiently resolved (Goo et al., 2013). Below are the peculiarities of some of the most intriguing mammalian RNases, and then we focus on human RNases, which clearly represent the most captivating and, at the same time, promising subject in terms of functional prospects in modern biology.

5.2.3 Eosinophil-associated RNases from rodents

Eosinophil-associated ribonucleases (EARs) are a subset of the RNaseA superfamily present in the granules of rodent eosinophils (Zhang et al., 2000). These proteins show similarity to human eosinophil-derived neurotoxin (EDN) and eosinophil cationic protein (ECP) and are thought to play a role in host defense, although their physiological functions remain unclear. At least 13 EARs have been identified in mice and 8 in rats (Cormier et al., 2001; Attery et al., 2017; Ishihara, 2003; Yamada et al., 2015). Some EARs have antimicrobial activity; for example, EAR-1 and EAR-2 proteins in rats inhibit the growth of Escherichia *coli* and Staphylococcus *aureus* at low concentrations (Ishihara, 2003). Antiviral activity has also been reported, with EAR-2 reducing the infectivity of murine pneumonia virus (PVM) in respiratory epithelial cells (Moreau et al., 2003). However, the level of EAR in the lungs of PVM-infected mice is reduced, suggesting that the virus may suppress the production of these proteins to evade the immune system. EARs may act as opsonins, binding to viral particles and cellular receptors to promote phagocytosis. Furthermore, EAR-2 has been shown to have chemotactic activity on human and murine dendrites, while EAR-6 increases its expression after *Schistosoma mansoni* infection, suggesting a possible role in defense against parasites (Nitto et al., 2004). However, the increase in EAR expression during infections could also be part of a more complex immune response. One study showed that EAR-11 is expressed in murine lungs only after ovalbumin stimulation and that its induction depends on CD4+ T lymphocytes. Furthermore, exposure to Th2-type cytokines (IL-4 or IL-13) increases EAR-11 expression in the lungs, suggesting its involvement in the type 2 immune response (Cormier et al., 2002). Overall, the physiological role of EARs remains uncertain. They may contribute to the maintenance of lung homeostasis by removing pathogenic RNAs or, during inflammation, promote the degradation of cellular RNA in infected cells, inducing cell death (Rosemberg, 2015). However, the precise mechanisms of their functions remain poorly understood.

5.2.4 Bovine RNases

RNaseA is the prototype of the RNaseA superfamily and at the same time the best-characterized member. It is a relatively small protein (124 residues) characterized as a two-layer $\alpha+\beta$ protein with a deep cleft for binding the RNA substrate. The first layer is composed of three alpha helices (residues 3–13, 24–34, and 50–60) from the N-terminal half of the protein. The second layer consists of three β-hairpins (residues 61–74, 79–104, and 105–124 from the C-terminal half) arranged in two β-sheets. The hairpins 61–74 and 105–124 form a four-stranded, antiparallel β-sheet that lies on helix 3 (residues 50–60). The longest β-hairpin 79–104 mates with a short β-strand (residues 42–45) to form a three-stranded, antiparallel β-sheet that lies on helix 2 (residues 24–34). RNaseA is stabilized by four disulfide bonds in its native state: Cys26–Cys84, Cys58–110, Cys40–95, and Cys65–72. The first two (26–84 and 58–110) are essential for conformational folding; each joins an alpha helix of the first layer to a beta sheet of the second layer, forming a small hydrophobic core in its vicinity. The latter two disulfide bonds (40–95 and 65–72) are less essential for folding; either one can be reduced (but not both) without affecting the native structure under physiological conditions. These disulfide bonds connect loop segments and are relatively exposed to solvent. The 65–72 disulfide bond has an extraordinarily high propensity to form, significantly more than would be expected from its loop entropy, both as a peptide and in the full-length protein. This suggests that the 61–74 β-hairpin has a high propensity to fold conformationally (Raines, 1998; Wlodawer et al., 1988; Abaturov et al., 2012; Graham et al., 2011). RNaseA is a cationic protein ($pI = 9.63$) with a high propensity to bind RNA (a poly-anion). RNaseA is unusually polar or, rather, unusually lacking in hydrophobic groups, especially aliphatic ones. This may account for its need for four disulfide bonds to stabilize its structure. The low hydrophobic content may also serve to reduce the physical repulsion between highly charged groups. The N-terminal α-helix of RNaseA (residues 3–13) is connected to the rest of RNaseA by a flexible linker (residues 16–23). Richards et al. showed that this linker may be cleaved by subtilisin between residues 20 and 21 without causing the N-terminal helix to dissociate from the rest of RNaseA. The peptide-protein complex is called "RNaseS" as the result of a peptide spanning from residues 1–20 (called "S-peptide") and the remaining protein body (residues 21–124) called "S-protein." The dissociation constant of the S-peptide for the S-protein is roughly 30 pM. The RNaseS model system is used for studying protein folding by coupling folding and association. The S-peptide was the first peptide from a native protein shown to have (flickering) secondary structure in isolation (Brown et al., 1969).

RNaseA cleaves specifically after pyrimidine nucleotides (Raines, 1998). Cleavage takes place in two steps: first, the 3',5'-phosphodiester bond is cleaved to generate a 2',3'-cyclic phosphodiester intermediate; second, the cyclic phosphodiester is hydrolyzed to a 3'-monophosphate. It can be inhibited by the ribonuclease inhibitor protein, by heavy metal ions, and by uridine-vanadate complexes (Raines, 1998). In some studies, RNaseA was shown to inhibit tumor growth in mice and rats at doses up to 1000 mg/kg, while in others, the absence of both cytotoxic and antitumor effects was observed, and the lack of antitumor effects was explained by the tight binding of RNaseA to intracellular ribonuclease inhibitor (RNH1) (Johnson et al., 2007). Nevertheless, despite the strong binding to RNH1, dimeric forms of RNaseA were shown to exhibit significant antitumor effects (Mohamed et al., 2022); thus, the question about the antitumor potential of RNaseA was raised again.

Bovine seminal ribonuclease (BS-RNase) is very similar to bovine pancreatic and brain RNases but is expressed exclusively in seminal fluid. It is the only known RNase that occurs naturally as a homodimer. Although it is present in an environment where host defense is crucial, its role in this context is poorly understood. BS-RNase has been shown to be cytotoxic to several tumor cell lines. Studies have shown that the substitution of three residues (G38K/K39G/G88R) increases its cytotoxic activity in K-562 cells, an effect that is even more marked in the monomeric variant of the protein (Lee et al., 2005). This is probably due to the lower affinity of the monomer for the ribonuclease inhibitor (RNH1), present in the cells. The cytotoxic activity seems to depend on the presence of cationic residues that favor the interaction with anionic cell membranes. In particular, the replacement of glycine 38 with lysine (G38K) strengthens the binding to the cell membrane, increasing the internalization of the protein by endocytosis (D'Errico, 2011). BS-RNase appears to induce apoptosis by degrading cellular RNA after entering the cells, although the exact mechanism is not yet fully understood. In addition to its cytotoxic activity, BS-RNase has immunosuppressive properties (Soucek, 1996). It has been shown to inhibit lymphocyte proliferation and reduce the expression of the α chain of the IL-2 receptor, suggesting that its immunosuppressive action occurs by interfering with the expression of this receptor in T lymphocytes. Furthermore, lymphocytes stimulated with phytohemagglutinin have been observed to undergo apoptosis in the presence of BS-RNase, showing typical signs such as DNA fragmentation, cell shrinkage, and mitochondrial alterations (Sinatra et al., 2000). BS-RNase is also selectively cytotoxic to thyroid carcinoma cells, inducing apoptosis through caspase-8- and caspase-9-dependent pathways (Kotchetkov et al., 2001). Cytotoxic efficacy varies among cell types, probably influenced by the mode of internalization of the protein. Studies on ONC, a similar RNase isolated from amphibians, suggest that the internalization pathway of RNases may determine their different cytotoxic activity in different cell types.

5.3 Human RNaseA homologs

5.3.1 Human RNase

In humans, 13 functional RNase genes have been identified (Cho et al., 2005). Of these, eight were discovered earlier and are collectively referred to as the canonical RNases. This group includes RNase1, RNase2 (EDN), RNase3 (ECP), RNase4, RNase5 (angiogenin), RNase6 (RNasek6), RNase7, and RNase8. These enzymes are characterized by the presence of conserved sequence motifs critical to RNase activity, such as the catalytic triad, composed of two histidines and one lysine, and the consensus signature "CKXXNTF."

Canonical RNases are associated with a broad-spectrum of physiological roles. These include the degradation of dietary RNA, induction of apoptosis, promotion of angiogenesis, facilitation of sperm maturation, and exertion of antimicrobial and antiviral effects (Barnard, 1969; Domachowske et al., 1998; Rosemberg, 2008; Sorrentino, 2010; Sorrentino et al., 1994).

5.3.2 Canonical RNase

Human RNases are a rapidly evolving family of proteins, with eight catalytically active members identified in humans called canonical RNases. These structurally diverse enzymes, which target various RNA substrates, are unique to vertebrates. Beyond RNA digestion, RNases have been linked to functions such as neurotoxicity, angiogenesis, immunosuppression, and antipathogen activity. Evolutionary studies suggest they originally evolved for host defense or angiogenesis. Human RNases are secretory proteins widely found across various organs, tissues, and body fluids, with some displaying multiple variants due to different glycosylation patterns. Based on structural, catalytic, and biological traits, these enzymes can be grouped into four types. Evolutionary analysis supports this classification, showing strong relationships between RNase2 and 3, as well as RNases7 and 8. RNases2, 3, 6, 7, and 8 form one group, distinct from RNases1, 4, and 5 (see Figure 1.3 in Chapter 1).

5.3.3 RNase1

RNase1 likely emerged from a common ancestral RNase gene through successive gene duplications and functional diversifications. In mammals, the remarkable conservation of the RNase1 gene underscores its pivotal role in critical physiological functions, including digestion, immune modulation, and tumor suppression.

RNase1 is the most active enzyme of the RNaseA superfamily in humans, exhibiting broad substrate specificity against single-stranded RNA, preferentially degrading poly(C) > poly(U) >> poly(A), as well as double-stranded RNA and RNA:DNA hybrids (Futami et al., 1997). RNase1 has a wide expression profile, with mRNA detectable across all human tissues (Weickmann et al., 1984).

In healthy individuals, RNase1 is present in the serum at concentrations of approximately 250 ng/mL, with vascular endothelial cells identified as the principal source of this circulating enzyme (Fischer, 2011). Although RNase1 is constitutively secreted, it can also accumulate in Weibel–Palade bodies within endothelial cells (Beintema, 1986).

Posttranslational modifications of RNase1 include N-linked glycosylation at three asparagine residues: Asn34, Asn76, and Asn88, the pattern of which varies depending on tissue origin (Beintema, 1986). While glycosylation at any of these sites attenuates enzymatic activity, it simultaneously enhances both thermal stability and resistance to proteolytic degradation. In addition to its catalytic function, RNase1 exhibits micromolar affinity for certain cell-surface glycans.

Initially studied as a potential biomarker for pancreatic cancer, subsequent studies revealed elevated RNase1 levels in various malignancies (Fischer, 2011; Maor et al., 1978). Furthermore, its levels have been found to be increased in many other pathological conditions, including infection, heart attack, smoking, trauma, myocardial infarction, aging, malnutrition, arthritis, and Alzheimer's disease (AD), making it unsuitable as an exclusive cancer biomarker. However, recent studies suggest that RNase1 may promote breast cancer initiation by activating the EphA4 receptor, indicating renewed interest in its biomarker potential role in cancer (Lee et al., 2021).

The association between cancer and RNases has been historically associated with ONC and BS-RNase, especially thanks to their ability to evade RNH1 inhibition. These peculiarities have inspired the creation of RNH1-evading variants of RNase1, some of which have entered clinical trials as potential anticancer agents (Hoang et al., 2018). The discovery of the presence of extracellular RNA (eRNA) has opened new perspectives on the biological role of RNase1 (Garnett et al., 2019). eRNA is involved in cell communication, inflammation, and blood clotting (Najem et al., 2024). RNase1 can degrade eRNA, modulating inflammation and reducing coagulation and macrophage recruitment in pathological settings such as atherosclerosis, infarction, and ischemia-reperfusion injury (Garnett et al., 2019). RNase1 has a complex role in human biology, with implications in cancer, inflammation, and coagulation. Its therapeutic potential continues to be explored, with the goal of developing new drugs based on its enzymatic properties. The negative effects of eRNA can be blocked by exogenous administration of RNase1, which has been shown to be more effective when administered prophylactically, since eRNA-induced damage becomes irreversible after the onset of injury (Garnett et al., 2019). Studies in animal models (Kleinert et al., 2016; Bedenbender et al., 2020) highlighted that: (1) in rats undergoing cardiac transplantation, treatment with RNase1 improved graft survival and reduced thrombus formation; (2) in a mouse model of arterial thrombosis, RNase1 reduced clot formation and delayed vascular occlusion. Human RNase1 modulates coagulation and inflammation by degrading eRNA, which would otherwise activate the coagulation cascade (FXII and FXI factors) and promote vasodilation via the release of bradykinin (Garnett et al., 2019). In addition, since eRNA can increase vascular permeability through VEGF and nitric oxide synthesis, RNase1 appears to play a key role in regulating inflammatory and cardiovascular processes, preventing excessive macrophage activation, blood clotting, and increased vascular permeability in response to tissue damage or disease (Garnett et al., 2019).

5.3.4 RNase2 (eosinophil-derived neurotoxin) and RNase3 (eosinophil cationic protein)

EDN (RNase2) is a major secretory protein present in the specific granules of human eosinophils (Rosemberg, 2008). Together with its more charged and cytotoxic paralogue, ECP (RNase3), it is released from eosinophils activated by cytokines and proinflammatory mediators (Kandikattu et al., 2019). Although EDN is also expressed in monocytes, dendritic cells, basophils, and neutrophils, it is best known for its role in eosinophils, where it is found in high concentrations (Tota et al., 2023). EDN is a biomarker for diseases such as asthma exacerbations, cow's milk allergy, eosinophilic esophagitis, and aberrant responses to vaccines (Bahbah et al., 2024). Orthologous sequences of EDN have been found in primates, with considerable interspecific divergence. In particular, EDN from New World primates (e.g., tamarin and night monkey) retains catalytic residues but has a drastically reduced enzymatic activity, attributed to variations in the C-terminal region (Rosemberg et al., 1997). The human genome has a single EDN pseudogene on 14q11.2, which appears to be inactive. In rodents, orthologs of EDN and ECP (mEars 1 and 2 in mice) are encoded in an expanded gene cluster that includes multiple functional genes and pseudogenes (Attery et al., 2017). This birth-death pattern of evolution is like that observed in MHC and immunoglobulin genes, suggesting a role in host defense. EDN is an EAR with important implications in inflammatory and immune diseases (Phipps et al., 2007). Its evolutionary role and interspecies variability make it an attractive target for understanding RNaseA functions in vertebrates.

ECP is a cytotoxic protein released by eosinophils with a key role in the immune response, contributing to host defense against pathogens and modulating inflammatory reactions (Venge et al., 1998). ECP is effective against bacteria, viruses, fungi, and parasites. Its cationic nature allows it to interact with negatively charged microbial surfaces, causing membrane permeabilization and microbial destruction (Pulido et al., 2016). ECP interacts with immune cells such as T, B, mast cells, and dendritic cells, modulating their function and cytokine production (Obeagu, 2024). It can influence the balance between Th1 and Th2 responses, with impacts on the adaptive immune system. ECP is often elevated in inflamed tissues, where it induces apoptosis and stimulates the release of pro-inflammatory cytokines. ECP participates in the pathogenesis of allergic diseases (asthma and allergic rhinitis) by contributing to the inflammatory response to allergens. Recent studies suggest an involvement of ECP in neuroinflammation and neuropsychiatric disorders such as schizophrenia (Obeagu, 2024). It may influence neuronal function, synaptic plasticity, and neuronal viability in the central nervous system (CNS). ECP may induce neurotoxic effects, activating glial cells and promoting the release of inflammatory cytokines (IL-6, TNF-α), amplifying neuroinflammatory processes. ECP may compromise the integrity of the blood-brain barrier, favoring the entry of inflammatory mediators and immune cells into the CNS (Obeagu, 2024). This alteration may contribute to neuroinflammation and influence the progression of neuropsychiatric disorders. Understanding the role of ECP in neuroinflammatory mechanisms could open the way for new therapeutic strategies to modulate inflammation and mitigate the progression of neuropsychiatric diseases.

5.4 RNases4 and 5

5.4.1 RNase4

RNase4 was first isolated from the conditioned medium of an adenocarcinoma cell line and then studied in several laboratories (Fett et al., 1985). However, it remains one of the lesser-studied members of the RNaseA superfamily. The mRNA encoding RNase4 is expressed across various human tissues, as well as in circulating immune cells (Rosemberg et al., 1995). RNase4 is recognized for its antimicrobial properties, particularly against uropathogenic *E. coli*, including multidrug-resistant strains (Sun et al., 2024). Its mRNA is widely expressed in various human tissues and circulating immune cells, with notable production in the kidney's collecting duct, where its expression is regulated by insulin receptor activation and the PI3K/AKT signaling pathway. In the urinary tract, RNase4 is constitutively secreted into urine, playing a critical role in defending against urinary tract infections (Becknell et al., 2019). Antibodies that neutralize the antimicrobial activity of RNase4 promote bacterial replication, further supporting its protective role (Schwartz et al., 2018). Additionally, it is plausible that monocytes and macrophages contribute to the production of RNase4 in the urinary tract.

Beyond the urinary tract, RNase4 has been identified as an important antimicrobial agent in the gut. It limits the growth of *Parasutterella*, a bacterium associated with various health disorders, including inflammatory bowel disease. RNase4 contributes to maintaining gut microbiota balance and has been shown to influence the gut's tryptophan metabolism, a process integral to the pathogenesis of colitis. In patients with inflammatory bowel disease, RNase4 levels are significantly reduced, while *Parasutterella* abundance is elevated, suggesting a link between RNase4 deficiency and dysbiosis in this type of disease (Sun et al., 2024). RNase4 may also play a role in modulating immune responses and maintaining intestinal epithelial cell integrity, although its specific regulatory mechanisms require further exploration. Interestingly, RNase4 is evolutionarily conserved across species, suggesting that its antimicrobial function is crucial for host defense (Schwartz et al., 2018). While its deficiency does not significantly alter the intestinal structure or major antimicrobial peptides, it may influence the maturation of secretory vesicles in Paneth cells, which are critical for gut immunity (Sun et al., 2024). Given its role in antimicrobial defense, RNase4 may have potential clinical applications as a biomarker for diagnosing disease and assessing therapeutic outcomes. Additionally, enhancing RNase4 expression or activity could offer new therapeutic strategies, as it has been identified in milk and shows promise as a functional food additive. Since the RNase4 gene is regulated by the same promoter as that for RNase5 (angiogenin), the question arose as to whether it could also be functionally involved in processes like those in which RNase5 is the main player. For this reason, several comparison studies between the two RNases were conducted. RNase4 shows significant angiogenic and neurogenic activity, which may have implications for neurodegenerative diseases such as amyotrophic lateral sclerosis (ALS) (Becknell et al., 2019). Angiogenesis and neurogenesis are closely linked, sharing common pathways and molecular signals. The angiogenic role of RNase was confirmed through multiple approaches, demonstrating its potential involvement in ALS pathogenesis. Unlike other RNases (e.g., RNaseA), RNase4 showed angiogenic activity, even with a 20-fold reduced catalytic activity in the K40A mutant. This suggests that RNase4 requires some enzymatic activity for angiogenesis but not full catalytic function (Becknell et al., 2019). RNase4, like angiogenin, undergoes nuclear translocation, a key mechanism for its biological effects (Becknell et al., 2019). However, unlike angiogenin, RNase4 does not stimulate rRNA transcription, suggesting different mechanisms of action. Both RNase4 and angiogenin promote neurogenesis by inducing neural differentiation and neurite outgrowth, particularly in P19 and ES cells (Li et al., 2013). RNase4 also offers neuroprotection, comparable to angiogenin, against stress-induced neurofilament damage. RNase4 and angiogenin share high sequence similarity and are regulated by the same promoter, although they have distinct tissue expression patterns. RNase4 is mainly expressed in organs involved in host defense, such as the liver, intestine, testes, and mammary gland (Sun et al., 2024). While it can suppress *Candida albicans* growth (Schwartz et al., 2018), its antimicrobial potency is weaker than angiogenin, and its broader immune role remains unclear. Both enzymes bind nucleic acids, suggesting an immunomodulatory function. In conclusion, although RNase4 and angiogenin share similarities, their distinct roles in immune defense, particularly through nucleic acid interaction, warrant further investigation for potential therapeutic applications.

5.4.2 RNase5 (angiogenin)

RNase5, also known as angiogenin, is a protein composed of 123 amino acids with a structure constituded by three α helices, seven β strands, and a 3_{10} helix and stabilized by three disulfide bridges (in a completely similar way to the above-described fish RNases). It contains three distinct functional sites: a catalytic site involved in tRNA cleavage, a nuclear localization sequence essential for nuclear transport of angiogenin, and a cellular binding site crucial for angiogenic activity (Sheng et al., 2015). Angiogenin is a potent stimulator of angiogenesis, acting mainly on endothelial cells, fibroblasts, and smooth muscle cells (Pyatibratov et al., 2012). Angiogenin promotes cell cycle progression, pushing cells from quiescence to active proliferation (Mao et al., 2024). Its nuclear translocation is essential for proliferation: blocking this process reduces cell growth, making it a key target in antitumor therapy. Under growth conditions, angiogenin interacts with follistatin, PLSCR1, and histone H3 to stimulate rRNA transcription and promote cell growth (Zhu et al., 2013; Gao et al., 2007). Angiogenin is an essential regulator of angiogenesis and cell proliferation, with a significant impact on vascular formation and tumor development (Miyake et al., 2015). Its critical role in rRNA transcription and cell cycle progression makes it a promising target for targeted therapeutic strategies, particularly in oncology. Angiogenin is a protein involved in numerous cellular processes, including proliferation, antiapoptosis, and regulation of oxidative stress. It can bind to p53 and Mdm2, influencing p53 ubiquitination and promoting cell proliferation (Sadagopan et al., 2012). Under normal conditions, its activity is tightly regulated, but in pathological conditions, such as cancer, abnormal levels of angiogenin can promote uncontrolled proliferation of tumor cells through the activation of pathways such as NF-κB, Wnt, and Notch (Cao et al., 2022). Angiogenin is also involved in the regulation of the cellular stress response, particularly in conditions of oxidative stress, where it promotes mitochondrial biogenesis and the antioxidant capacity of cells (Pizzo et al., 2013; Yamasaki et al., 2009). In addition, it plays an important role in the prevention of apoptosis through various mechanisms, including activation of the ILK/PI3K/AKT and Nf-κB pathways, regulation of Bcl-2 proteins, and inhibition of caspase-3 (Sheng et al., 2015; Sadagopan et al., 2012; Saikia et al., 2014). Angiogenin contributes to angiogenesis and inflammation regulation, influencing cell survival and the tumor microenvironment. Its ability to modulate transcription and DNA repair suggests a crucial role in the cellular response to damage and the development of diseases, including cancer and neurodegenerative disorders. Angiogenin plays a crucial role in the regulation of inflammation and neuronal growth. During inflammatory processes, angiogenin modulates the immune response by inhibiting the expression of pro-inflammatory cytokines (IL-6, IL-8, and TNF-α) and promoting antiinflammatory ones (IL-4 and IL-10) (Lee et al., 2014). In addition, it reduces nuclear translocation of NF-κB and contributes to the maintenance of intestinal microbial homeostasis, mitigating gastrointestinal inflammation and facilitating the resolution of the inflammatory process through angiogenesis (Yao et al., 2024). In the nervous system, angiogenin is essential for the development and survival of neurons, influencing the growth of axons and dendrites and improving synaptic connectivity. It can be regulated by neurotrophic factors such as BDNF and intervenes in the response to cellular stress by, among others, the generation of tiRNAs (Yamasaki et al., 2009; Mori et al., 2018). Furthermore, its involvement in neurogenesis and neuronal protection suggests a possible therapeutic role for neurodegenerative diseases such as ALS. Angiogenin plays a critical role in maintaining synaptic stability, which is crucial for the treatment of neurodegenerative diseases (Zhang et al., 2024). Mutations in angiogenin can cause translation dysregulation, affecting synaptic protein dynamics and neuronal activity (Wang et al., 2023). Angiogenin is widely expressed in the nervous system, particularly in neurons of the brain and spinal cord, where it supports neuronal growth, synapse formation, and neuroprotection (Kishikawa et al., 2008). Astrocytes also uptake angiogenin, contributing to neuronal metabolism, waste clearance, and tissue repair (Halpern et al., 2019). In the spinal cord, angiogenin is strongly expressed in glial cells and motor neurons, suggesting a potential role in spinal cord-related neurological diseases. The distribution and function of angiogenin vary across different regions of the nervous system, indicating its diverse regulatory mechanisms. Mutations in the angiogenin gene are linked to neurodegenerative diseases such as ALS, Parkinson's disease (PD), AD, and frontotemporal dementia (FTD) (Kishikawa et al., 2008; Subramanian et al., 2007). These mutations can disrupt its angiogenic activity, impair nuclear translocation, and alter rRNA translation, leading to neuronal damage. Reduced angiogenin levels have been observed in PD patients, contributing to axonal degeneration and neuronal apoptosis (Wu et al., 2007). Studies suggest that angiogenin supplementation can protect neurons from excitotoxicity, endoplasmic reticulum stress, and nutrient deprivation, potentially slowing disease progression (Niedzielska et al., 2016). Animal models have allowed researchers to detect that angiogenin enhances neuronal regeneration, adaptability, and resilience against neurodegenerative damage, highlighting its therapeutic potential for treating neurodegenerative diseases.

Angiogenin is instrumental in maintaining the delicate equilibrium between pro-inflammatory and antiinflammatory cytokines, thereby mitigating inflammation and providing tissue protection. Furthermore, angiogenin inhibits the nuclear translocation of NF-κB, a key mediator of inflammation, while also contributing to the preservation of intestinal microbial homeostasis (Lee et al., 2014).

Genetic mutations in the angiogenin gene have been implicated in the onset of ALS, PD, and FTD, underscoring its vital role in maintaining neuronal integrity (van Es et al., 2011). Angiogenin further protects motor neurons from stress-induced damage and delays neurodegenerative progression through molecular pathways such as the PI3K/Akt signaling cascade (Zhang et al., 2024). By mitigating oxidative stress and inflammation, angiogenin enhances cerebrovascular health, providing a therapeutic benefit in the context of vascular injuries and dysfunction. Also noteworthy is its key role in the cellular stress response and restoration of homeostasis. This function is linked to the ability of angiogenin to cleave tRNAs during stress, generating stress-induced tRNAs (tiRNAs) that regulate cellular processes such as translation repression, inhibition of apoptosis, and stress granule formation (Culurciello et al., 2022). tiRNAs help prioritize gene translation in response to stress while suppressing housekeeping genes (Barros et al., 2023).

The activity of angiogenin is tightly controlled by RNH1, but during stress, angiogenin accumulates in the cytoplasm, where it cleaves tRNAs and contributes to the assembly of stress granules (Buchan et al., 2009). Stress granules, which are formed when translation stops during stress, serve as hubs for mRNA regulation and cell-fate decisions (Emara et al., 2010). The role of angiogenin in stress granule formation and restoration of homeostasis is essential for the adaptive response to stress (Ivanov et al., 2011), and it is well known in several cell lines (Buchan et al., 2009; Culurciello et al., 2024). Recently, angiogenin's involvement in the cell stress response was also highlighted in keratinocytes (Culurciello et al., 2022, 2024), opening a new line of study in an organ (the skin) that until recently was incredibly unexplored regarding the functional properties of angiogenin. The role of angiogenin in the keratinocytes' stress response is suggestive of both therapeutic and diagnostic applications in skin diseases.

In summary, angiogenin is a key regulator of cellular stress responses, promoting survival through translational reprogramming and stress granule formation, with broad implications in cancer, neurodegenerative diseases, immune responses, and skin health.

5.5 RNases6, 7, and 8

5.5.1 RNase6

Human RNase6 is a cationic antimicrobial protein that plays a critical role in defending the urinary tract, especially against uropathogenic *E. coli*, a common cause of urinary tract infections (Ruiz-Rosado et al., 2023). RNase6 exhibits potent bactericidal activity against both Gram-positive and Gram-negative bacteria at micromolar concentrations

(Becknell et al., 2015). The antibacterial activity of RNase6 is likely due to its high cationicity (pI of 9.49), which facilitates binding to the negatively charged microbial membranes, leading to membrane permeabilization and subsequent cell death (Anguita et al., 2024; Lee et al., 2019). RNase6 acts as a novel leukocytic antimicrobial peptide, particularly active during infection when the body's first-line defenses are compromised. RNase6 shares similar antimicrobial properties with RNase7, but unlike RNase7, its expression is notably upregulated during pyelonephritis, an infection of the kidney (Becknell et al., 2015).

RNase6 is generally localized within cytoplasmic granular structures of myeloid cells, such as neutrophils and macrophages (Ruiz-Rosado et al., 2023). This subcellular localization suggests that it may function in intracellular compartments like the phagosome or lysosome, where it may degrade microbial RNA or directly contribute to pathogen destruction, particularly within phagocytosing macrophages (Lee et al., 2019).

When compared to other antimicrobial RNases, such as RNase3 and RNase7, RNase6 shares similar microbicidal activity against *E. coli*. However, RNase6 and other RNases do not exhibit synergistic effects, indicating their activities may be additive rather than cooperative (Becknell et al., 2015).

5.5.2 RNase7

RNase7 was initially isolated from human skin (stratum corneum) and is highly cationic (Rademacher, 2016). It is expressed in keratinocytes and other tissues, including the heart, liver, kidney, pharynx, and tonsils (Rademacher et al., 2019). Constitutively expressed in skin and present in fetal skin, suggesting a role in early microbial defense and development (Schuster et al., 2013). RNase7 expression is induced by proinflammatory cytokines (IL-17A, IFN-γ, and IL-1β) via pathways like STAT3, TAK-1, and MAPK (Simanski et al., 2013). Its expression increases in response to pathogens (*S. aureus, Pseudomonas aeruginosa, Enterococcus faecium*, and *Trichophyton rubrum*) (Rademacher et al., 2019). Commensal bacteria (*Staphylococcus epidermidis, Corynebacterium amycolatum*) also induce RNase7, enhancing skin defense (Rademacher et al., 2021). Epidermal growth factor receptor (EGFR) is crucial for RNase7 induction, explaining susceptibility to infections in patients on anti-EGFR therapy (Walter et al., 2017). *Borrelia burgdorferi* induces RNase7, but tick saliva proteins can inhibit this response, aiding bacterial evasion (Anderson et al., 2021). UV-B radiation and niacinamide also upregulate RNase7 expression, contributing to skin immunity (Gläser et al., 2009). RNase7 has broad-spectrum antimicrobial activity against Gram-positive and Gram-negative bacteria, fungi, and yeast. Functional studies show that blocking RNase7 increases pathogen outgrowth, emphasizing its role in skin defense (Spencer et al., 2013). High baseline levels of RNase7 in healthy individuals correlate with protection against *S. aureus* infections. RNase7 binds to RNH1, which reduces its antimicrobial activity by preventing cell wall binding (Sun et al., 2022). This inhibition is likely due to steric interference rather than blocking its RNase function. RNH1 is present in suprabasal layers but absent in the stratum corneum (Rademacher et al., 2019). RNH1 is degraded in the stratum corneum, which allows RNase7 to function freely as an antimicrobial agent (Spencer et al., 2014). RNase7 binds negatively charged bacterial membranes and disrupts them via electrostatic interactions. Moreover, it binds lipopolysaccharide and peptidoglycan in *S. aureus* and *E. coli* and recognizes OprI, an outer membrane protein in *P. aeruginosa*, facilitating bacterial membrane permeability (Rademacher, 2016). RNase7 presents a dual action against *C. albicans*: first, it acts as a membrane-lytic protein and then degrades cellular RNA, contributing to fungal killing (Salazar et al., 2016). Immunomodulatory functions of RNase7 comprise enhancing IFN-α response in plasmacytoid dendritic cells, triggering a potent interferon-α response stronger than LL-37 or hBD-2/-3, and providing protection against HSV-1 infection despite no direct antiviral activity (Rademacher et al., 2019). RNase7 reduces Th2 cytokine release (IL-4, IL-5, and IL-13) by inhibiting GATA3 transcription factor activation (Kopfnagel et al., 2017). RNase7 binds self-DNA, promoting CXCL10 (IP-10) release and an IFN-β-dependent antiviral response (Kopfnagel et al., 2020). It may act as an alarm, detecting skin barrier disruptions and triggering innate immune responses. RNase7 is upregulated in psoriasis and atopic dermatitis, suggesting involvement in skin inflammation (Dong et al., 2024). It may degrade host RNA from damaged cells, helping to control RNA-mediated inflammation during skin injury (Rademacher et al., 2019).

5.5.3 RNase8

RNase8 is uniquely expressed in the human placenta, with no detectable expression in any of the other tissues examined, a feature that sets it apart from most other RNaseA genes, which are typically expressed in multiple tissues (Zhang et al., 2002). Phylogenetic analysis suggests that RNase8 is most closely related to RNase7, with the two genes likely having arisen from a recent gene duplication event in primates. This duplication event appears to have occurred approximately 44 million years ago, shortly before the divergence of New World and Old World monkeys (Zhang et al., 1998). Recombinant human RNase8 exhibits very low ribonucleolytic activity compared to other RNaseA superfamily members and lacks the antiviral and antibacterial activities characteristic of many RNaseA homologs (Sun et al., 2022). Studies reveal that several primate orthologs of RNase8 have been deactivated by frameshift deletions or functionally disruptive point mutations at crucial structural or catalytic residues (Zhang et al., 2002). This species-specific deactivation suggests that the function of RNase8 may be more specialized and limited to humans, potentially fulfilling a role that is not yet fully understood in evolutionary terms. Tissue-specific expression of RNase8 in the placenta suggests that RNase8 may play a specialized role in placental function, possibly as an angiogenic factor or in the regulation of placental development. Interestingly, the gene encoding RNase8 contains a codon encoding a cysteine at position 66, which may play a role in the enzyme's structure. This cysteine is atypical for RNaseA family members, as it may not participate in the usual disulfide bond formation seen in many RNaseA enzymes. It is possible that this cysteine can form interchain disulfide bonds, potentially with the cysteine at position 23, contributing to the structural uniqueness of RNase8. However, the precise structural features of RNase8 await further investigation, including crystallographic analysis to clarify its three-dimensional conformation.

In summary, human RNase8 represents a unique member of the superfamily with distinct characteristics. Its exclusive expression in the placenta, rapid evolutionary rate, and the presence of atypical structural features suggest that RNase8 fulfills a specialized function, potentially related to placental development or other tissue-specific processes (Zhang et al., 2002).

RNase8 is also expressed in the male genital reproductive tract, where it could play a role in reproductive and immune functions.

5.5.4 Noncanonical human RNase

Five additional RNase genes—RNase9 through RNase13—have been identified within the human genome (Cho et al., 2005). The physiological functions of the noncanonical RNases remain largely unexplored. From the few reports available, it has emerged that RNase9 and RNase10 are expressed in the human epididymis, suggesting a role in the male reproductive tract (Cheng et al., 2009). Recombinant human RNase9 has demonstrated potential antibacterial properties, and RNase10 (in mice) seems to be essential for sperm maturation (Cheng et al., 2009). In contrast, the biological functions of RNase11, RNase12, and RNase13 remain widely unknown.

Notably, all 13 functional RNase genes are clustered within a ~500-kilobase region on the long arm of chromosome 14 (14q11.2) in the human genome (Cho et al., 2005).

Noncanonical RNases share key features with canonical ones, like conserved disulfide bonds and a secretion signal, but lack the catalytic triad signature and likely lack ribonucleolytic activity due to mutations or insertions. Despite this, some, such as hRNase9, exhibit bactericidal activity against *E. coli* through membrane destabilization, independent of ribonucleolytic action (Liu et al., 2008). Noncanonical RNases are primarily expressed in the male reproductive tract, with hRNase9 and hRNase10 playing roles in sperm maturation and being regulated by androgens (Liu et al., 2008; Krutskikh et al., 2012). Though little is known about their full physiological roles, these RNases may fulfill functions beyond ribonucleolytic activity. Evidence suggests that ribonucleolytic activity-independent functions are common in the RNaseA superfamily. One such example is hRNase5/ANG, which acts as an EGFR ligand and a serum biomarker for predicting clinical responses to EGFR-targeted therapies (Wang et al., 2018).

RNases9–13 share the common feature of lacking ribonucleolytic activity due to mutations or alterations in the catalytic triad, but still play significant roles in immune defense and reproductive processes (Lee et al., 2019).

RNase9 exhibits bactericidal activity against *E. coli* by destabilizing bacterial membranes, independent of ribonucleolytic activity. It is androgen-regulated and contributes to sperm maturation, as shown by impaired sperm maturation in RNase9-knockout mice (Westmuckett et al., 2014). RNase10 shares similar functions, also being androgen-regulated and associated with sperm maturation and immune defense (Shaffer et al., 2024). Its precise functions are still under investigation, but it is thought to play a role in the male reproductive system. RNase11 is expressed in the testis and is believed to contribute to immune defense and reproductive functions, including sperm maturation. Like the others, its activity is likely regulated by androgens, but further research is needed to elucidate its roles. RNase12 is less characterized but is believed to function similarly to other noncanonical RNases, with involvement in immune and reproductive processes, particularly within the testis and epididymis. It is also androgen-regulated (Shaffer et al., 2024). RNase13 is expressed in the male reproductive tract and likely has nonribonucleolytic roles, contributing to immune defense and reproductive health (Lee et al., 2019). However, its specific functions remain to be fully defined.

In summary, while these RNases lack catalytic activity, they are involved in various immune and reproductive functions, suggesting they may play nonclassical roles in antimicrobial defense and male reproductive health. The 3D models of five non-canonical RNases are shown in Fig. 5.3 and compared with those of angiogenin and RNase1.

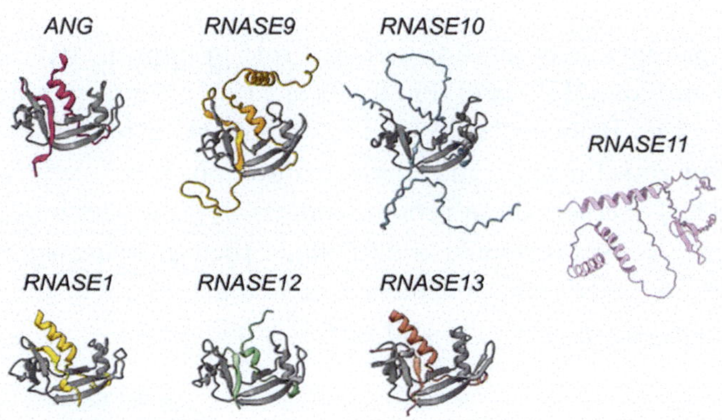

Figure 5.3 3D fold prediction for the five noncanonical human RNases. Noncanonical fold prediction for the five noncanonical human RNases (RNase9_P60153, RNase10_Q5GAN6, RNase11_Q8TAA1, RNase12_Q5GAN4, RNase13_Q5GAN3), human angiogenin (ANG), and human RNase1. RNAse11, the only one not adopting a canonical RNase fold, is on the right (*RNase*, ribonuclease).

5.6 Conclusions

The RNaseA superfamily represents a remarkable instance of evolutionary innovation and functional diversification within vertebrates. Initially identified in the bovine pancreas, RNaseA has long served as a cornerstone in the field of protein biochemistry, providing critical insights into enzyme structure, folding, and catalytic mechanisms. RNaseA homologs subsequently discovered across a wide range of species, including fish, amphibians, reptiles, birds, and mammals, gave the superfamily a remarkable adaptive flexibility. Its functional repertoire spans a diverse array of physiological processes, including angiogenesis, host defense, selective cytotoxicity, and immunoregulation.

Structural variations among members of the RNaseA superfamily, such as the presence or absence of disulfide bonds, as well as differences in catalytic activity and inhibitor sensitivity, reflect evolutionary adaptations driven by environmental pressures and organismal physiology. From foundational research to potential therapeutic applications, the RNaseA superfamily continues to provide invaluable insights into molecular evolution, protein function, and the development of biomedical therapies. The clinical exploration of ONC, for instance, underscores its promise in therapeutic contexts.

RNaseA homologs from humans, both canonical and noncanonical enzymes, exhibit distinct, specialized functions beyond their role in RNA degradation. Canonical RNases (Jones, 1920; Dubos, 1937; Dubos and Thompson, 1938; Kunitz, 1940; Anfinsen, 1959; Groves, 1966; Scheraga, 1967; Smyth et al., 1963) are critical to processes such as host defense, inflammation, angiogenesis, neuroprotection, and immune modulation, with distinct patterns of tissue-specific expression and activity. RNase1, for example, plays a pivotal role in regulating inflammation and coagulation, while RNases2 and 3 are integral to eosinophil-mediated immunity. RNases4 and 5 contribute to neurogenesis and vascular growth, whereas RNases6 and 7 act as antimicrobial agents within the urinary tract and skin. RNase8, which is uniquely expressed in the placenta, may serve an essential role in reproductive functions.

In contrast, noncanonical RNases (Findlay et al., 1961; Hantgan et al., 1974; Raines, 1998; Beintema, 1998; Beintema et al., n.d.), which lack enzymatic activity, still play key roles in antimicrobial defense and sperm maturation, particularly within the male reproductive tract. Together, these proteins underscore the extraordinary functional versatility of the RNaseA superfamily and highlight their growing significance in areas such as immunity, reproduction, and disease pathology. As such, they offer exciting prospects for therapeutic intervention and further exploration.

References

Abaturov, L.V., et al., 2012. Hydrogen exchange and proteolytic degradation of ribonuclease A. The local splitting of the native structure and the conformation of loop segments. Biophysics 57 (1), 35–54. Available at https://doi.org/10.1134/S0006350912010022.

Anderson, C., et al., 2021. The brilliance of *Borrelia:* mechanisms of host immune evasion by lyme disease-causing spirochetes. Pathogens 10 (3), 281. Available at https://doi.org/10.3390/pathogens10030281.

Anfinsen, C.B.P.I., 1959. Crystalline pancreatic ribonuclease. Introductory remarks. Annals of the New York Academy of Sciences 81, 513–514. Available at https://doi.org/10.1111/j.1749-6632.1959.tb49332.x.

Anguita, R., et al., 2024. A common polymorphism in *RNASE6* impacts its antimicrobial activity toward uropathogenic *Escherichia coli*. International Journal of Molecular Sciences 25 (1), 604. Available at https://doi.org/10.3390/ijms25010604.

Ardelt, B., et al., 2007. Cytostatic and cytotoxic properties of amphinase: a novel cytotoxic ribonuclease from *Rana pipiens* oocytes. Cell Cycle (Georgetown, Tex.) 6, 3097–3102.

Ardelt, W., et al., 1991. Amino acid sequence of an anti-tumor protein from *Rana pipiens* oocytes and early embryos. Homology to pancreatic ribonucleases. The Journal of Biological Chemistry 266, 245–251.

Arnold, U., 2014. Stability and folding of amphibian ribonuclease a superfamily members in comparison with mammalian homologues. The FEBS Journal 281 (16), 3559–3575. Available at https://doi.org/10.1111/febs.12891.

Attery, A., et al., 2017. Mouse eosinophil associated ribonucleases: mechanism of cytotoxic, antibacterial and antiparasitic activities. International Journal of Biological Macromolecules 94 (Pt A), 445–450. Available at https://doi.org/10.1016/j.ijbiomac.2016.10.041.

Bahbah, W.A., et al., 2024. Serum eosinophil-derived neurotoxin: a new promising biomarker for cow's milk allergy diagnosis. Pediatric Research 96 (7), 1812–1821. Available at https://doi.org/10.1038/s41390-024-03260-x.

Barnard, E.A., 1969. Biological function of pancreatic ribonuclease. Nature 221 (5178), 340–344. Available at https://doi.org/10.1038/221340a0.

Barros, G.C., et al., 2023. The central role of translation elongation in response to stress. Biochemical Society Transactions 51 (3), 959–969. Available at https://doi.org/10.1042/BST20220584.

Becknell, B., et al., 2019. The responses of the ribonuclease a superfamily to urinary tract infection. Frontiers in Immunology 10, 2786. Available at https://doi.org/10.3389/fimmu.2019.02786.

Becknell, B., et al., 2015. Ribonucleases 6 and 7 have antimicrobial function in the human and murine urinary tract. Kidney International 87 (1), 151–161. Available at https://doi.org/10.1038/ki.2014.268.

Bedenbender, K., et al., 2020. Endothelial ribonuclease 1 in cardiovascular and systemic inflammation. Frontiers in Cell and Developmental Biology 8, 576491. Available at https://doi.org/10.3389/fcell.2020.576491.

Beintema, JJ, Gaastra, W, Lenstra, JA, Welling, GW, Fitch, WM, 1977. The molecular evolution of pancreatic ribonuclease. J Mol Evol 10 (1), 49–71. Available at 10.1007/BF01796134.

Beintema, J.J. et al., 1997. Evolution of vertebrate ribonucleases: ribonuclease A superfamily. In: D'Alessio, G., Riordan, J.F. (Eds.), Ribonucleases: Structures and Functions. Academic Press, San Diego, pp. 245–269.

Beintema, J.J., et al., 1977. The molecular evolution of pancreatic ribonuclease. Journal of Molecular Evolution, 10, 49–71.

Beintema, J.J., 1986. Evolutionary role of posttranslational modifications of proteins, as illustrated by the glycosylation characteristics of the digestive enzyme pancreatic ribonuclease. Journal of Molecular Evolution 24, 118–120.

Beintema, J.J., 1998. The ribonuclease a superfamily: General discussion. Cellular and Molecular Life Sciences CMLS 54, 825–832. Available at https://doi.org/10.1007/s000180050211.

Brachet, J., 1947. The metabolism of nucleic acids during embryonic development. Cold Spring Harbor Symposia on Quantitative Biology 12, 18–27.

Brown, J.E., et al., 1969. Conformational studies of a series of overlapping peptides from ribonuclease and their relationship to the protein structure. Biochemistry 8 (7), 2876–2879. Available at https://doi.org/10.1021/bi00835a027.

Buchan, J.R., et al., 2009. Eukaryotic stress granules: the ins and outs of translation. Molecular Cell 36 (6), 932–941. Available at https://doi.org/10.1016/j.molcel.2009.11.020.

Cao, M., et al., 2022. Classical angiogenic signaling pathways and novel anti-angiogenic strategies for colorectal cancer. Current Issues in Molecular Biology 44 (10), 4447–4471. Available at https://doi.org/10.3390/cimb44100305.

Cheng, G.Z., et al., 2009. Human ribonuclease 9, a member of ribonuclease a superfamily, specifically expressed in epididymis, is a novel sperm-binding protein. Asian Journal of Andrology 11 (2), 240–251. Available at https://doi.org/10.1038/aja.2008.30.

Cho, S., et al., 2005. The ribonuclease a superfamily of mammals and birds: identifying new members and tracing evolutionary histories. Genomics 85 (2), 208–220. Available at https://doi.org/10.1016/j.ygeno.2004.10.008.

Cho, S., et al., 2007. Zebrafish ribonucleases are bactericidal: implications for the origin of the vertebrate RNase a superfamily. Molecular Biology and Evolution 24 (5), 1259–1268. Available at https://doi.org/10.1093/molbev/msm047.

Cormier, S.A., et al., 2001. Mouse eosinophil-associated ribonucleases: a unique subfamily expressed during hematopoiesis. Mammalian Genome Official Journal of the International Mammalian Genome Society 12 (5), 352–361. Available at https://doi.org/10.1007/s003350020007.

Cormier, S.A., et al., 2002. T(H)2-mediated pulmonary inflammation leads to the differential expression of ribonuclease genes by alveolar macrophages. American Journal of Respiratory Cell and Molecular Biology 27 (6), 678–687. Available at https://doi.org/10.1165/rcmb.4882.

Culurciello, R., et al., 2022. Protective effects of recombinant human angiogenin in keratinocytes: new insights on oxidative stress response mediated by RNases. International Journal of Molecular Sciences 23 (15), 8781. Available at https://doi.org/10.3390/ijms23158781.

Culurciello, R., et al., 2024. Tailoring the stress response of human skin cells by substantially limiting the nuclear localization of angiogenin. Heliyon 10 (3), e24556. Available at https://doi.org/10.1016/j.heliyon.2024.e24556.

D'Errico, G., 2011. Enforcing the positive charge of N-termini enhances membrane interaction and antitumor activity of bovine seminal ribonuclease. Biochimica et Biophysica Acta 1808 (12), 3007–3015. Available at https://doi.org/10.1016/j.bbamem.2011.08.009.

Darzynkiewicz, Z., et al., 1988. Cytostatic and cytotoxic effects of Pannon (P-30 protein), a novel anticancer agent. Cell and Tissue Kinetics 21, 169–182. Available at https://doi.org/10.1111/j.1365-2184.1988.tb00855.x.

Domachowske, J.B., et al., 1998. Recombinant human eosinophil-derived neurotoxin/RNase 2 functions as an effective antiviral agent against respiratory syncytial virus. The Journal of Infectious Diseases 177 (6), 1458–1464. Available at https://doi.org/10.1086/515322.

Dong, S., et al., 2024. Skin barrier-inflammatory pathway is a driver of the psoriasis-atopic dermatitis transition. Frontiers in Medicine (Lausanne) 11, 1335551. Available at https://doi.org/10.3389/fmed.2024.1335551.

Dubos, R.J., 1937. The decomposition of yeast nucleic acid by a heat resistant enzyme. Science (New York, N.Y.) 85 (2214), 549–550. Available at https://doi.org/10.1126/science.85.2214.549.

Dubos, R.J., Thompson, R.H.S., 1938. The decomposition of yeast nucleic acid by a heat-resistant enzyme. The Journal of Biological Chemistry 124 (2), 501–510.

Emara, M.M., et al., 2010. Angiogenin-induced tRNA-derived stress-induced RNAs promote stress-induced stress granule assembly. The Journal of Biological Chemistry 285 (14), 10959–10968. Available at https://doi.org/10.1074/jbc.M109.077560.

Ferguson, R., 2019. The catalytic activity and secretion of zebrafish RNases are essential for their in vivo function in motor neurons and vasculature. Scientific Reports 9 (1), 1107. Available at https://doi.org/10.1038/s41598-018-37140-2.

Fett, J.W., et al., 1985. Isolation and characterization of angiogenin, an angiogenic protein from human carcinoma cells. Biochemistry 24, 5480–5486. Available at https://doi.org/10.1021/bi00341a030.

Finamore, F.J., 1955. A study of ribonuclease and desoxyribonuclease during the early developmental stages of *Rana pipiens*. Experimental Cell Research 8, 533–542.

Findlay, D., et al., 1961. The active site and mechanism of action of bovine pancreatic ribonuclease. Nature 27 (190), 781–784. Available at https://doi.org/10.1038/190781a0.

Fischer, S., 2011. Expression and localisation of vascular ribonucleases in endothelial cells. 105(2), 345-355. doi: 10.1160/TH10-06-0345.
Futami, J., et al., 1997. Tissue-specific expression of pancreatic-type RNases and RNase inhibitor in humans. DNA and Cell Biology 16, 413–419.
Gao, X., et al., 2007. Identification and characterization of follistatin as a novel angiogenin-binding protein. FEBS Letters 581 (28), 5505–5510. Available at https://doi.org/10.1016/j.febslet.2007.10.059.
Garnett, E.R., et al., 2019. Phenotype of ribonuclease 1 deficiency in mice. RNA (New York, N.Y.) 25 (8), 921–934. Available at https://doi.org/10.1261/rna.070433.119.
Gläser, R., et al., 2009. UV-B radiation induces the expression of antimicrobial peptides in human keratinocytes in vitro and in vivo. The Journal of Allergy and Clinical Immunology 123 (5), 1117–1123. Available at https://doi.org/10.1016/j.jaci.2009.01.043.
Goo, S.M., et al., 2013. The expansion and functional diversification of the mammalian ribonuclease a superfamily epitomizes the efficiency of multigene families at generating biological novelty. Genome Biology and Evolution 5 (11), 2124–2140. Available at https://doi.org/10.1093/gbe/evt161.
Goparaju, C.M., Blasberg, J.D., Volinia, S., Palatini, J., Ivanov, S., Donington, J.S., et al., 2011. Onconase mediated NFKbeta downregulation in malignant pleural mesothelioma. Oncogene 30, 2767–2777.
Graham, D.J., et al., 2011. On the information expressed in enzyme structure: more lessons from ribonuclease A. Molecular Diversity 15 (3), 769–779. Available at https://doi.org/10.1007/s11030-011-9307-4.
Groves, M.L., 1966. Crystalline ribonuclease from bovine milk. Journal of Dairy Science 49 (2), 204–205. Available at https://doi.org/10.3168/jds.S0022-0302(66)87829-3.
Hacke, M., et al., 2013. Consequences of proline-to-alanine substitutions for the stability and refolding of onconase. The FEBS Journal 280, 4454–4462.
Haigis, M.C., Kurten, E.L., Raines, R.T., 2003. Ribonuclease inhibitor as an intracellular sentry. Nucleic Acids Research 31, 1024–1032.
Halpern, M., et al., 2019. Examining the relationship between astrocyte dysfunction and neurodegeneration in ALS using hiPSCs. Neurobiology of Disease 132, 104562. Available at https://doi.org/10.1016/j.nbd.2019.104562.
Hantgan, R.R., et al., 1974. Pathways of folding of reduced bovine pancreatic ribonuclease. Biochemistry 13 (17), 3421–3431. Available at https://doi.org/10.1021/bi00714a0.
Hayano, K., et al., 1993. Characterization of poly-C-preferential ribonuclease from chicken liver. Journal of Biochemistry 114, 156–162.
Hoang, T.T., et al., 2018. A human ribonuclease variant and ERK-pathway inhibitors exhibit highly synergistic toxicity for cancer cells. Molecular Cancer Therapeutics 17 (12), 2622–2632. Available at https://doi.org/10.1158/1535-7163.MCT-18-0724.
Ishihara, K., 2003. Preparation of recombinant rat eosinophil-associated ribonuclease-1 and -2 and analysis of their biological activities. Biochimica et Biophysica Acta 1638 (2), 164–172. Available at https://doi.org/10.1016/s0925-4439(03)00077-2.
Ivanov, P., Emara, M.M., Villen, J., Gygi, S.P., Anderson, P., 2011. Angiogenin-induced tRNA fragments inhibit translation initiation. Molecular Cell 43 (4), 613–623. Available at https://doi.org/10.1016/j.molcel.2011.06.022.
Johnson, R.J., et al., 2007. Inhibition of human pancreatic ribonuclease by the human ribonuclease inhibitor protein. Journal of Molecular Biology 368, 434–449. Available at https://doi.org/10.1016/j.jmb.2007.02.005.
Jones, W., 1920. The action of boiled pancreas extract on yeast nucleic acid. Journal of American Physiology 52, 203–207. Available at https://doi.org/10.1152/ajplegacy.1920.52.1.203.
Kandikattu, H.K., et al., 2019. Synergy of interleukin (IL)-5 and IL-18 in eosinophil mediated pathogenesis of allergic diseases. Cytokine & Growth Factor Reviews 47, 83–98. Available at https://doi.org/10.1016/j.cytogfr.2019.05.003.
Katekaew, S., et al., 2010. Structure of the newly found green turtle egg-white ribonuclease. Acta Crystallographica. Section F, Structural Biology and Crystallization Communications 66 (Pt 7), 755–759. Available at https://doi.org/10.1107/S1744309110016751.

Kishikawa, H., et al., 2008. Targeting angiogenin in therapy of amyotropic lateral sclerosis. Expert Opinion on Therapeutic Targets 12 (10), 1229–1242. Available at https://doi.org/10.1517/14728222.12.10.1229.

Kleinert, E., et al., 2016. Ribonuclease (RNase) prolongs survival of grafts in experimental heart transplantation. Journal of the American Heart Association 5 (5), e003429. Available at https://doi.org/10.1161/JAHA.116.003429.

Klenova, E.M., et al., 1992. Isolation of a cDNA clone encoding the RNasesuperfamily-related gene highly expressed in chicken bone-marrow cells. Biochemical and Biophysical Research Communications 185, 231–239.

Kopfnagel, V., et al., 2017. RNase 7 downregulates TH2 cytokine production by activated human T cells. Allergy 72 (11), 1694–1703. Available at https://doi.org/10.1111/all.13173.

Kopfnagel, V., et al., 2020. RNase 7 promotes sensing of self-DNA by human keratinocytes and activates an antiviral immune response. The Journal of Investigative Dermatology 140 (8), 1589–1598. Available at https://doi.org/10.1016/j.jid.2019.09.029 e3.

Kotchetkov, R., et al., 2001. Selective activity of BS-RNase against anaplastic thyroid cancer. Anticancer Research 21 (2A), 1035–1042.

Krutskikh, A., et al., 2012. Epididymal protein Rnase10 is required for post-testicular sperm maturation and male fertility. The FASEB Journal 26 (10), 4198–4209. Available at https://doi.org/10.1096/fj.12-205211.

Kunitz, M., 1940. Crystalline ribonuclease. The Journal of General Physiology 24 (1), 15–32. Available at https://doi.org/10.1085/jgp.24.1.15.

Lee, F.S., et al., 1989. Tight-binding inhibition of angiogenin and ribonuclease a by placental ribonuclease inhibitor. Biochemistry 28 (1), 225–230. Available at https://doi.org/10.1021/bi00427a031.

Lee, H.H., et al., 2021. Human ribonuclease 1 serves as a secretory ligand of ephrin A4 receptor and induces breast tumor initiation. Nat Commun 12 (1), 2788. Available at https://doi.org/10.1038/s41467-021-23075-2.

Lee, H.H., et al., 2019. Functional roles of the human ribonuclease a superfamily in RNA metabolism and membrane receptor biology. Molecular Aspects of Medicine 70, 106–116. Available at https://doi.org/10.1016/j.mam.2019.03.003.

Lee, J.E., et al., 2005. Cytotoxicity of bovine seminal ribonuclease: monomer versus dimer. Biochemistry 44 (48), 15760–15767. Available at https://doi.org/10.1021/bi051668z.

Lee, S.H., et al., 2014. Angiogenin reduces immune inflammation via inhibition of TANK-binding kinase 1 expression in human corneal fibroblast cells. Mediators of Inflammation 2014, 861435. Available at https://doi.org/10.1155/2014/861435.

Leland, P.A., et al., 1998. Ribonuclease a variants with potent cytotoxic activity. Proceedings of the National Academy of Sciences of the United States of America 95, 10407–10412.

Levy, C.C., et al., 1980. The purification and properties of chicken liver RNase: an enzyme which is useful in distinguishing between cytidylic and uridylic acid residues. The Journal of Biological Chemistry 255 (5), 2153–2159.

Li, S., et al., 2013. Ribonuclease 4 protects neuron degeneration by promoting angiogenesis, neurogenesis, and neuronal survival under stress. Angiogenesis 16 (2), 387–404. Available at https://doi.org/10.1007/s10456-012-9322-9.

Liu, J., et al., 2008. Cloning, expression and location of RNase9 in human epididymis. BMC Reserach Notes 1, 111. Available at https://doi.org/10.1186/1756-0500-1-111.

Malicka-Błaszkiewicz, M., Kubicz, A., 1979. Partial purification and some properties of a liver alkaline ribonuclease from the frog *Rana esculenta*. Acta Biochimica Polonica 26, 275–283.

Mao, M., et al., 2024. Research progress on the structure, function, and use of angiogenin in malignant tumours. Heliyon 10 (9), e30654. Available at https://doi.org/10.1016/j.heliyon.2024.e30654.

Maor, D., et al., 1978. Alteration of human serum ribonuclease activity in malignancy. CRC Critical Reviews in Clinical Laboratory Sciences 10, 89–111.

Mikulski, S.M., et al., 2002. Phase II trial of a single weekly intravenous dose of ranpirnase in patients with unresectable malignant mesothelioma. Journal of Clinical Oncology Official Journal of the American Society of Clinical Oncology 20, 274–281.

Miyake, M., et al., 2015. Angiogenin promotes tumoral growth and angiogenesis by regulating matrix metallopeptidase-2 expression via the ERK1/2 pathway. Oncogene 34 (7), 890–901. Available at https://doi.org/10.1038/onc.2014.2.

Mohamed, I.S.E., et al., 2022. Bovine pancreatic RNase A: an insight into the mechanism of antitumor activity in vitro and in vivo. Pharmaceutics 14 (6), 1173. Available at https://doi.org/10.3390/pharmaceutics14061173.

Monti, D.M., et al., 2009. Characterization of the angiogenic activity of zebrafish ribonucleases. The FEBS Journal 276 (15), 4077–4090. Available at https://doi.org/10.1111/j.1742-4658.2009.07115.x.

Moreau, J.M., et al., 2003. Diminished expression of an antiviral ribonuclease in response to pneumovirus infection in vivo. Antiviral Research 59 (3), 181–191. Available at https://doi.org/10.1016/s0166-3542(03)00111-6.

Mori, A., et al., 2018. Brain-derived neurotrophic factor induces angiogenin secretion and nuclear translocation in human umbilical vein endothelial cells. Pathology, Research and Practice 214 (4), 521–526. Available at https://doi.org/10.1016/j.prp.2018.02.013.

Mosimann, S.C., et al., 1994. Refined 1.7A X-ray crystallographic structure of P-30 protein, an amphibian ribonuclease with anti-tumor activity. Journal of Molecular Biology 236 (4), 1141–1153. Available at https://doi.org/10.1016/0022-2836(94)90017-5.

Nagano, H., Kiuchi, H., Abe, Y., Shukuya, R., 1976. Purification and properties of an alkaline ribonuclease from the hepatic cytosol fraction of bullfrog, *Rana catesbeiana*. Journal of Biochemistry 80, 19–26.

Najem, M.Y., et al., 2024. Extracellular RNA induces neutrophil recruitment via toll-like receptor 3 during venous thrombosis after vascular injury. Journal of the American Heart Association 13 (15), e034492. Available at https://doi.org/10.1161/JAHA.124.034492.

Nakano, T., et al., 1992. Identification of genes differentially expressed in two types of v-myb-transformed avian myelomonocytic cells. Oncogene 7 (3), 527–534.

Newton, D.L., et al., 1998. Single amino acid substitutions at the N-terminus of a recombinant cytotoxic ribonuclease markedly influence biochemical and biological properties. Biochemistry 37 (15), 5173–5183. Available at https://doi.org/10.1021/bi972147h.

Niedzielska, E., et al., 2016. Oxidative stress in neurodegenerative diseases. Molecular Neurobiology 53 (6), 4094–4125. Available at https://doi.org/10.1007/s12035-015-9337-5.

Nitto, T., et al., 2005. Characterization of a ribonuclease gene and encoded protein from the reptile, Iguana iguana. Gene 352, 36–44. Available at https://doi.org/10.1016/j.gene.2005.03.002.

Nitto, T., et al., 2004. Characterization of the divergent eosinophil ribonuclease, mEar 6, and its expression in response to *Schistosoma mansoni* infection in vivo. Genes and Immunity 5 (8), 668–674. Available at https://doi.org/10.1038/sj.gene.6364143.

Nitto, T., et al., 2006. Evolution and function of leukocyte RNase A ribonucleases of the avian species, *Gallus gallus*. The Journal of Biological Chemistry 281 (35), 25622–25634. Available at https://doi.org/10.1074/jbc.M604313200.

Obeagu, E.I., 2024. Exploring the role of eosinophil cationic protein (ECP) in schizophrenia: insights and implications. Medicine 103 (22), e38380. Available at https://doi.org/10.1097/MD.0000000000038380.

Phipps, S., et al., 2007. Eosinophils contribute to innate antiviral immunity and promote clearance of respiratory syncytial virus. Blood 110 (5), 1578–1586. Available at https://doi.org/10.1182/blood-2007-01-071340.

Pizzo, E., et al., 2011. A new RNase sheds light on the RNase/angiogenin subfamily from zebrafish. The Biochemical Journal 433 (2), 345–355. Available at https://doi.org/10.1042/BJ20100892.

Pizzo, E., et al., 2008. Ribonucleases with angiogenic and bactericidal activities from the Atlantic salmon. The FEBS Journal 275 (6), 1283–1295. Available at https://doi.org/10.1111/j.1742-4658.2008.06289.x.

Pizzo, E., et al., 2006. Ribonucleases and angiogenins from fish. The Journal of Biological Chemistry 281 (37), 27454–27460. Available at https://doi.org/10.1074/jbc.M605505200.

Pizzo, E., et al., 2007. The success of the RNase scaffold in the advance of biosciences and in evolution. Gene 406 (1-2), 8–12. Available at https://doi.org/10.1016/j.gene.2007.05.006.

Pizzo, E., et al., 2013. Ribonuclease/angiogenin inhibitor 1 regulates stress-induced subcellular localization of angiogenin to control growth and survival. Journal of Cell Science 126 (Pt 18), 4308–4319. Available at https://doi.org/10.1242/jcs.134551.

Pulido, D., et al., 2016. A novel RNase 3/ECP peptide for pseudomonas aeruginosa biofilm eradication that combines antimicrobial, lipopolysaccharide binding, and cell-agglutinating activities. Antimicrobial Agents and Chemotherapy 60 (10), 6313–6325. Available at https://doi.org/10.1128/AAC.00830-16.

Pyatibratov, M.G., et al., 2012. New insights into the role of angiogenin in actin polymerization. International Review of Cell and Molecular Biology 295, 175–198. Available at https://doi.org/10.1016/B978-0-12-394306-4.00011-3.

Qiao, M., et al., 2012. Onconase downregulates microRNA expression through targeting microRNA precursors. Cell Research 22, 1199–1202.

Rademacher, F., et al., 2021. Antimicrobial peptides and proteins: interaction with the skin microbiota. Experimental Dermatology 30 (10), 1496–1508. Available at https://doi.org/10.1111/exd.14433.

Rademacher, F., et al., 2019. The antimicrobial and immunomodulatory function of RNase 7 in skin. Frontiers in Immunology 10, 2553. Available at https://doi.org/10.3389/fimmu.2019.02553.

Rademacher, F., 2016. et. RNase 7 in cutaneous defense. International Journal of Molecular Sciences 17 (4), 560. Available at https://doi.org/10.3390/ijms17040560.

Raines, R.T., 1998. Ribonuclease A. Chemical Reviews 98 (3), 1045–1066. Available at https://doi.org/10.1021/cr960427h.

Rosemberg, H.F., 2008. Eosinophil-derived neurotoxin / RNase 2: connecting the past, the present and the future. Current Pharmaceutical Biotechnology 9 (3), 135–140. Available at https://doi.org/10.2174/138920108784567236 2008.

Rosemberg, H.F., 2015. Eosinophil-derived neurotoxin (EDN/RNase 2) and the mouse eosinophil-associated RNases (mEars): expanding roles in promoting host defense. International Journal of Molecular Sciences 16 (7), 15442–15455. Available at https://doi.org/10.3390/ijms160715442.

Rosemberg, H.F., et al., 1997. Diversity among the primate eosinophil-derived neurotoxin genes: a specific C-terminal sequence is necessary for enhanced ribonuclease activity. Nucleic Acids Research 25 (17), 3532–3536. Available at https://doi.org/10.1093/nar/25.17.3532.

Rosemberg, H.F., et al., 1995. Human ribonuclease 4 (RNase 4): coding sequence, chromosomal localization and identification of two distinct transcripts in human somatic tissues. Nucleic Acids Research 23 (21), 4290–4295. Available at https://doi.org/10.1093/nar/23.21.4290.

Ruiz-Rosado, J.D., et al., 2023. Human ribonuclease 6 has a protective role during experimental urinary tract infection. Journal of Innate Immunity 15 (1), 865–875. Available at https://doi.org/10.1159/000534736.

Sadagopan, S., et al., 2012. Angiogenin functionally interacts with p53 and regulates p53 mediated apoptosis and cell survival. Oncogene 31 (46), 4835–4847. Available at https://doi.org/10.1038/onc.2011.648.

Saikia, M., et al., 2014. Angiogenin-cleaved tRNA halves interact with cytochrome c, protecting cells from apoptosis during osmotic stress. Molecular and Cellular Biology 34 (13), 2450–2463. Available at https://doi.org/10.1128/MCB.00136-14.

Salazar, V.A., et al., 2016. Exploring the mechanisms of action of human secretory RNase 3 and RNase 7 against Candida albicans. Microbiologyopen 5 (5), 830–845. Available at https://doi.org/10.1002/mbo3.373.

Saxena, S.K., et al., 2002. Entry into cells and selective degradation of tRNAs by a cytotoxic member of the RNase a family. The Journal of Biological Chemistry 277, 15142–15146.

Scheraga, H.A., 1967. Structural studies of pancreatic ribonuclease. Federation Proceedings 26 (5), 1380–1387.

Schuster, C., et al., 2013. Prenatal human skin expresses the antimicrobial peptide RNase 7. Archives for Dermatological Research. Archiv fur Dermatologische Forschung 305 (6), 545–549. Available at https://doi.org/10.1007/s00403-013-1340-y Epub 2013 Apr 2.

Schwartz, L., et al., 2018. The immunomodulatory and antimicrobial properties of the vertebrate ribonuclease a superfamily. Vaccines (Basel) 6 (4), 76. Available at https://doi.org/10.3390/vaccines6040076.

Shaffer, J.F., et al., 2024. Epididymis-specific RNase a family genes regulate fertility and small RNA processing. The Journal of Biological Chemistry 300 (12), 107933. Available at https://doi.org/10.1016/j.jbc.2024.107933.
Sheng, J., et al., 2015. Three decades of research on angiogenin: a review and perspective. Acta Biochimica et Biophysica Sinica 48 (5), 399–410. Available at https://doi.org/10.1093/abbs/gmv131.
Sica, F., et al., 2021. The structural features of an ancient ribonuclease from Salmo salar reveal an intriguing case of auto-inhibition. International Journal of Biological Macromolecules 182, 659–668. Available at https://doi.org/10.1016/j.ijbiomac.2021.04.041.
Simanski, M., et al., 2013. IL-17A and IFN-γ synergistically induce RNase 7 expression via STAT3 in primary keratinocytes. PLoS One 8 (3), e59531. Available at https://doi.org/10.1371/journal.pone.0059531.
Sinatra, F., et al., 2000. Bovine seminal RNase induces apoptosis in normal proliferating lymphocytes. International Journal of Clinical & Laboratory Research 30 (4), 191–196. Available at https://doi.org/10.1007/s005990070006.
Singh, U.P., et al., 2007. Enzymatic and structural characterisation of amphinase, a novel cytotoxic ribonuclease from *Rana pipiens* oocytes. Journal of Molecular Biology 371, 93–111.
Smyth, D.G., et al., 1963. The sequence of amino acid residues in bovine pancreatic ribonuclease: revisions and confirmations. The Journal of Biological Chemistry 238, 227–234.
Sorrentino, S., et al., 1994. Human pancreatic-type and nonpancreatic-type ribonucleases: a direct side-by-side comparison of their catalytic properties. Archives of Biochemistry and Biophysics 312, 340–348.
Sorrentino, S., 2010. The eight human "canonical" ribonucleases: molecular diversity, catalytic properties, and special biological actions of the enzyme proteins. FEBS Letters 584 (11), 2194–2200. Available at https://doi.org/10.1016/j.febslet.2010.04.018.
Soucek, J., 1996. Immunosuppressive activity of bovine seminal ribonuclease and its mode of action. Immunobiology 195 (3), 271–285. Available at https://doi.org/10.1016/S0171-2985(96)80045-3.
Spencer, J.D., et al., 2014. An endogenous ribonuclease inhibitor regulates the antimicrobial activity of ribonuclease 7 in the human urinary tract. Kidney International 85 (5), 1179–1191. Available at https://doi.org/10.1038/ki.2013.395.
Spencer, J.D., et al., 2013. Ribonuclease 7, an antimicrobial peptide upregulated during infection, contributes to microbial defense of the human urinary tract. Kidney International 83 (4), 615–625. Available at https://doi.org/10.1038/ki.2012.410.
Subramanian, V., et al., 2007. A new role for angiogenin in neurite growth and pathfinding: implications for amyotrophic lateral sclerosis. Human Molecular Genetics 16 (12), 1445–1453. Available at https://doi.org/10.1093/hmg/ddm095.
Sun, D., et al., 2022. The role of human ribonuclease a family in health and diseases: A systematic review. IScience 25 (11), 105284. Available at https://doi.org/10.1016/j.isci.2022.105284.
Sun, J., et al., 2024. Ribonuclease 4 functions as an intestinal antimicrobial protein to maintain gut microbiota and metabolite homeostasis. Nature Communications 15, 5778. Available at https://doi.org/10.1038/s41467-024-50223-1.
Tota, M., 2023. The role of eosinophil-derived neurotoxin and vascular endothelial growth factor in the pathogenesis of eosinophilic asthma. Cells 12 (9), 1326. Available at https://doi.org/10.3390/cells12091326.
Turcotte, R.F., et al., 2008. Interaction of onconase with the human ribonuclease inhibitor protein. Biochemical and Biophysical Research Communications 377 (2), 512–514. Available at https://doi.org/10.1016/j.bbrc.2008.10.032.
van Es, M.A., et al., 2011. Angiogenin variants in Parkinson disease and amyotrophic lateral sclerosis. Annals of Neurology 70 (6), 964–973. Available at https://doi.org/10.1002/ana.22611.
Van Nguyen, T., et al., 2017. A study of ribonuclease activity in venom of vietnam cobra. Journal of Animal Science and Technology 59, 20. Available at https://doi.org/10.1186/s40781-017-0145-5.
Venge, P., et al., 1998. Eosinophil cationic protein (ECP). The International Journal of Biochemistry & Cell Biology 30 (4), 433–437. Available at https://doi.org/10.1016/s1357-2725(97)00127-1.
Walter, S., et al., 2017. RNase 7 participates in cutaneous innate control of Corynebacterium amycolatum. Scientific Reports 7 (1), 13862. Available at https://doi.org/10.1038/s41598-017-14383-z.

Wang, S., et al., 2023. Translation dysregulation in neurodegenerative diseases: a focus on ALS. Molecular Neurodegeneration 18 (1), 58. Available at https://doi.org/10.1186/s13024-023-00642-3.

Wang, Y.N., et al., 2018. Angiogenin/ribonuclease 5 Is an EGFR ligand and a serum biomarker for erlotinib sensitivity in pancreatic cancer. Cancer Cell 33 (4), 752–769.e8. Available at https://doi.org/10.1016/j.ccell.2018.02.012.

Weickmann, J.L., et al., 1984. Immunological assay of pancreatic ribonuclease in serum as an indicator of pancreatic cancer. Cancer Research 44, 1682–1687.

Westmuckett, A.D., et al., 2014. Impaired sperm maturation in RNASE9 knockout mice. Biology of Reproduction 90 (6), 120. Available at https://doi.org/10.1095/biolreprod.113.116863.

Wlodawer, A., et al., 1988. Structure of phosphate-free ribonuclease A refined at 1.26 A. Biochemistry 27 (8), 2705–2717.

Wu, D., et al., 2007. Angiogenin loss-of-function mutations in amyotrophic lateral sclerosis. Annals of Neurology 62 (6), 609–617. Available at https://doi.org/10.1002/ana.21221.

Wu, Y., et al., 1993. A cytotoxic ribonuclease. Study of the mechanism of onconase cytotoxicity. The Journal of Biological Chemistry 268 (14), 10686–10693.

Yamada, K.J., et al., 2015. Eosinophil-associated ribonuclease 11 Is a macrophage chemoattractant. The Journal of Biological Chemistry 290 (14), 8863–8875. Available at https://doi.org/10.1074/jbc.M114.626648.

Yamasaki, S., et al., 2009. Angiogenin cleaves tRNA and promotes stress-induced translational repression. The Journal of Cell Biology 185 (1), 35–42. Available at https://doi.org/10.1083/jcb.200811106.

Yao, Z., et al., 2024. Activation of angiogenin expression in macrophages by lipopolysaccharide via the TLR4/NF-κB pathway in colitis. Acta Biochimica et Biophysica Sinica 56 (6), 857–865. Available at https://doi.org/10.3724/abbs.2024013.

Zanfardino, A., et al., 2010. The bactericidal action on *Escherichia coli* of ZF-RNase-3 is triggered by the suicidal action of the bacterium OmpT protease. The FEBS Journal 277 (8), 1921–1928. Available at https://doi.org/10.1111/j.1742-4658.2010.07614.x.

Zhang, A., et al., 2024. Function of angiogenin and its role in neurological diseases. Authorea. Available at https://doi.org/10.22541/au.172227700.03344408/v1.

Zhang, J., et al., 2000. Evolution of the rodent eosinophil-associated RNase gene family by rapid gene sorting and positive selection. Proceedings of the National Academy of Sciences of the United States of America 97, 4701–4706.

Zhang, J., et al., 1998. Positive Darwinian selection after gene duplication in primate ribonuclease genes. Proceedings of the National Academy of Sciences of the United States of America 95 (7), 3708–3713. Available at https://doi.org/10.1073/pnas.95.7.3708.

Zhang, J., et al., 2002. RNase 8, a novel RNase A superfamily ribonuclease expressed uniquely in placenta. Nucleic Acids Research 30 (5), 1169–1175. Available at https://doi.org/10.1093/nar/30.5.1169.

Zhu, J., et al., 2013. Phospholipid scramblase 1 functionally interacts with angiogenin and regulates angiogenin-enhanced rRNA transcription. Cellular Physiology and Biochemistry International Journal of Experimental Cellular Physiology, Biochemistry, and Pharmacology 32 (6), 1695–1706. Available at https://doi.org/10.1159/000356604.

CHAPTER 6

From monomers to multimers

Giovanni Gotte
Biological Chemistry Section of the Department of Neuroscience, Biomedicine and Movement, University of Verona, Verona, Italy

6.1 Introduction

A large number of ribonucleolytic enzymes (ribonucleases, or RNases) exist in nature, and their role is to cleave many different RNA types. RNases act in many different contexts under variable conditions, and for this reason they are divided into different families. In particular, their classification involves the mammalian, amphibian, and bacterial ones, as well as endo- or exoribonucleases, or the secretory and nonsecretory RNases.

The last classification, in particular the one relative to secretory RNases, includes enzymes that can undergo self-association, a phenomenon that will be the main object of this chapter. The prototype of this family is the bovine pancreatic RNaseA (Barnard, 1969), an enzyme abundantly expressed by the *Bos Taurus* pancreas and deeply studied in the second part of the last century (Raines, 1998) (Fig. 6.1A). These studies permitted the assignment of three Nobel Prizes in the last century, in particular for the protein sequence by Sir S. Moore and W.H. Stein (Smyth et al., 1963), for the protein folding studied by Anfinsen (1973), and for the solid-phase protein synthesis based on the sequence of RNaseA by Merrifield (1986).

6.2 Bovine pancreatic RNaseA and secretory pancreatic-type (pt)-RNases

RNaseA is a 13.7 kDa basic enzyme formed by 124 AA residues (Fig. 6.1A), with a p.I. of 9.3 (Ui, 1971) and a catalytic triad composed of His12 located in the N-terminus, Lys41 in the protein core, and His119 in the C-terminus (Kartha et al., 1967; Thompson and Raines, 1994). Its structure is V-shaped, with a central cleft accommodating the substrate. Albeit being small, RNaseA presents four disulfides involving all its eight Cys residues, precisely 26–84, 40–95, 58–110, and 65–72. Hence, RNaseA is a highly stable enzyme, with a T_m of about 67°C.

RNaseA displays three α-helices, the first comprising residues 3–13 of the 1–20 N-terminal domain that faces into the catalytic cleft responsible for the V-shaped conformation of the enzyme. The protein also shows four antiparallel β-strands, and

RNases in Health and Diseases
DOI: https://doi.org/10.1016/B978-0-443-27312-4.00011-0

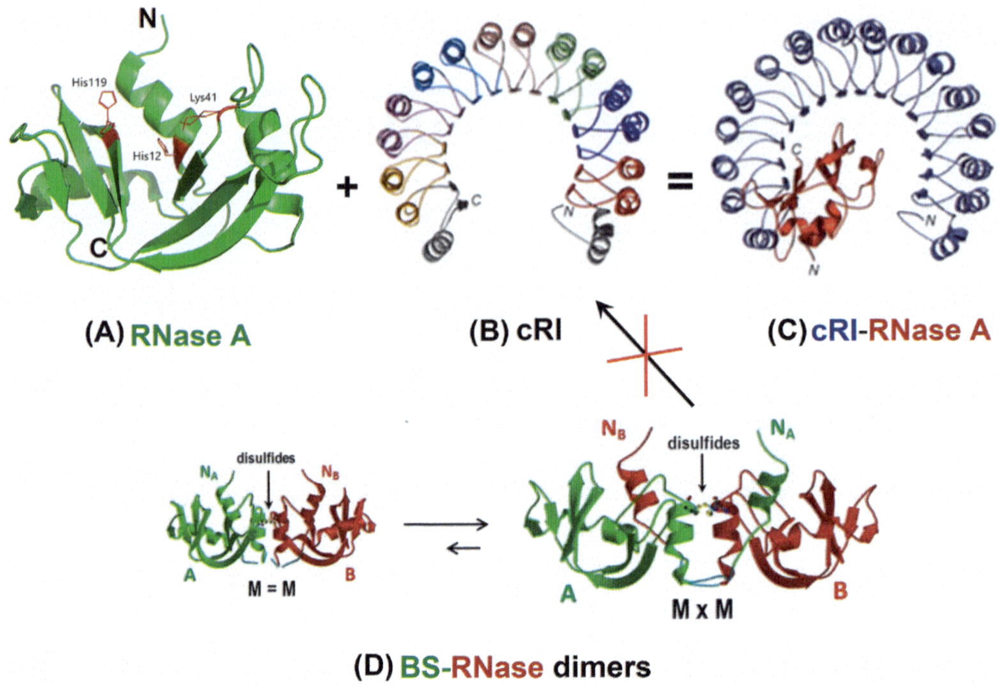

Figure 6.1 Structures of RNaseA, BS-RNase, and of the ubiquitous cellular RNases' inhibitor (cRI). (A) RNaseA, (B) cRI, (C) RNase A-cRI complex, and (D) BS-RNase native dimer isoforms (M = M and M × M).

all secondary structure elements are connected by unstructured loops of variable length (Kartha et al., 1967).

The main targets of RNaseA are single-stranded (ss)-RNA substrates that are cleaved through a two-step mechanism, each involving one catalytic His residue: in detail, a first cleavage of the $P–O^{5'}$ bond of the RNA strand is catalyzed by His12, and the subsequent hydrolysis of the $P–O^{2'}$ bond forming the nucleoside 2',3'-cyclic phosphodiester intermediate by His119, while Lys41 stabilizes the transition state of the first step in which the phosphorous (P1 site) is transiently pentacovalently coordinated (Raines, 1998). The enzyme is active on the 3'-side of a pyrimidine, preferring cytosines over uridines (Sorrentino and Libonati, 1994). Other residues, defined as catalytic subsites, are crucial for RNaseA activity (Nogués et al., 1995): in particular, the phosphates labeled P0 and P2, and the nitrogen bases labeled B1, B2, and B3 are located near the P1 phosphate and represent the actual enzyme active site, in turn facing the enzymatic cleft (Boix et al., 1994; delCardayré and Raines, 1994; delCardayré and Raines, 1995; Raines, 1998). Notably, RNaseA can also bind and destabilize double-stranded (ds)-DNA substrates, although in this case the result is enzymatically unproductive (Felsenfeld et al., 1963; Libonati and Sorrentino, 1992).

Importantly, RNaseA can enter cells through endocytosis (Leich et al., 2007), thereby making its enzymatic activity potentially harmful to cells and the individuals involved. Hence, evolution has engineered a cell ubiquitous RNase inhibitor (cRI, Fig. 6.1B) (Kobe and Deisenhofer, 1993, 1996), sometimes recently named RNH1 (Neu et al., 2024), that is able to block it upon forming an almost irreversible complex (Fig. 6.1C) (Kobe and Deisenhofer, 1996).

The enzyme can be glycosylated at the Ans34 residue, becoming the so-called RNaseB (Plummer and Hirs, 1963; Liang et al., 1980), while the more heterogeneously glycosylated RNaseC and D are less abundant (Plummer, 1968). These glycosylation events sterically affect the enzyme moiety and dynamics (Joao and Dwek, 1993; Solis et al., 2001), although they do not directly interact with the catalytic residues.

RNaseA is highly resistant to many proteases, but its N-terminal domain can be cleaved at the 20–21 peptide bond by the action of bacterial subtilisin, in this way forming the so-called peptide-S (residues 1–20) and residual protein-S (residues 21–124) (Richards and Vithayathil, 1959). These derivatives can electrostatically re-associate to form an enzymatically active adduct called RNaseS (Kim et al., 1992). However, the 20–21 bond cleavage was thereafter considered the structural basis for the formation of RNaseA aggregates when the protein was purified from the tissue of bovine pancreas under acidic conditions (Crestfield et al., 1962). Indeed, the RNaseA noncovalent aggregates, as well as the oligomers obtainable through chemical cross-linking (Gotte and Menegazzi, 2019), or upon the interaction of the enzyme with metal derivatives (Picone et al., 2015), would permit it to sterically evade the interaction with the cellular RI (cRI) and make the enzyme active, preferably against malignant tumor cells, as it will be discussed later.

Hence, the oligomerization events involving RNaseA will be the topic of this chapter. Then, the oligomerization of other members included in the "pancreatic-type" (pt)-RNases superfamily (Sorrentino and Libonati, 1994), so-called because they share crucial structural and functional features of bovine pancreatic RNaseA, will be described as well, besides the oligomerization tendency of the nonmammalian RNases, like the amphibian ones.

6.3 Natively dimeric bovine seminal RNase

6.3.1 Covalent dimerization through intersubunit disulfides

Bovine seminal ribonuclease (BS-RNase), highly expressed in the *Bos Taurus* seminal plasma (D'Alessio et al., 1972a), is the RNase variant that displays the highest sequence identity (about 82%) with RNaseA (Suzuki et al., 1987). Both are bovine enzymes, although deriving from different organs, and the most important difference distinguishing them is that BS-RNase is natively homo-dimeric (D'Alessio et al., 1972b), while RNaseA is monomeric (M). The dimericity of BS-RNase is warranted by the Cys31 and 32 residues, which are Lys31 and Ser32 in RNaseA, or however different from Cys in all the other secretory pt-RNases. The two Cys residues form in BS-RNase two antiparallel disulfide bonds that covalently link the two monomeric subunits (M=M) of the seminal enzyme variant, as shown in Fig. 6.1D.

6.3.2 Additional stabilization of the native dimer through the swapping of the N-terminal ends

Interestingly, BS-RNase is actually the result of an equilibrium between two isoforms (Fig. 6.1D), the first already called M = M, while the other is named M × M (panel B): beyond the mentioned 31–32 and 32–31 disulfides, this latter isoform displays the peculiar ability to reciprocally exchange the N-termini of the two subunits (Piccoli et al., 1992). This event enhances the stability of the enzyme, with an M = M / M × M equilibrium ratio of about 30/70, and the crystal structure of the M x M BS-RNase isoform confirms the presence of the N-termini intertwining (Mazzarella et al., 1993).

The property to reciprocally interchange a protein domain was defined as three-dimensional domain swapping (3D-DS) (Bennett et al., 1995) by Prof. D. Eisenberg when, in 1994, the reciprocal "entanglement" of the subunits was detected in the diphtheria toxin dimer (Bennett et al., 1994) and has been thereafter discovered in more than 100 protein structures formed upon undergoing self-association.

6.3.3 Enzymatic and antitumor activity of bovine seminal-RNases

BS-RNase is active versus many substrates against which monomeric RNaseA is instead inactive; indeed, BS-RNase is remarkably active versus viral and other artificial dsRNA substrates (Libonati and Floridi, 1969). These early results were quickly paralleled by the discovery that the BS variant displays, together with other remarkable biological properties, a high antitumor activity (Matoušek and Grozdanovič, 1973; Vescia et al., 1980; Mastronicola et al., 1995). Indeed, the dimeric nature of this enzyme is crucial to exerting this activity in the cells, and in particular, only the domain-swapped M × M isoform can do this (Cafaro et al., 1995): in fact, once entering cells, the cytosolic reducing environment is sufficient to reduce the two exposed 31–32 and 32–31 disulfides. Therefore only the swapped isoform can remain dimeric for a time sufficient to exert its activity, while the unswapped M = M counterpart quickly dissociates to M and is thereby sequestered by cRI (Kim et al., 1995). Later, Prof. A. Merlino and colleagues revealed that P19 and L28 are essential for BS-RNase to maintain the reciprocal orientation of the two M × M subunits also after reducing the two aforementioned disulfide bonds; indeed, upon inserting both P19A and L28Q mutations, hence making BS-RNase similar to RNaseA, the M × M isoform also becomes inactive in the cytosol of cancer cells because the conformation of this variant changes dramatically. Consequently, upon losing the two antiparallel disulfides in the cytosol, the dimeric P19A + L28Q BS-RNase can become susceptible to cRI (Merlino et al., 2008).

6.4 Covalent oligomers of RNaseA: structural, catalytic, and biological properties

6.4.1 "BS-RNase-like" covalent dimerization of RNaseA

An elegant approach to merge the advantageous features of BS-RNase and RNaseA was to obtain an engineered RNaseA dimer by producing BS-RNaseA chimeras; to this end, the two Cys31 and 32, allowing BS-RNase to be natively dimeric, were introduced in RNaseA. In addition, the presence of A19P and Q28L mutations seems to be mandatory to make this RNaseA derivative active against cancer cells. This so-called PLCC-RNaseA is cytotoxic because the reciprocal orientation of the two subunits is maintained in a BS-like form capable of escaping the cRI interaction only in the presence of both P19 and L28 residues (Di Donato et al., 1994; Cafaro et al., 1998; Ercole et al., 2009). The Pro-Leu-Cys-Cys mutations were also introduced in human pancreatic RNase1 (i.e., A19P-Q28L-R31C-R32C), therefore making it a dimeric PLCC-RNase1 variant capable of evading the inhibitor and of exerting a remarkable activity against cancer cells (Piccoli et al., 1999).

6.4.2 Oligomerization of RNaseA induced by chemical cross-linkers

The production of enzymatically active RNaseA oligomers similar to the BS-RNase covalent dimer triggered efforts devoted to producing stable and active RNaseA derivatives already since the late 60s of the last century. To do so, the bifunctional diimidoesters, which are molecules reactive with lysine residues, in particular dimethyl-adipimidate, permitted the obtaining of covalently fixed RNaseA dimers and oligomers (Hartman and Wold, 1967). Then, the choice of longer spacers provided by additional (-CH_2-)$_n$ groups between the two reactive imido-ester groups helped to increase the yield versus modified lysines ratio of the covalently fixed oligomers (Wang et al., 1976). In particular, dimethyl-suberimidate was also successful in augmenting the enzymatic activity versus dsRNA, with the RNaseA dimer resulting in higher activity than the monomer (Bartholeyns and Moore, 1974). The limitation was to find the ideal conditions and reagent ratios useful to avoid the modification of an excessive number of lysines, in particular Lys41 of the RNaseA active site. However, the lysine's side chain is modified, and this affects the basic charge distribution even without decreasing the basic charge of the derivative. Hence, the enzymatic activity of the RNaseA cross-linked dimers was found slightly lower than that of the corresponding noncovalent dimers, and this difference was also present within the trimers (Gotte et al., 1997). Nevertheless, either the covalent RNaseA dimers or, even more, the covalent trimers displayed activity against tumor cells and also in vivo, in mice (Bartholeyns and Baudhuin, 1976; Tarnowski et al., 1976; Gotte et al., 1997). Instead, the use of glutaraldehyde (Richards and Knowles, 1968), another common bifunctional cross-linker that causes the deletion of the Lys basic charge, was abandoned in the case of RNaseA (Jansen et al., 1971). In summary, a detailed list of linkers useful to produce covalent oligomers of RNases can be found in Gotte and Menegazzi (2019).

6.4.3 Covalent "zero-length" dimerization induced by heating RNaseA under vacuum

This experimental approach permits obtaining covalently fixed dimers and possibly larger oligomers without using chemicals: the so-called "zero-length" RNaseA dimers can be obtained through an under-vacuum heating treatment of the enzyme (Simons et al., 2007; Vottariello et al., 2010). These dimers, however, were active as antitumor agents only upon their cationization with poly-ethylenimine (Futami et al., 2005; Vottariello et al., 2010). Indeed, a previous work showed that the addition of polyamines, in particular of poly-spermine, to the suberimidate-derived dimeric RNaseA allows it to extremely increase its enzymatic activity against many ss- and ds-RNA substrates (Wang and Moore, 1977).

6.4.4 RNaseA covalent oligomerization induced by metalation

In the last decades, the interactions of many proteins, including RNaseA, with metals and metal-derived drugs have been extensively studied, especially by the two research groups led by L. Messori at the University of Florence and by A. Merlino of the "Federico II" University of Naples (Messori and Merlino, 2017). These studies shed light on the competitive role that proteins exert against the interactions of nucleic acids with metal drugs such as cis-platin and many of its derivatives: indeed, proteins interacting with metal drugs induce a dramatic lowering of the efficacy toward the DNA desired target (Pinato et al., 2013). Interestingly, the two mentioned research groups registered that many proteins, including RNaseA, can form strong complexes with cis-platin or similar drug derivatives (Ferraro et al., 2021, 2025). In addition, RNaseA can form aggregates thanks to metalation (Picone et al., 2015), but this ability resulted in quantitative differences with diverse platinum derivatives (Picone et al., 2017). Notably, the percentages of RNaseA oligomers formed with cis-platin were the highest ever registered for this enzyme, even higher than the ones obtained through noncovalent aggregation through the domain swapping mechanism (see next paragraphs). However, these metal derivatives were enzymatically and biologically inactive because the oligomers' formation directly affects the His119 residue that is part of the RNaseA catalytic triad (Picone et al., 2015).

6.5 Noncovalent self-association of RNaseA

6.5.1 RNaseA oligomerization through three-dimensional domain swapping

The first studies of the structural and functional features of RNaseA were performed on the enzyme purified from bovine pancreatic tissues during the '50s of the last century (Ledoux, 1955; Richards and Vithayathil, 1959). Large amounts could be extracted and purified, following protocols including the use of acidic solutions. Purification and analyses performed in the first '60s through size-exclusion chromatography (SEC) showed the presence of species larger than native RNaseA. These were RNaseA aggregates, enzymatically active versus RNA substrates, hence confirming they derived from the native enzyme (Crestfield et al., 1962). Sir S. Moore and coworkers hypothesized that the N-terminal arm could be displaced from the protein core interaction (Crestfield et al., 1962) (Fig. 6.2A), linking this event to the subtilisin susceptibility of the 20–21 peptide bond inserted in the loop connecting the α-helix N-terminus to the protein core (Richards and Vithayathil, 1959). The opening of the N-terminus would allow it to be re-associated to another protomer that behaves identically (Crestfield et al., 1962), thereby permitting the formation of a noncovalent RNaseA dimer. However, this hypothesis did not explain why larger aggregates of the enzyme were produced as well (Crestfield et al., 1962). In addition, RNaseA was shown to form two different dimeric derivatives (Fruchter and Crestfield, 1965), but no further structural data emerged until D. Eisenberg and colleagues solved in 1998 the crystal structure of an RNaseA dimer that was shown to be actually formed through the swapping of the N-termini of the two subunits (Liu et al., 1998). This dimer was then called N-dimer, or N_D (Fig. 6.2A). Hence, this evidence confirmed what was hypothesized 35 years before in the S. Moore labs (Crestfield et al., 1962). The two dimers were partially resolved with SEC (Gotte and Libonati, 1998) and completely purified from one to the other with cation-exchange chromatography (Gotte et al., 1999). Thereafter, the crystal structure of the second RNaseA dimer was also solved, displaying it to be formed by the swapping of its C-termini (C-dimer, or C_D, Fig. 6.2B) (Liu et al., 2001) and therefore being structurally different from N_D. The two dimers are differently susceptible to subtilisin proteolytic cleavage because of the different conformation of the loop containing the Ala20-Ser21 peptide bond susceptible to the protease (Nenci et al., 2001a). Then, the C-dimer exposes a definitely higher basic charge density than the N-swapped one (Gotte et al., 1999; Nenci et al., 2001a and 2001b), hence permitting it to exert a higher enzymatic activity than N_D versus ds-

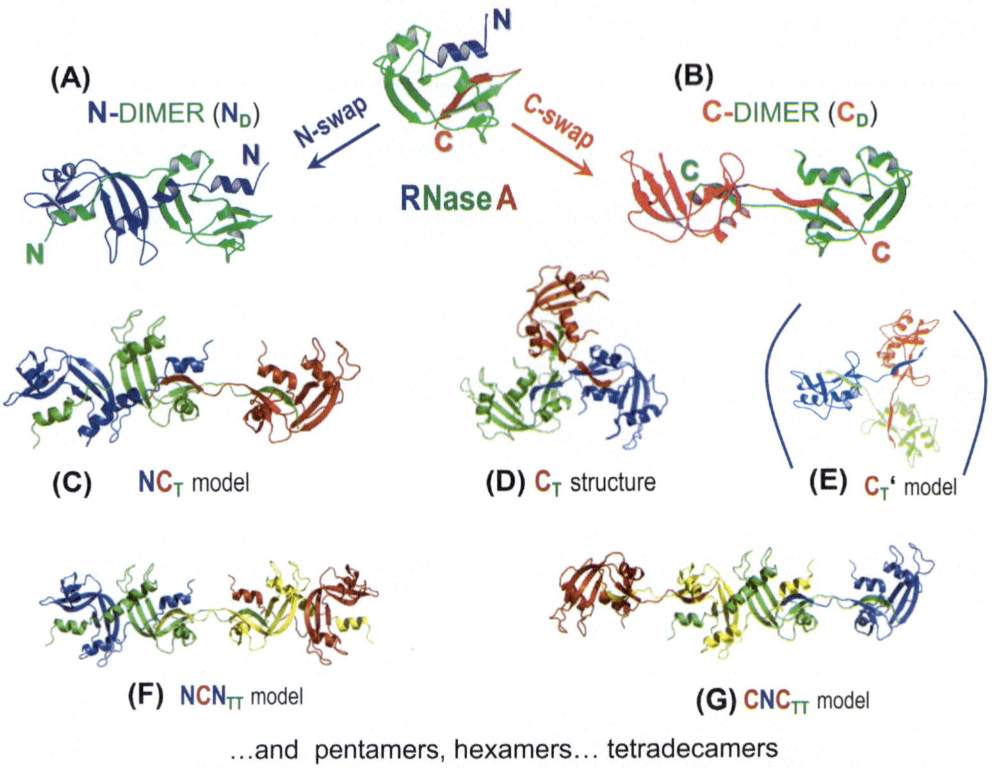

Figure 6.2 Self-association of RNaseA through 3D-domain swapping. (A and B) Crystal structures of the N-swapped and C-swapped dimers (pdb codes: N_D, 1A2W (Liu et al., 1998); C_D, 1F0V (Liu et al., 2001)); (C) model of the N+C swapped linear trimer; (D) crystal structure of the cyclic C_T (pdb code 1JS0 (Liu et al., 2002)), compared with (E) the flower-like C_T' model in brackets; and (F and G) models of two RNaseA N+C+N and C+N+C tetramers.

RNA substrates, as well as versus viral RNAs (Gotte and Libonati, 1998; Gotte et al., 1999).

Cation-exchange chromatography permitted the characterization of oligomers larger than dimers, revealing, above all, the formation of two different trimers (Gotte et al., 1999). Starting from experimental data, one trimer was modeled as resulting from an N + C-swapping event occurring in the central subunit, thereby embracing an N-swapping monomer on one side and a C-swapping subunit on the other (N + C-trimer, or NC_T, Fig. 6.2C) (Liu et al., 2001; Nenci et al., 2001b; Liu et al., 2002). Instead, the crystal structure of the second trimer was solved as a symmetric cyclic derivative formed by the swapping of the enzyme C-terminal β-strand (C-trimer, or C_T, Fig. 6.2D) (Liu et al., 2002). Interestingly, the orientation of the subunits was different from that visible in the model obtained from the C-dimer structure (C_T^I, Fig. 6.2E). This suggests that a hinge loop can find more than one energy minimum in turn, corresponding to different conformations characterizing an oligomer forming through 3D-DS: indeed, although both the C_T crystal structure and C_T^I model are related by a threefold axis, the former is similar to a helical propeller (Fig. 6.2D) and displays a buried surface more than the less compact flower-like model (Fig. 6.2E).

Beyond these oligomers, at least four RNaseA tetramers have been partially characterized, suggesting other structural models reported in Gotte et al. (1999), Nenci et al. (2001b), Liu and Eisenberg (2002), Gotte et al. (2006a), and Cozza et al. (2008), two of which are visible in Fig. 6.2F and G. Also, many pentamers, hexamers, and larger multimers were found and modeled as shown in Gotte et al. (2006a) and Cozza et al. (2008), while species up to traces of tetradecamers have also been detected (López-Alonso et al., 2009a). All these multimers are characterized by the simultaneous presence of both N- and C-swapped termini.

Notably, the same oligomers can be produced by heating the protein dissolved at very high concentrations in various solvents; in this way, however, the relative amounts of the oligomers can vary, or invert, depending on the particular conditions and/or solvent used (Gotte et al., 2003a).

All RNaseA oligomers displayed cytotoxicity either in vitro, against HL-60 and ML2 leukemia cells, or in vivo, by augmenting the activity from dimers to trimers, tetramers, and so on. Again, the higher the basic charge density exposed (C-swapping > N-swapping), the higher the cytotoxic effect measured (Matousek et al., 2003). More recently, the RNaseA C-dimer was detected to be active also against A375 cells, a very aggressive human melanoma cell line (Montioli et al., 2021).

These latter results suggest again that the oligomerization of RNaseA, and possibly of pt-RNases in general, may represent a good strategy to obtain species accompanied by a therapeutic potential against malignant cells and tissues. The possibility to escape cRI and to augment the basic charge density upon their self-association may explain the remarkable activity exerted by RNaseA oligomers. Their higher basic charge density could help them to be internalized in the malignant cells (Chao et al., 2010; Sundlass et al., 2013), thereby resulting in more efficacy than the monomer versus ds-RNAs, viral RNAs, or also versus DNA:RNA hybrids (Libonati and Sorrentino, 1992; Sorrentino and Libonati, 1994; Libonati and Gotte, 2004).

6.5.2 Stability of the RNaseA domain-swapped oligomers

All RNaseA oligomers are metastable, but their dissociation rate can be braked by storing them at low temperature, 4°C (Gotte et al., 1999; Libonati and Gotte, 2004), and by a buffer containing phosphate: indeed, phosphate forms an electrostatic bridge between the two His12 and His119 residues belonging to each of the N- or C-swapping domains, respectively (López-Alonso et al., 2006). The polarity of the two domains can partially affect the yield and the stability of the oligomers (Gotte et al., 2006b; Ercole et al., 2009), while the H-bond through Ser18 and Ser80 can affect the tendency of the N-terminus to be detached from the protein core and undergo the swapping event (Ercole et al., 2007; Vottariello et al., 2011). Then, also glycosylation (RNaseB) and/or deamidation events affecting RNaseA can reduce its ability to self-associate through 3D-DS (Gotte et al., 2003b; Fagagnini et al., 2017a).

The most important limitation for possible therapeutic applications is indeed the not very long-term stability of the 3D-DS oligomers. Hence, the covalent fixation occurring after their production could be a successful approach, provided that excessive modification is avoided. In this context, the antitumor activity of an RNaseA C-dimer covalently fixed with EDC (López-Alonso et al., 2009b) has not been tested so far, but this approach could help to measure the antitumor activity potential of other pt-RNases that might undergo a 3D-DS self-association.

6.5.3 RNase A-like oligomerization of bovine seminal RNase through three-dimensional domain swapping

The acidic treatment followed by lyophilization and initially used to purify RNaseA from pancreas tissues (Crestfield et al., 1962, 1963) was also applied to BS-RNase: upon re-dissolving the lyophilized powder in a phosphate buffer, aggregates characterized by MW values doubling the 28 kDa corresponding to natively dimeric BS-RNase were detected (Adinolfi et al., 1996). Hence, an RNase A-like tetramer, that is, a dimer of the natively homo-dimeric enzyme, was produced. This derivative was hypothesized to derive from a propagative swapping of the N-terminal ends of each subunit, thereby forming a cyclically closed structure (Adinolfi et al., 1996). Some years later, the swapping of the C-terminal end of at least one of the dimer subunits was found: indeed, BS-RNase can form, beside the mentioned N-swapped tetramer, an N+C+N-swapped tetramer (i.e., two different dimers of the dimer) that should resemble the one formed by RNaseA shown in Fig. 6.2F, and at least two hexamers (trimers of the dimer). As well, traces of larger multimers are formed by BS-RNase (Gotte et al., 2012). Importantly, the oligomeric aggregates of BS-RNase displayed an increased antitumor activity against human pancreatic cancer cells with respect to the already active native dimer (Gotte et al., 2012; Fiorini et al., 2014). The presence of the C-terminal swapping was confirmed by the K113N-BS RNase artificial mutant (Gotte et al., 2012) and by the ability of the BS-RNase monomeric derivative to form an RNase A-like C-swapped dimer (Sica et al., 2013).

6.5.4 Is it actually possible for RNaseA and its oligomers to undergo fibrillation?

Up to a few years ago, wt-RNaseA was considered totally unable to undergo fibrillation, although six or seven short peptides forming the enzyme were found able to form amyloid-like fibrils (Goldschmidt et al., 2010; Teng et al., 2012) similar to the ones formed by the amyloid-prone proteins (Sawaya et al., 2007). This inability was ascribed to the ability of the protein to act as "auto-chaperone" (Goldschmidt et al., 2010; Teng et al., 2012), thereby avoiding an amyloidogenic destiny. This could sound somewhat contradictory considering that RNaseA is able to form many 3D-DS multimers. Indeed, only RNaseA mutants displaying two or more glycine residue insertions or a poly-Q spacer inside the hinge loops connecting the protein core to the N- or C-terminal ends, respectively, were able to form amyloid-like derivatives. These fibrils were in turn modeled as structures formed through 3D-DS of RNaseA elongated N- or C-termini (Teng et al., 2012; Sambashivan et al., 2005).

However, and notably, M. Noji et al. reported in 2021 that, when wt-RNaseA breaks the supersaturation barrier, it can undergo fibrillation (Noji et al., 2021), although this tendency is inhibited by sugar osmolytes (Siraj et al., 2023). Recently, it was also registered that the RNaseA oligomers characterized by the presence of at least one N-term-swapping (i.e., N_D, NC_T, NCN_{TT}, CNC_{TT}, and so on, see Fig. 6.2), while not exclusively C-term-swapped ones, are able to slowly promote the formation of large amorphous so-called "super-aggregates" (SAs) (Gotte et al., 2022) that structurally resemble the protofibrils detected in crowding scenarios (Siraj et al., 2023). However, the tendency of these N-swapped oligomers to slowly evolve toward SAs is actually limited by their contemporary dissociation to the native monomer (Gotte et al., 2022). Nevertheless, the native RNaseA monomer also fails to produce SAs (Gotte et al., 2022), while the N-terminal 20-residues-long S-peptide is known to produce fibrils once detached from the residual S-protein (Liu et al., 1998). Hence, a controlled N-swapped oligomerization of RNaseA may be envisaged as a preliminary step toward massive aggregation and potential fibrillation, although its auto-chaperone action impedes this amyloid-like evolution (Goldschmidt et al., 2010; Teng et al., 2012). By the way, some mutants of the amyloidogenic cystatin-like monellin have recently been suggested to form a dimer through 3D-DS, which could be protective against fibrillation (Lucignano et al., 2025). Considering, however, that some proteins capable of undergoing 3D-DS self-association are fibrillogenic (Bennett et al., 2006), the role of 3D-DS oligomerization may be considered as protective against, or instead favorable to evolving toward amyloidosis as related to each specific protein involved.

6.6 Oligomerization of other mammalian secretory pancreatic-type (pt)-RNases

6.6.1 Features necessary to be included in the secretory pt-RNase superfamily

Secretory RNases can belong to the RNase A-like, pt-RNase superfamily if they share the following features: (1) a V-shaped folding (Fig. 6.1A), accompanied by four disulfide bonds, except RNase5 (human angiogenin (ANG)) that maintains only three disulfides;

(2) a His-Lys-His catalytic triad, H12-K41-H119 in RNaseA (Raines, 1998), with the two His residues located in the N- and C-termini, respectively, and Lys in the protein core; (3) a preferential activity exerted on pyrimidines located on the 3' side of the RNA substrate, and with a two-step transphosphorylation/hydrolysis mechanism involving the two His residues alternatively acting as acid or base (Raines, 1998); and (4) a C**K**xxNTF consensus sequence, comprising the catalytic Lys, and another PV**H** triplet located at the enzyme C-terminus and comprising the second catalytic His residue (Smith and Raines, 2006). The most important RNases that can undergo self-association are reported and described in the next paragraphs and shown in Fig. 6.3.

6.6.2 Dimers and oligomers of human pancreatic RNase1

The human pancreatic RNase (HP-RNase), also called RNase1 (Fig. 6.3A), is the 128 AA-residue human homolog of RNaseA (Beintema et al., 1984). Indeed, it shares the same shape and catalytic residues as it, and is actually expressed in the human pancreas as well as in other organs, like, for example, the brain or kidneys (Yasuda et al., 1993; Mizuta et al., 1990). Differently glycosylated RNase1 derivatives (Ressler and Raines, 2019) display enzymatic activity values that increase with their glycosylation degree (Libonati and Sorrentino, 1992). Notably, RNase1 displays a disordered ^{125}EDST^{128} C-terminus elongation, not present in RNaseA. Then, its crystal structure has been solved only when complexed with cRI (Johnson et al., 2007) and not as the enzyme alone, while only variants displaying a mutated N-terminus, or a shortened loop connecting it with the protein core, have been shown in the past to constitutively form N-swapped dimers (Russo et al., 2000; Canals et al., 2001; Pica et al., 2013). Instead, wt-RNase1 has recently been shown to extensively oligomerize, even more than RNaseA, when it is subjected to the same acidic lyophilization treatment (Noro et al., 2023). Moreover, a truncated des(125–128)-RNase1 variant lacking the four C-terminal ^{125}EDST^{128} tail, like RNaseA, is able to oligomerize even more abundantly: the lowest percentage of monomer can be recovered with this shortened variant, which in turn forms relevant amounts of trimer(s), tetramers, and so on, beyond the highest yield of dimers (Noro et al., 2023). The oligomers formed by both RNase1 variants are enzymatically active, although their activity does not increase with respect to the already highly active relative monomers (Libonati and Sorrentino, 1992; Bal and Batra, 1997; Noro et al., 2023). SEC chromatography profiles, as well as electrophoretic and spectroscopic data collected for the oligomers, strongly suggest that the RNase1 oligomers structurally resemble the ones formed by RNaseA; indeed, a couple of dimers are formed, one N-term-swapped and the other C-swapped. These dimers and one C-trimer have been modeled as similar to the corresponding RNaseA ones (Noro et al., 2023) (Fig. 6.3B–D), or, alternatively, for the N-dimer, as similar to the RNase1 mutant that previously showed to be constitutively an N-swapped dimer (Fig. 6.3E) (Canals et al., 2001; Noro et al., 2023).

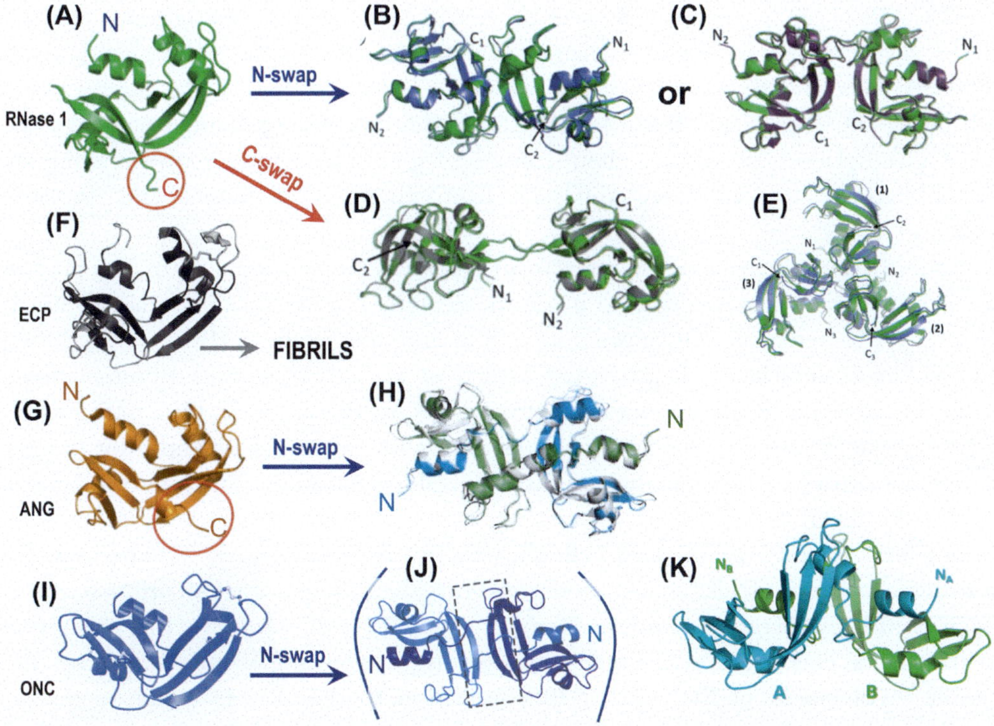

Figure 6.3 Self-association of four pt-RNases different from RNaseA. (A)-RNase1 (HP-RNase), (F)-RNase3 (ECP), (G)-RNase5 (ANG), and (I)-onconase (ONC). (B)-model of the RNase1 N-dimer (green) overlapped with the RNaseA N-dimer structure (blue, pdb code 1A2W (Liu et al., 1998)); (C)-model of the RNase1 N-dimer (green) overlapped with the structure of N_D PM8-RNase1 (violet, pdb 1H8X (Canals et al., 2001)); (D)-model of the RNase1 C-dimer (green) overlapped with the RNaseA C-swapped dimer structure (gray, pdb 1F0V (Liu et al., 2001)); (E)-model of the RNase1 C-trimer (green) overlapped with RNaseA cyclic C-trimer structure (gray, pdb 1JS0 (Liu et al., 2002)); (H) model of the ANG N-dimer (green-cyan) overlapped with the RNaseA N-dimer structure (grey, pdb code 1A2W); (J)- ONC N-dimer modeled starting from the RNaseA N-dimer structure: the dashed box highlights the open interface (Bennett et al., 1995) existing associating the two subunits, like in RNaseA N_D (pdb code 1A2W); and (K)-crystal structure of the ONC N-dimer (pdb code 7OR6, (Gotte et al., 2021)), with a different conformation from the model visible in panel J.

All RNase1 oligomers were proposed as potential antitumoral compared with the known inactivity of the monomer, since the latter is irreversibly sequestered by cRI complexation (Johnson et al., 2007).

6.6.3 Aggregation of RNase3, the eosinophil cationic protein

RNase3, also known as eosinophil cationic protein (ECP) (Fig. 6.3F), is another secretory RNase A-like variant of about 16 kDa and formed by 133 AA residues with an H15-K38-H128 active site (Boix et al., 1999). However, many residues that are catalytically important in all mammalian pt-RNases (K7, R10, K66, and F120) (Raines, 1998) are replaced with nonconservative ones for a sequence identity of 32% and a very high pI of 10.8 (Venge and Bystrom, 1998). In addition, three potential sites for Asn-linked glycosylation, as well as their natural N-glycan structures, contain complex oligosaccharides that are very different from those of RNaseA and RNase1 (Sorrentino, 1998).

RNase3's substrate preference differs from RNase1's (Sorrentino, 2010). The protein exerts neurotoxic potency linked to its ribonucleolytic activity (Gleich et al., 1986). It also displays bactericidal, antiparasitic, as well as antiviral activities (Boix et al., 2008), and inhibits mammalian cell growth (Carreras et al., 2003), but its ribonucleolytic activity seems to be not necessary for cytotoxicity (Rosenberg, 1995). Data indicate that both cell toxicity and bactericidal action correlate with ECP's ability to disrupt the cell membranes thanks to the aromatic and cationic residues located on its surface (Carreras et al., 2003, 2005).

Notably, RNase3/ECP displayed the ability to form at 37°C amyloid-like protofibrils under particular acidic conditions, revealing the presence of an N-terminal region prone to undergo amyloidosis (Torrent et al., 2010). Considering that eosinophil infiltration was detected in Alzheimer's disease tests (Wegiel and Wisniewski, 1994), the possible involvement of ECP in, or its connection to this disease could be further investigated. However, RNase3 is the unique secretory member of the pt-RNase superfamily that to date has been shown to undergo amyloid-like aggregation, although proto-fibril formation does not follow the 3D-DS mechanism (Torrent et al., 2010).

6.6.4 Dimerization of RNase5, the human angiogenin

First characterized in 1985 from human carcinoma cells, another important member of the RNase A-like superfamily is human ANG (Fig. 6.3G), a 14.1 kDa RNase member involved in the neo-formation of blood vessels (Fett et al., 1985). Although its AA sequence reaches only 35% identity with RNaseA (Strydom, 1998), its catalytic (H13-K40-H114) and key residues permit to include it in the pt-RNase superfamily as RNase5 (Sorrentino, 2010). However, ANG is definitely less active than RNaseA and RNase1 (Leland et al., 2002), but its low activity, exerted especially versus tRNAs (Saxena et al., 1992), is mandatory to maintain its biological activity (Shapiro and Vallee, 1989). ANG shows a nuclear localization R31-R32-R33 sequence (Moroianu and Riordan, 1994) that permits it to exert several other biological functions progressively discovered in more recent years (Garnett and Raines, 2022); indeed, ANG displays neuroprotective activity in astrocytes (Skorupa et al., 2012), also upon Nrf2 antioxidant activation (Hoang et al., 2019). This protective action is a consequence of the cleavage of tRNAs that, in turn, inhibits translation. Then, the 5'tiRNA fragment derivatives are responsible for the formation of G-quadruplexes and stress granules (Ivanov et al., 2014). Consequently, many loss-of-function RNase5 mutants have been displayed to be pathogenic, in particular in neurodegenerative contexts such as Parkinson's disease (PD) and amyotrophic lateral sclerosis (ALS) (Greenway et al., 2006; Van Es et al., 2011). Considering that these diseases are multifactorial but certainly linked to uncontrolled protein fibrillation, in particular of α-synuclein (Polymeropoulos et al., 1997), SOD-1 (Rosen et al., 1993), and TDP-43 (Neumann et al., 2006) mutants, RNase5/ANG has also been tested in its tendency to self-associate; the results obtained, however, show that it can dimerize, again through the 3D-DS mechanism, but neither forms large aggregates nor generates amyloid-like fibrils (Fasoli et al., 2021). However, both S28N- and R121C-ANG, two pathogenic mutants found in ALS patients, can dimerize to a greater extent than wt-ANG, while structural analyses suggest that only one dimer can be formed through an N-terminal domain swapping (Fig. 6.3H). In fact, the 3_{10}-helix tract (Acharya et al., 1994) might somehow hinder the C-terminus swapping, hence avoiding the formation of a C-dimer and large multimers (Fasoli et al., 2021).

Interestingly, and finally, an ANG dimer has been characterized as a complex with ds-RNA, but in this case formed through interactions not involving the 3D-DS mechanism. However, these interactions, although registered only in the crystals, might hinder the coupling of the enzyme with the RNA substrate and somehow affect important biological events (Sievers and Ficner, 2022).

The biological features of ANG dimerization, although not being immanent in this RNase type, could be further analyzed, above all in the numerous pathogenic mutants involved in the previously mentioned neurodegenerative diseases (Greenway et al., 2006; Van Es et al., 2011; Gotte, 2024).

6.7 Dimerization of the antitumor amphibian RNase onconase

The amphibian RNase secreted by *Leopard Frog* oocytes or early embryos is the 11.8 kDa ranpirnase, or, more commonly named, onconase (ONC, Fig. 6.3I) (Ardelt et al., 1991). Notwithstanding being formed by only 104 AA residues and hence definitely smaller than all other RNase A-like variants, ONC satisfies the four structural and functional requisites, listed in Section 6.1 and mandatory for being included in the pt-RNase superfamily. Notably, ONC lacks 20 AA residues out of the 124 of RNaseA and shares only about 30% sequence identity with it. However, it displays an RNase A-like folding and an H10-K31-H97 catalytic triad resembling that of the prototype (Ardelt et al., 1991). The smaller length is principally related to the absence of many flexible loops characterizing all other pt-RNases. One of these loops lacks some key residues that would permit the interaction with cRI; hence, ONC is able to evade cRI (Turcotte and Raines, 2008; Rutkoski and Raines, 2008) and exerts its action versus many substrates, like tRNAs and miRNAs (Suhasini and Sirdeshmukh, 2007; De Tomi et al., 2022; Menegazzi and Gotte, 2022), thereby resulting in selective activity versus cancer cells. This preferential action is related to the high number of negative charges (Chao et al., 2010) provided, for example, by sialic acids (Titani et al., 1987) and by the acidic tumor microenvironment (Estrella et al., 2013), which is more abundant on cancer cells' membranes than on normal cells.

Data suggest that ONC induces cell death through apoptosis (Halicka et al., 2007), while the antitumor activity of ONC is enhanced by combining it with radiation, interferon, or other active factors (Rybak et al., 1996; Mikulski et al., 1992; Shen et al., 2016; Menegazzi and Gotte, 2022).

This evidence permitted to test ONC in vivo by directly injecting it into the circulating system, thereby providing encouraging results (Mikulski et al., 1993, 1995), especially when administered after "standard" chemotherapy in patients affected by mesothelioma or lung cancers (Mikulski et al., 1995; Costanzi et al., 2005). Nevertheless, an undesired renal accumulation, although reversible upon stopping treatments (Vasandani et al., 1999), obliged the clinicians to suspend these promising tests.

However, ONC was also recently revealed to be able to dimerize through the swapping of its N-terminal end alone, since the C-terminus is blocked by the 87–104 disulfide involving the last cysteine residue of the enzyme (Fagagnini et al., 2017b). Initially modeled as similar to the RNaseA N-dimer (Fig. 6.3J), the crystal structure subsequently solved confirmed the formation of the ONC dimer through the swapping of the protein N-terminus, but with a conformation different from the RNaseA N-dimer (Fig. 6.3K) (Gotte et al., 2021). This evidence confirms that proteins that are very similar can form supramolecular structures different from each other and different from models previously resulting in energetically reliable, as was found for the cyclic RNaseA C-trimer (Liu et al., 2002).

Beside the dimer, low amounts of a larger species, possibly an ONC trimer or tetramer, appear in SEC (Gotte et al., 2021). This species has not been characterized further, but this suggests that a large ONC derivative not bypassing the glomerular barrier, and therefore avoiding its accumulation in the kidneys, could be possible to form. The unlock of the C-terminus, however, produced worse results (Noro et al., 2023), in line with the previous statement asserting that the C87–C104 disulfide is mandatory for ONC stability (Leland et al., 2000). It has to be mentioned, however, that a cyclic ONC cross-linked trimer covalently produced with maleimides by a thiol-reactive S72C-ONC mutant was less cytotoxic than the ONC monomer (Rutkoski et al., 2010). Nevertheless, the structure of an ONC trimer, or tetramer formed through 3D-DS could result in a different reciprocal orientation of the subunits, promoting easier interaction with the cell membrane than the cross-linked S72C-ONC trimer (Sundlass et al., 2013).

6.8 Therapeutic potential of the RNases oligomers

All data reported about the self-association of the secretory pt-RNases can suggest that the controlled oligomerization of these enzymes can become advantageous in terms of their possible use as anticancer agents.

First, the basic nature of these enzymes can ensure their ability to penetrate the cell membrane selectively for cancer cells (Leich et al., 2007): indeed, being the normal cells less rich in acidic derivatives than cancerous ones (Titani et al., 1987; Estrella et al., 2013), the increase in basic charge density provided by the self-association could actually help RNases to penetrate the cells even more than their native monomers. However, since the interaction with the cell membrane requires a precise reciprocal orientation of charged key residues (Sundlass et al., 2013), it seems to be crucial that the actual structure of the oligomers should be useful to favor a productive interaction of these residues with the membrane. The structure of the ONC dimer, different from the RNaseA N-dimer previously modeled (Fagagnini et al., 2017b), may support this argument (Gotte et al., 2021).

Once inside cells (Rodríguez et al., 2007), the activity of RNases can be blocked by cRI (Kobe and Deisenhofer, 1996; Rutkoski and Raines, 2008). Certainly, either the ability to evade cRI permits the RNases to be harmful to cells (Turcotte and Raines, 2008; Rutkoski and Raines, 2008), or the cytotoxicity exerted by ANG only in the absence of cRI confirms this possibility (Thomas et al., 2018). However, it has to be recalled that RNases' oligomerization limited to dimerization could not be sufficient to ensure cRI evasion: indeed, although being debated, this could be the case of the RNaseA domain-swapped dimers interacting with cRI (Naddeo et al., 2005), and not of the BS-RNase native dimer evading it (Kim et al., 1995). Indeed, also the cross-linked covalently fixed dimer(s) displayed to be definitely less cytotoxic than the corresponding trimers (Gotte et al., 1997), somehow confirming that dimerization is not "per se" sufficient to exert cytotoxicity.

Once avoiding the cRI interaction, the augmentation of basic charge density due to self-association might likely allow RNases' oligomers to be productive against substrates like ds-RNAs or DNA:RNA hybrids that are not susceptible to RNaseA (Gotte et al., 1999; Libonati and Gotte, 2004). In addition, the discovery of the action of ONC on miRNAs and of ANG on tRNAs certainly opens new scenarios to better control the determinants governing the activity of these enzymes and, in this way, explore their possible therapeutic use.

The effects registered with ONC on both pancreatic cancer and malignant melanoma cells were found to be additive together with classic chemotherapeutic (Fiorini et al., 2015; Raineri et al., 2019a), and above all, active versus melanoma drug-resistant cells (Raineri et al., 2019b; Passarini et al., 2026).

However, the antitumor activity of a generic RNase can be paralleled by an undesired renal accumulation, as it occurred with ONC (Vasandani et al., 1999). Hence, the enlargement of the dimensions of the RNase(s) moiety proportional to the oligomerization degree of the enzyme(s) could delete, or at least drastically reduce, the glomerular filtration of the kidneys. Moreover, this would be paralleled by the augmentation of the circulating half-life of the oligomers, thereby increasing their beneficial antitumor effects, considering that ONC was easily administered intravenously in clinical trials (Mikulski et al., 1993; Vogelzang et al., 2001).

6.9 Concluding remarks

From all reported here, the future efforts in the research of the effects of RNase oligomerization should be principally devoted to increasing the tendency of these RNases to self-associate even to a larger extent than dimers. The experimental data collected indicate that this possibility is strongly related to the ability of the RNase enzymes to swap both their N- and C-terminals. Where not possible, possible strategies to exploit specific mutations to ease this event or to alternatively use appropriate cross-linkers, could drive toward successful results.

However, and finally, the possibility to solve the structures of all the oligomers formed, and to precisely characterize the key residues active in the interactions with cell membranes (Sundlass et al., 2013) and/or with cRI (Rutkoski and Raines, 2008), as well as also with the actual RNA targets involved, may help to unveil which could be the oligomers more active versus malignant cells and tissues.

Last, but not least, these results could also clarify the best conditions useful to induce the lowest undesired side-effects in case of using these species as therapeutic agents.

References

Acharya, K.R., Shapiro, R., Allen, S.C., Riordan, J.F., Vallee, B.L., 1994. Crystal structure of human angiogenin reveals the structural basis for its functional divergence from ribonuclease. Proceedings of the National Academy of Sciences of the United States of America 91 (8), 2915–2919. Available at https://doi.org/10.1073/pnas.91.8.2915, www.pnas.org.

Adinolfi, S., Piccoli, R., Sica, F., Mazzarella, L., 1996. BS-RNase tetramers: an example of domain-swapped oligomers. FEBS Letters 398 (2–3), 326–332. Available at https://doi.org/10.1016/S0014-5793(96)01034-4.

Anfinsen, C.B., 1973. Principles that govern the folding of protein chains. Science (New York, N.Y.) 181 (4096), 223–230. Available at https://doi.org/10.1126/science.181.4096.223.

Ardelt, W., Mikulski, S.M., Shogen, K., 1991. Amino acid sequence of an anti-tumor protein from Rana pipiens oocytes and early embryos. Homology to pancreatic ribonucleases. Journal of Biological Chemistry 266 (1), 245–251. Available at https://doi.org/10.1016/s0021-9258(18)52427-3.

Bal, H.P., Batra, J.K., 1997. Human pancreatic ribonuclease. European Journal of Biochemistry 245 (2), 465–469. Available at https://doi.org/10.1111/j.1432-1033.1997.t01-1-00465.x, www.blackwell-science.com/ejb.

Barnard, E.A., 1969. Biological function of pancreatic ribonuclease. Nature 221 (5178), 340–344. Available at https://doi.org/10.1038/221340a0.

Bartholeyns, J., Baudhuin, P., 1976. Inhibition of tumor cell proliferation by dimerized ribonuclease. Proceedings of the National Academy of Sciences 73 (2), 573–576. Available at https://doi.org/10.1073/pnas.73.2.573.

Bartholeyns, J., Moore, S., 1974. Pancreatic ribonuclease: enzymic and physiological properties of a cross-linked dimer. Science (New York, N.Y.) 186 (4162), 444–445. Available at https://doi.org/10.1126/science.186.4162.444.

Beintema, J.J., Wietzes, P., Weickmann, J.L., Glitz, D.G., 1984. The amino acid sequence of human pancreatic ribonuclease. Analytical Biochemistry 136 (1), 48–64. Available at https://doi.org/10.1016/0003-2697(84)90306-3.

Bennett, M.J., Choe, S., Eisenberg, D., 1994. Domain swapping: entangling alliances between proteins. Proceedings of the National Academy of Sciences 91 (8), 3127–3131. Available at https://doi.org/10.1073/pnas.91.8.3127.

Bennett, M.J., Sawaya, M.R., Eisenberg, D., 2006. Deposition diseases and 3D domain swapping. Structure (London, England 1993) 14 (5), 811–824. Available at https://doi.org/10.1016/j.str.2006.03.011.

Bennett, M.J., Schlunegger, M.P., Eisenberg, D., 1995. 3D domain swapping: a mechanism for oligomer assembly. Protein Science 4 (12), 2455–2468. Available at https://doi.org/10.1002/pro.5560041202.

Boix, E., Leonidas, D.D., Nikolovski, Z., Nogués, M.V., Cuchillo, C.M., Acharya, K.R., 1999. Crystal structure of eosinophil cationic protein at 2.4A resolution. Biochemistry 38 (51), 16794–16801. Available at https://doi.org/10.1021/bi9919145.

Boix, E., Nogués, M.V., Schein, C.H., Benner, S.A., Cuchillo, C.M., 1994. Reverse transphosphorylation by ribonuclease a needs an intact p2-binding site. Point mutations at Lys-7 and Arg-10 alter the catalytic properties of the enzyme. Journal of Biological Chemistry 269 (4), 2529–2534. Available at https://doi.org/10.1016/s0021-9258(17)41977-6.

Boix, E., Torrent, M., Sanchez, D., Nogues, M., 2008. The antipathogen activities of eosinophil cationic protein. Current Pharmaceutical Biotechnology 9 (3), 141–152. Available at https://doi.org/10.2174/138920108784567353.

Cafaro, V., Bracale, A., Di Maro, A., Sorrentino, S., D'Alessio, G., Di Donato, A., 1998. New muteins of RNase A with enhanced antitumor action. FEBS Letters 437 (1-2), 149–152. Available at https://doi.org/10.1016/S0014-5793(98)01221-6.

Cafaro, V., De Lorenzo, C., Piccoli, R., Bracale, A., Mastronicola, M.R., Di Donato, A., et al., 1995. The antitumor action of seminal ribonuclease and its quaternary conformations. FEBS Letters 359 (1), 31–34. Available at https://doi.org/10.1016/0014-5793(94)01450-F.

Canals, A., Pous, J., Guasch, A:cia, Benito, A., Ribó, M., Vilanova, M., et al., 2001. The structure of an engineered domain-swapped ribonuclease dimer and its implications for the evolution of proteins toward oligomerization. Structure (London, England 1993) 9 (10), 967–976. Available at https://doi.org/10.1016/s0969-2126(01)00659-1.

Carreras, E., Boix, E., Rosenberg, H.F., Cuchillo, C.M., Nogués, M.V., 2003. Both aromatic and cationic residues contribute to the membrane-lytic and bactericidal activity of eosinophil cationic protein. Biochemistry 42 (22), 6636–6644. Available at https://doi.org/10.1021/bi0273011.

Carreras, E., Boix, E., Navarro, S., Rosenberg, H.F., Cuchillo, C.M., Nogués, M.V., 2005. Surface-exposed amino acids of eosinophil cationic protein play a critical role in the inhibition of mammalian cell proliferation. Molecular and Cellular Biochemistry 272 (1-2), 1–7. Available at https://doi.org/10.1007/s11010-005-4777-2.

Chao, T.Y., Lavis, L.D., Raines, R.T., 2010. Cellular uptake of ribonuclease a relies on anionic glycans. Biochemistry 49 (50), 10666–10673. Available at https://doi.org/10.1021/bi1013485.

Costanzi, J., Sidransky, D., Navon, A., Goldsweig, H., 2005. Ribonucleases as a novel pro-apoptotic anticancer strategy: review of the preclinical and clinical data for ranpirnase. Cancer Investigation 23 (7), 643–650. Available at https://doi.org/10.1080/07357900500283143.

Cozza, G., Moro, S., Gotte, G., 2008. Elucidation of the ribonuclease a aggregation process mediated by 3D domain swapping: a computational approach reveals possible new multimeric structures. Biopolymers 89 (1), 26–39. Available at https://doi.org/10.1002/bip.20833.

Crestfield, A.M., Stein, W.H., Moore, S., 1962. On the aggregation of bovine pancreatic ribonuclease. Archives of Biochemistry and Biophysics 1, 217–222.

Crestfield, A.M., Stein, W.H., Moore, S., 1963. On the preparation of bovine pancreatic ribonuclease A. The Journal of Biological Chemistry 238, 618–621. Available at https://doi.org/10.1016/s0021-9258(18)81307-2.

D'Alessio, G., Flobidi, A., Prisco, R.D., Pignero, A., Leone, E., 1972a. Bull semen ribonucleases: 1. Purification and physico-chemical properties of the major component. European Journal of Biochemistry 26 (2), 153–161. Available at https://doi.org/10.1111/j.1432-1033.1972.tb01751.x.

D'Alessio, G., Parente, A., Guida, C., Leone, E., 1972b. Dimeric structure of seminal ribonuclease. FEBS Letters 27 (2), 285–288. Available at https://doi.org/10.1016/0014-5793(72)80642-2.

De Tomi, E., Campagnari, R., Orlandi, E., Cardile, A., Zanrè, V., Menegazzi, M., et al., 2022. Upregulation of miR-34a-5p, miR-20a-3p and miR-29a-3p by Onconase in A375 melanoma cells correlates with the downregulation of specific onco-proteins. International Journal of Molecular Sciences 23 (3). Available at https://doi.org/10.3390/ijms23031647, https://www.mdpi.com/1422-0067/23/3/1647/pdf.

delCardayré, S.B., Raines, R.T., 1994. Structural determinants of enzymic processivity. Biochemistry 33 (20), 6031–6037. Available at https://doi.org/10.1021/bi00186a001.

delCardayré, S.B., Raines, R.T., 1995. A residue to residue hydrogen bond mediates the nucleotide specificity of ribonuclease A. Journal of Molecular Biology 252 (3), 328–336. Available at https://doi.org/10.1006/jmbi.1995.0500.

Di Donato, A., Cafaro, V., D'Alessio, G., 1994. Ribonuclease A can be transformed into a dimeric ribonuclease with antitumor activity. Journal of Biological Chemistry 269 (26), 17394–17396. Available at https://doi.org/10.1016/s0021-9258(17)32451-1.

Ercole, C., Spadaccini, R., Alfano, C., Tancredi, T., Picone, D., 2007. A new mutant of bovine seminal ribonuclease with a reversed swapping propensity. Biochemistry 46 (8), 2227–2232. Available at https://doi.org/10.1021/bi0613630.

Ercole, C., Colamarino, R.A., Pizzo, E., Fogolari, F., Spadaccini, R., Picone, D., 2009. Comparison of the structural and functional properties of RNase A and BS-RNase: a stepwise mutagenesis approach. Biopolymers 91 (12), 1009–1017. Available at https://doi.org/10.1002/bip.21176.

Estrella, V., Chen, T., Lloyd, M., Wojtkowiak, J., Cornnell, H.H., Ibrahim-Hashim, A., et al., 2013. Acidity generated by the tumor microenvironment drives local invasion. Cancer Research 73 (5), 1524–1535. Available at https://doi.org/10.1158/0008-5472.CAN-12-2796, http://cancerres.aacrjournals.org/content/73/5/1524.full.pdf+html.

Fagagnini, A., Montioli, R., Caloiu, A., Ribó, M., Laurents, D.V., Gotte, G., 2017a. Extensive deamidation of RNase A inhibits its oligomerization through 3D domain swapping. Biochimica et Biophysica Acta - Proteins and Proteomics 1865 (1), 76–87. Available at https://doi.org/10.1016/j.bbapap.2016.10.008, http://www.elsevier.com/locate/bbapap.

Fagagnini, A., Pica, A., Fasoli, S., Montioli, R., Donadelli, M., Cordani, M., et al., 2017b. Onconase dimerization through 3D domain swapping: structural investigations and increase in the apoptotic effect in cancer cells* Biochemical Journal 474 (22), 3767–3781. Available at https://doi.org/10.1042/BCJ20170541, http://www.biochemj.org/content/474/22/3767.full-text.pdf.

Fasoli, S., Bettin, I., Montioli, R., Fagagnini, A., Peterle, D., Laurents, D.V., et al., 2021. Dimerization of human angiogenin and of variants involved in neurodegenerative diseases. International Journal of Molecular Sciences 22 (18). Available at https://doi.org/10.3390/ijms221810068, https://www.mdpi.com/1422-0067/22/18/10068/pdf.

Felsenfeld, G., Sandeen, G., Vonhippel, P.H., 1963. The Destabilizing Effect Of Ribonuclease On The Helical Dna Structure. Proceedings of the National Academy of Sciences of the United States of 50, 644–651. Available at https://doi.org/10.1073/pnas.50.4.644.

Ferraro, G., Loreto, D., Merlino, A., 2021. Interaction of platinum-based drugs with proteins: an overview of representative crystallographic studies. Current Topics in Medicinal Chemistry 21 (1), 6–27. Available at https://doi.org/10.2174/1568026620666200624162213, http://www.eurekaselect.com/615/journal/current-topics-medicinal-chemistry.

Ferraro, G., Pracharova, J., Gotte, G., Massai, L., Berecka, M., Starha, P., et al., 2025. Cytotoxicity and binding to DNA, lysozyme, ribonuclease a, and human serum albumin of the diiodido analog of picoplatin. Inorganic Chemistry 64 (18), 8895–8905. Available at https://doi.org/10.1021/acs.inorgchem.4c05424, http://pubs.acs.org/journal/inocaj.

Fett, J.W., Strydom, D.J., Lobb, R.R., Alderman, E.M., Bethune, J.L., Riordan, J.F., et al., 1985. Isolation and characterization of angiogenin, an angiogenic protein from human carcinoma cells. Biochemistry 24 (20), 5480–5486. Available at https://doi.org/10.1021/bi00341a030.

Fiorini, C., Gotte, G., Donnarumma, F., Picone, D., Donadelli, M., 2014. Bovine seminal ribonuclease triggers Beclin1-mediated autophagic cell death in pancreatic cancer cells. Biochimica et Biophysica Acta - Molecular Cell Research 1843 (5), 976–984. Available at https://doi.org/10.1016/j.bbamcr.2014.01.025.

Fiorini, C., Cordani, M., Gotte, G., Picone, D., Donadelli, M., 2015. Onconase induces autophagy sensitizing pancreatic cancer cells to gemcitabine and activates Akt/mTOR pathway in a ROS-dependent manner. Biochimica et Biophysica Acta - Molecular Cell Research 1853 (3), 549–560. Available at https://doi.org/10.1016/j.bbamcr.2014.12.016, www.elsevier.com/locate/bbamcr.

Fruchter, R.G., Crestfield, A.M., 1965. Preparation and properties of two active forms of ribonuclease dimer. The Journal of Biological Chemistry 240 (10), 3868–3874. Available at https://doi.org/10.1016/s0021-9258(18)97123-1.

Futami, J., Kitazoe, M., Maeda, T., Nukui, E., Sakaguchi, M., Kosaka, J., et al., 2005. Intracellular delivery of proteins into mammalian living cells by polyethylenimine-cationization. Journal of Bioscience and Bioengineering 99 (2), 95–103. Available at https://doi.org/10.1263/jbb.99.95, http://www.elsevier.com/wps/find/journaldescription.cws_home/505516/description#description.

Garnett, E.R., Raines, R.T., 2022. Emerging biological functions of ribonuclease 1 and angiogenin. Critical Reviews in Biochemistry and Molecular Biology 57 (3), 244–260. Available at https://doi.org/10.1080/10409238.2021.2004577, http://www.tandfonline.com/loi/ibmg20.

Gleich, G.J., Loegering, D.A., Bell, M.P., Checkel, J.L., Ackerman, S.J., McKean, D.J., 1986. Biochemical and functional similarities between human eosinophil-derived neurotoxin and eosinophil cationic protein: homology with ribonuclease. Proceedings of the National Academy of Sciences of the United States of America 83 (10), 3146–3150. Available at https://doi.org/10.1073/pnas.83.10.3146.

Goldschmidt, L., Teng, P.K., Riek, R., Eisenberg, D., 2010. Identifying the amylome, proteins capable of forming amyloid-like fibrils. Proceedings of the National Academy of Sciences of the United States of America 107 (8), 3487–3492. Available at https://doi.org/10.1073/pnas.0915166107, http://www.pnas.org/content/107/8/3487.full.pdf.

Gotte, G., Testolin, L., Costanzo, C., Sorrentino, S., Armato, U., Libonati, M., 1997. Cross-linked trimers of bovine ribonuclease A: activity on double-stranded RNA and antitumor action. FEBS Letters 415 (3), 308–312. Available at https://doi.org/10.1016/S0014-5793(97)01147-2.

Gotte, G., Libonati, M., 1998. Two different forms of aggregated dimers of ribonuclease A. Biochimica et Biophysica Acta - Protein Structure and Molecular Enzymology 1386 (1), 106–112. Available at https://doi.org/10.1016/S0167-4838(98)00087-9.

Gotte, G., Bertoldi, M., Libonati, M., 1999. Structural versatility of bovine ribonuclease A. Distinct conformers of trimeric and tetrameric aggregates of the enzyme. European Journal of Biochemistry 265 (2), 680–687. Available at https://doi.org/10.1046/j.1432-1327.1999.00761.x.

Gotte, G., Vottariello, F., Libonati, M., 2003a. Thermal aggregation of ribonuclease A. A contribution to the understanding of the role of 3D domain swapping in protein aggregation. Journal of Biological Chemistry 278 (12), 10763–10769. Available at https://doi.org/10.1074/jbc.M213146200.

Gotte, G., Libonati, M., Laurents, D.V., 2003b. Glycosylation and specific deamidation of ribonuclease B affect the formation of three-dimensional domain-swapped oligomers. Journal of Biological Chemistry 278 (47), 46241–46251. Available at https://doi.org/10.1074/jbc.M308470200.

Gotte, G., Laurents, D.V., Libonati, M., 2006a. Three-dimensional domain-swapped oligomers of ribonuclease A: identification of a fifth tetramer, pentamers and hexamers, and detection of trace heptameric, octameric and nonameric species. Biochimica et Biophysica Acta - Proteins and Proteomics 1764 (1), 44–54. Available at https://doi.org/10.1016/j.bbapap.2005.10.011.

Gotte, G., Donadelli, M., Laurents, D.V., Vottariello, F., Morbio, M., Libonati, M., 2006b. Increase of RNase A N-terminus polarity or C-terminus apolarity changes the two domains' propensity to swap and form the two dimeric conformers of the protein. Biochemistry 45 (36), 10795–10806. Available at https://doi.org/10.1021/bi060933t.

Gotte, G., Helmy, A.M., Ercole, C., Spadaccini, R., Laurents, D.V., Donadelli, M., et al., 2012. Double domain swapping in bovine seminal RNase: formation of distinct N- and C-swapped tetramers and multimers with increasing biological activities. PLoS One 7 (10), e46804. Available at https://doi.org/10.1371/journal.pone.0046804.

Gotte, G., Menegazzi, M., 2019. Biological activities of secretory RNases: focus on their oligomerization to design antitumor drugs. Frontiers in Immunology 10. Available at https://doi.org/10.3389/fimmu.2019.02626, https://www.frontiersin.org/journals/immunology#.

Gotte, G., Campagnari, R., Loreto, D., Bettin, I., Calzetti, F., Menegazzi, M., et al., 2021. The crystal structure of the domain-swapped dimer of onconase highlights some catalytic and antitumor activity features of the enzyme. International Journal of Biological Macromolecules 191, 560–571. Available at https://doi.org/10.1016/j.ijbiomac.2021.09.095, www.elsevier.com/locate/ijbiomac.

Gotte, G., Butturini, E., Bettin, I., Noro, I., Mahmoud Helmy, A., Fagagnini, A., et al., 2022. Slow evolution toward "super-aggregation" of the oligomers formed through the swapping of RNase A N-termini: a wish for amyloidosis? International Journal of Molecular Sciences 23 (19). Available at https://doi.org/10.3390/ijms231911192, http://www.mdpi.com/journal/ijms.

Gotte, G., 2024. Effects of pathogenic mutants of the neuroprotective RNase 5-angiogenin in amyotrophic lateral sclerosis (ALS). Genes 15 (6). Available at https://doi.org/10.3390/genes15060738, http://www.mdpi.com/journal/genes/.

Greenway, M.J., Andersen, P.M., Russ, G., Ennis, S., Cashman, S., Donaghy, C., et al., 2006. ANG mutations segregate with familial and 'sporadic' amyotrophic lateral sclerosis. Nature Genetics 38 (4), 411–413. Available at https://doi.org/10.1038/ng1742.

Halicka, H.D., Ardelt, B., Shogen, K., Darzynkiewicz, Z., 2007. Mild hyperthermia predisposes tumor cells to undergo apoptosis upon treatment with onconase. International Journal of Oncology 30 (4), 841–847. Available at https://doi.org/10.3892/ijo.30.4.841, http://www.spandidos-publications.com/serveFile/ijo_30_4_841_PDF.pdf?type=article&article_id=ijo_30_4_841&item=PDF.

Hartman, F.C., Wold, F., 1967. Cross-linking of bovine pancreatic ribonuclease a with dimethyl adipimidate. Biochemistry 6 (8), 2439–2448. Available at https://doi.org/10.1021/bi00860a021.

Hoang, T.T., Johnson, D.A., Raines, R.T., Johnson, J.A., 2019. Angiogenin activates the astrocytic Nrf2/antioxidant-response element pathway and thereby protects murine neurons from oxidative stress. Journal of Biological Chemistry 294 (41), 15095–15103. Available at https://doi.org/10.1074/jbc.RA119.008491, http://www.jbc.org/content/294/41/15095.full.pdf.

Ivanov, P., O'Day, E., Emara, M.M., Wagner, G., Lieberman, J., Anderson, P., 2014. G-quadruplex structures contribute to the neuroprotective effects of angiogenin-induced tRNA fragments. Proceedings of the National Academy of Sciences of the United States of America 111 (51), 18201–18206. Available at https://doi.org/10.1073/pnas.1407361111, http://www.pnas.org/content/111/51/18201.full.pdf+html.

Jansen, E.F., Tomimatsu, Y., Olson, A.C., 1971. Cross-linking of α-chymotrypsin and other proteins by reaction with glutaraldehyde. Archives of Biochemistry and Biophysics 144 (1), 394–400. Available at https://doi.org/10.1016/0003-9861(71)90492-9.

Joao, H.C., Dwek, R.A., 1993. Effects of glycosylation on protein structure and dynamics in ribonuclease B and some of its individual glycoforms. European Journal of Biochemistry 218 (1), 239–244. Available at https://doi.org/10.1111/j.1432-1033.1993.tb18370.x.

Johnson, R.J., McCoy, J.G., Bingman, C.A., Phillips, G.N., Raines, R.T., 2007. Inhibition of human pancreatic ribonuclease by the human ribonuclease inhibitor protein. Journal of Molecular Biology 368 (2), 434–449. Available at https://doi.org/10.1016/j.jmb.2007.02.005, https://www.journals.elsevier.com/journal-of-molecular-biology.

Kartha, G., Bello, J., Harker, D., 1967. Tertiary structure of ribonuclease. Nature 213 (5079), 862–865. Available at https://doi.org/10.1038/213862a0.

Kim, E.E., Wyckoff, H.W., Richards, F.M., Varadarajan, R., 1992. Refinement of the crystal structure of ribonuclease S. Comparison with and between the various ribonuclease A structures. Biochemistry 31 (49), 12304–12314. Available at https://doi.org/10.1021/bi00164a004.

Kim, J.-S., Soucek, J., Matousek, J., Raines, R.T., 1995. Mechanism of ribonuclease cytotoxicity. Journal of Biological Chemistry 270 (52), 31097–31102. Available at https://doi.org/10.1074/jbc.270.52.31097.

Kobe, B., Deisenhofer, J., 1993. Crystal structure of porcine ribonuclease inhibitor, a protein with leucine-rich repeats. Nature 366 (6457), 751–756. Available at https://doi.org/10.1038/366751a0.

Kobe, B., Deisenhofer, J., 1996. Mechanism of ribonuclease inhibition by ribonuclease inhibitor protein based on the crystal structure of its complex with ribonuclease A. Journal of Molecular Biology 264 (5), 1028–1043. Available at https://doi.org/10.1006/jmbi.1996.0694, https://www.journals.elsevier.com/journal-of-molecular-biology.

Ledoux, L., 1955. Action of ribonuclease on certain ascites tumours. Nature 175 (4449), 258–259. Available at https://doi.org/10.1038/175258b0.

Leich, F., Stöhr, N., Rietz, A., Ulbrich-Hofmann, R., Arnold, U., 2007. Endocytotic internalization as a crucial factor for the cytotoxicity of ribonucleases. Journal of Biological Chemistry 282 (38), 27640–27646. Available at https://doi.org/10.1074/jbc.M702240200Germany, http://www.jbc.org/cgi/reprint/282/38/27640.

Leland, P.A., Staniszewski, K.E., Kim, B.M., Raines, R.T., 2000. A synapomorphic disulfide bond is critical for the conformational stability and cytotoxicity of an amphibian ribonuclease. FEBS Letters 477 (3), 203–207. Available at https://doi.org/10.1016/S0014-5793(00)01804-4.

Leland, P.A., Staniszewski, K.E., Park, C., Kelemen, B.R., Raines, R.T., 2002. The ribonucleolytic activity of angiogenin. Biochemistry 41 (4), 1343–1350. Available at https://doi.org/10.1021/bi0117899.

Liang, C.J., Yamashita, K., Kobata, A., 1980. Structural study of the carbohydrate moiety of bovine pancreatic ribonuclease B1. Journal of Biochemistry 88 (1), 51–58.

Libonati, M., Floridi, A., 1969. Breakdown of double-stranded RNA by bull semen ribonuclease. European Journal of Biochemistry 8 (1), 81–87. Available at https://doi.org/10.1111/j.1432-1033.1969.tb00498.x.

Libonati, M., Gotte, G., 2004. Oligomerization of bovine ribonuclease A: structural and functional features of its multimers. Biochemical Journal 380 (2), 311–327. Available at https://doi.org/10.1042/BJ20031922.

Libonati, M., Sorrentino, S., 1992. Revisiting the action of bovine ribonuclease A and pancreatic-type ribonucleases on double-stranded RNA. Molecular and Cellular Biochemistry 117 (2), 139–151. Available at https://doi.org/10.1007/BF00230753.

Liu, Y., Hart, P.J., Schlunegger, M.P., Eisenberg, D., 1998. The crystal structure of a 3D domain-swapped dimer of RNase A at a 2.1-Å resolution. Proceedings of the National Academy of Sciences of the United States of America 95 (7), 3437–3442. Available at https://doi.org/10.1073/pnas.95.7.3437, www.pnas.org.

Liu, Y., Gotte, G., Libonati, M., Eisenberg, D., 2001. A domain-swapped RNase A dimer with implications for amyloid formation. Nature Structural Biology 8 (3), 211–214. Available at https://doi.org/10.1038/84941.

Liu, Y., Gotte, G., Libonati, M., Eisenberg, D., 2002. Structures of the two 3D domain-swapped RNase A trimers. Protein Science 11 (2), 371–380. Available at https://doi.org/10.1110/ps.36602.

Liu, Y., Eisenberg, D., 2002. 3D domain swapping: as domains continue to swap. Protein Science 11 (6), 1285–1299. Available at https://doi.org/10.1110/ps.0201402.

López-Alonso, J.P., Bruix, M., Font, J., Ribó, M., Vilanova, M., Rico, M., et al., 2006. Formation, structure, and dissociation of the ribonuclease S three-dimensional domain-swapped dimer. Journal of Biological Chemistry 281 (14), 9400–9406. Available at https://doi.org/10.1074/jbc.M510491200Spain, http://www.jbc.org/cgi/reprint/281/14/9400.

López-Alonso, J.P., Gotte, G., Laurents, D.V., 2009a. Kinetic analysis provides insight into the mechanism of Ribonuclease A oligomer formation. Archives of Biochemistry and Biophysics 489 (1-2), 41–47. Available at https://doi.org/10.1016/j.abb.2009.07.013.

López-Alonso, J.P., Diez-García, F., Font, J., Ribó, M., Vilanova, M., Scholtz, J.M., et al., 2009b. Carbodiimide EDC induces cross-links that stabilize RNase a C-dimer against dissociation: EDC adducts can affect protein net charge, conformation, and activity. Bioconjugate Chemistry 20 (8), 1459–1473. Available at https://doi.org/10.1021/bc9001486, http://pubs.acs.org/doi/pdfplus/10.1021/bc9001486.

Lucignano, R., Bologna, A., Gramazio, S., Wang, P.H., Taxis, C., Essen, L.O., et al., 2025. Unravelling the amyloid aggregation mechanism of the sweet protein Monellin: insights from circular permutated mutants. International Journal of Biological Macromolecules 308. Available at https://doi.org/10.1016/j.ijbiomac.2025.142239, https://www.sciencedirect.com/science/journal/01418130.

Mastronicola, M.R., Piccoli, R., D'Alessio, G., 1995. Key extracellular and intracellular steps in the antitumor action of seminal ribonuclease. European Journal of Biochemistry 230 (1), 242–249. Available at https://doi.org/10.1111/j.1432-1033.1995.tb20557.x.

Matoušek, J., Grozdanovič, J., 1973. Specific effect of bull seminal ribonuclease (AS RNase) on cell systems in mice. Comparative Biochemistry and Physiology Part A: Physiology 46 (2), 241–248. Available at https://doi.org/10.1016/0300-9629(73)90415-5.

Matousek, J., Gotte, G., Pouckova, P., Soucek, J., Slavik, T., Vottariello, F., et al., 2003. Antitumor activity and other biological actions of oligomers of ribonuclease A. Journal of Biological Chemistry 278 (26), 23817–23822. Available at https://doi.org/10.1074/jbc.M302711200.

Mazzarella, L., Capasso, S., Demasi, D., Di Lorenzo, G., Mattia, C.A., Zagari, A., 1993. Bovine seminal ribonuclease: structure at 1.9Å resolution. Acta Crystallographica. Section D, Biological Crystallography 49 (4), 389–402. Available at https://doi.org/10.1107/s0907444993003403.

Menegazzi, M., Gotte, G., 2022. Role of the ribonuclease ONCONASE in miRNA biogenesis and tRNA processing: focus on cancer and viral infections. International Journal of Molecular Sciences 23 (12). Available at https://doi.org/10.3390/ijms23126556, https://www.mdpi.com/1422-0067/23/12/6556/pdf?version=1655017334.

Merlino, A., Ercole, C., Picone, D., Pizzo, E., Mazzarella, L., Sica, F., 2008. The buried diversity of bovine seminal ribonuclease: shape and cytotoxicity of the swapped non-covalent form of the enzyme. Journal of Molecular Biology 376 (2), 427–437. Available at https://doi.org/10.1016/j.jmb.2007.11.008, https://www.journals.elsevier.com/journal-of-molecular-biology.

Merrifield, B., 1986. Solid phase synthesis. Science (New York, N.Y.) 232 (4748), 341–347. Available at https://doi.org/10.1126/science.3961484.

Messori, L., Merlino, A., 2017. Protein metalation by metal-based drugs: X-ray crystallography and mass spectrometry studies. Chem. Commun. 53 (85), 11622–11633. Available at https://doi.org/10.1039/C7CC06442J.

Mikulski, S., Viera, A., Shogen, K., 1992. Invitro synergism between a novel amphibian oocytic ribonuclease (onconase(r)) and tamoxifen, lovastatin and Cisplatin, in human ovcar-3 ovarian-carcinoma cell-line. International Journal of Oncology 1 (7), 779–785.

Mikulski, S.M., Grossman, A.M., Carter, P.W., Shogen, K., Costanzi, J.J., 1993. Phase I human clinical trial of ONCONASE® (P-30 protein) administered intravenously on a weekly schedule in cancer patients with solid tumors. International Journal of Oncology 3 (1), 57–64.

Mikulski, S.M., Chun, H.G., Mittelman, A., Panella, T., Puccio, C.A., Shogen, K., et al., 1995. Relationship between response rate and median survival in patients with advanced non-small cell lung cancer: comparison of ONCONASE(®) with other anticancer agents. International Journal of Oncology 6 (4), 889–897.

Mizuta, K., Awazu, S., Yasuda, T., Kishi, K., 1990. Purification and characterization of three ribonucleases from human kidney: comparison with urine ribonucleases. Archives of Biochemistry and Biophysics 281 (1), 144–151. Available at https://doi.org/10.1016/0003-9861(90)90424-W.

Montioli, R., Campagnari, R., Fasoli, S., Fagagnini, A., Caloiu, A., Smania, M., et al., 2021. Rnase a domain-swapped dimers produced through different methods: structure–catalytic properties and antitumor activity. Life (Chicago, Ill. 1978) 11 (2), 1–19. Available at https://doi.org/10.3390/life11020168, https://www.mdpi.com/2075-1729/11/2/168/pdf.

Moroianu, J., Riordan, J.F., 1994. Identification of the nucleolar targeting signal of human angiogenin. Biochemical and Biophysical Research Communications 203 (3), 1765–1772. Available at https://doi.org/10.1006/bbrc.1994.2391.

Naddeo, M., Vitagliano, L., Russo, A., Gotte, G., D'Alessio, G., Sorrentino, S., 2005. Interactions of the cytotoxic RNase A dimers with the cytosolic ribonuclease inhibitor. FEBS Letters 579 (12), 2663–2668. Available at https://doi.org/10.1016/j.febslet.2005.03.087, www.elsevier.com/locate/febslet.

Nenci, A., Gotte, G., Maras, B., Libonati, M., 2001a. Different susceptibility of the two dimers of ribonuclease A to subtilisin. Implications for their structure. Biochimica et Biophysica Acta - Protein Structure and Molecular Enzymology 1545 (1-2), 255–262. Available at https://doi.org/10.1016/S0167-4838(00)00287-9.

Nenci, A., Gotte, G., Bertoldi, M., Libonati, M., 2001b. Structural properties of trimers and tetramers of ribonuclease A. Protein Science 10 (10), 2017–2027. Available at https://doi.org/10.1110/ps.14101.

Neu, C., Beckers, C., Frank, N., Thomas, K., Bartneck, M., Simon, T.P., et al., 2024. Ribonuclease inhibitor 1 emerges as a potential biomarker and modulates inflammation and iron homeostasis in sepsis. Scientific Reports 14 (1). Available at https://doi.org/10.1038/s41598-024-65778-8, https://www.nature.com/srep/.

Neumann, M., Sampathu, D.M., Kwong, L.K., Truax, A.C., Micsenyi, M.C., Chou, T.T., et al., 2006. Ubiquitinated TDP-43 in frontotemporal lobar degeneration and amyotrophic lateral sclerosis. Science (New York, N.Y.) 314 (5796), 130–133. Available at https://doi.org/10.1126/science.1134108.

Nogués, M.V., Vilanova, M., Cuchillo, C.M., 1995. Bovine pancreatic ribonuclease A as a model of an enzyme with multiple substrate binding sites. Biochimica et Biophysica Acta (BBA)/Protein Structure and Molecular 1253 (1), 16–24. Available at https://doi.org/10.1016/0167-4838(95)00138-K.

Noji, M., Samejima, T., Yamaguchi, K., So, M., Yuzu, K., Chatani, E., et al., 2021. Breakdown of supersaturation barrier links protein folding to amyloid formation. Communications Biology 4 (1). Available at https://doi.org/10.1038/s42003-020-01641-6, https://www.springer.com/journal/11852.

Noro, I., Bettin, I., Fasoli, S., Smania, M., Lunardi, L., Giannini, M., et al., 2023. Human RNase 1 can extensively oligomerize through 3D domain swapping thanks to the crucial contribution of its C-terminus. International Journal of Biological Macromolecules 249, 126110. Available at https://doi.org/10.1016/j.ijbiomac.2023.126110.

Passarini, C., Cardile, A., Zuanetti, F., Zanrè, V., Pacchiana, R., Celesia, A., Danzi, F., Fiore, A., Gotte, G., Menegazzi, M., 2026. Onconase induces apoptosis in dabrafenib-resistant melanoma cell lines through dysregulation of ROS homeostasis, antioxidant protein expression, and mitochondrial dynamics. International Journal of Molecular Sciences 27 (4). Available at 10.3390/ijms27041638.

Pica, A., Merlino, A., Buell, A.K., Knowles, T.P.J., Pizzo, E., D'Alessio, G., et al., 2013. Three-dimensional domain swapping and supramolecular protein assembly: insights from the X-ray structure of a dimeric swapped variant of human pancreatic RNase. Acta Crystallographica Section D: Biological Crystallography 69 (10), 2116–2123. Available at https://doi.org/10.1107/S0907444913020507.

Piccoli, R., Di Gaetano, S., De Lorenzo, C., Grauso, M., Monaco, C., Spalletti-Cernia, D., et al., 1999. A dimeric mutant of human pancreatic ribonuclease with selective cytotoxicity toward malignant cells. Proceedings of the National Academy of Sciences of the United States of America 96 (14), 7768–7773. Available at https://doi.org/10.1073/pnas.96.14.7768, www.pnas.org.

Piccoli, R., Tamburrini, M., Piccialli, G., Di Donato, A., Parente, A., D'Alessio, G., 1992. The dual-mode quaternary structure of seminal RNase. NProceedings of the National Academy of Sciences of the United States of America 89 (5), 1870–1874. Available at https://doi.org/10.1073/pnas.89.5.1870, www.pnas.org.

Picone, D., Donnarumma, F., Ferraro, G., Gotte, G., Fagagnini, A., Butera, G., et al., 2017. A comparison study on RNase A oligomerization induced by cisplatin, carboplatin and oxaliplatin. Journal of Inorganic Biochemistry 173, 105–112. Available at https://doi.org/10.1016/j.jinorgbio.2017.05.005, www.elsevier.com/locate/jinorgbio.

Picone, D., Donnarumma, F., Ferraro, G., Russo Krauss, I., Fagagnini, A., Gotte, G., et al., 2015. Platinated oligomers of bovine pancreatic ribonuclease: structure and stability. Journal of Inorganic Biochemistry 146, 37–43. Available at https://doi.org/10.1016/j.jinorgbio.2015.02.011, www.elsevier.com/locate/jinorgbio.

Pinato, O., Musetti, C., Farrell, N.P., Sissi, C., 2013. Platinum-based drugs and proteins: reactivity and relevance to DNA adduct formation. Journal of Inorganic Biochemistry 122, 27–37. Available at https://doi.org/10.1016/j.jinorgbio.2013.01.007.

Plummer, T.H., 1968. Glycoproteins of bovine pancreatic juice. Isolation of Ribonucleases C and D.. The Journal of Biological Chemistry 243 (22), 5961–5966. Available at https://doi.org/10.1016/s0021-9258(18)94514-x.

Plummer, T.H., Hirs, C.H., 1963. The isolation of ribounclease B, a glycoprotein, from bovine pancreatic juice. The Journal of Biological Chemistry 238, 1396–1401. Available at https://doi.org/10.1016/s0021-9258(18)81195-4.

Polymeropoulos, M.H., Lavedan, C., Leroy, E., Ide, S.E., Dehejia, A., Dutra, A., et al., 1997. Mutation in the α-synuclein gene identified in families with Parkinson's disease. Science (New York, N.Y.) 276 (5321), 2045–2047. Available at https://doi.org/10.1126/science.276.5321.2045.

Raineri, A., Prodomini, S., Fasoli, S., Gotte, G., Menegazzi, M., 2019a. Influence of onconase in the therapeutic potential of PARP inhibitors in A375 malignant melanoma cells. Biochemical Pharmacology 167, 173–181. Available at https://doi.org/10.1016/j.bcp.2019.06.006, www.elsevier.com/locate/biochempharm.

Raineri, A., Fasoli, S., Campagnari, R., Gotte, G., Menegazzi, M., 2019b. Onconase restores cytotoxicity in dabrafenib-resistant A375 human melanoma cells and affects cell migration, invasion and colony formation capability. International Journal of Molecular Sciences 20 (23). Available at https://doi.org/10.3390/ijms20235980, https://www.mdpi.com/1422-0067/20/23/5980/pdf.

Raines, R.T., 1998. Ribonuclease A. Chemical Reviews 98 (3), 1045–1066. Available at https://doi.org/10.1021/cr960427h.

Ressler, V.T., Raines, R.T., 2019. Consequences of the endogenous N-glycosylation of human ribonuclease 1. Biochemistry 58 (7), 987–996. Available at https://doi.org/10.1021/acs.biochem.8b01246, http://pubs.acs.org/journal/bichaw.

Richards, F.M., Knowles, J.R., 1968. Glutaraldehyde as a protein cross-linking reagent. Journal of Molecular Biology 37 (1), 231–233. Available at https://doi.org/10.1016/0022-2836(68)90086-7.

Richards, F.M., Vithayathil, P.J., 1959. The preparation of subtilisn-modified ribonuclease and the separation of the peptide and protein components. The Journal of Biological Chemistry 234 (6), 1459–1465.

Rodríguez, M., Torrent, G., Bosch, M., Rayne, F., Dubremetz, J.F., Ribó, M., et al., 2007. Intracellular pathway of onconase that enables its delivery to the cytosol. Journal of Cell Science 120 (8), 1405–1411. Available at https://doi.org/10.1242/jcs.03427.

Rosen, D.R., Siddique, T., Patterson, D., Figlewicz, D.A., Sapp, P., Hentati, A., et al., 1993. Mutations in Cu/Zn superoxide dismutase gene are associated with familial amyotrophic lateral sclerosis. Nature 362 (6415), 59–62. Available at https://doi.org/10.1038/362059a0.

Rosenberg, H.F., 1995. Recombinant human eosinophil cationic protein: ribonuclease activity is not essential for cytotoxicity. Journal of Biological Chemistry 270 (14), 7876–7881. Available at https://doi.org/10.1074/jbc.270.14.7876, http://www.jbc.org/content/by/year.

Russo, N., Antignani, A., D'Alessio, G., 2000. In vitro evolution of a dimeric variant of human pancreatic ribonuclease. Biochemistry 39 (13), 3585–3591. Available at https://doi.org/10.1021/bi992367q.

Rutkoski, T.J., Kink, J.A., Strong, L.E., Schilling, C.I., Raines, R.T., 2010. Antitumor activity of ribonuclease multimers created by site-specific covalent tethering. Bioconjugate Chemistry 21 (9), 1691–1702. Available at https://doi.org/10.1021/bc100292x, http://pubs.acs.org/journal/bcches.

Rutkoski, T.J., Raines, R.T., 2008. Evasion of ribonuclease inhibitor as a determinant of ribonuclease cytotoxicity. Current Pharmaceutical Biotechnology 9 (3), 185–199. Available at https://doi.org/10.2174/138920108784567344, http://www.ingentaconnect.com/content/ben/cpb/2008/00000009/00000003/art00008.

Rybak, S.M., Pearson, J.W., Fogler, W.E., Volker, K., Spence, S.E., Newton, D.L., et al., 1996. Enhancement of vincristine cytotoxicity in drug-resistant cells by simultaneous treatment with onconase, an antitumor ribonuclease. Journal of the National Cancer Institute 88 (11), 747–753. Available at https://doi.org/10.1093/jnci/88.11.747.

Sambashivan, S., Liu, Y., Sawaya, M.R., Gingery, M., Eisenberg, D., 2005. Amyloid-like fibrils of ribonuclease a with three-dimensional domain-swapped and native-like structure. Nature 437 (7056), 266–269. Available at https://doi.org/10.1038/nature03916.

Sawaya, M.R., Sambashivan, S., Nelson, R., Ivanova, M.I., Sievers, S.A., Apostol, M.I., et al., 2007. Atomic structures of amyloid cross-β spines reveal varied steric zippers. Nature 447 (7143), 453–457. Available at https://doi.org/10.1038/nature05695, http://www.nature.com/nature/index.html.

Saxena, S.K., Rybak, S.M., Davey, R.T., Youle, R.J., Ackerman, E.J., 1992. Angiogenin is a cytotoxic, tRNA-specific ribonuclease in the RNase A superfamily. Journal of Biological Chemistry 267 (30), 21982–21986. Available at https://doi.org/10.1016/s0021-9258(19)36710-9.

Shapiro, R., Vallee, B.L., 1989. Site-directed mutagenesis of histidine-13 and histidine-114 of human angiogenin. alanine derivatives inhibit angiogenin-induced angiogenesis. Biochemistry 28 (18), 7401–7408. Available at https://doi.org/10.1021/bi00444a038.

Shen, R., Li, J., Ye, D., Wang, Q., Fei, J., 2016. Combination of onconase and dihydroartemisinin synergistically suppresses growth and angiogenesis of non-small-cell lung carcinoma and malignant mesothelioma. Acta Biochimica et Biophysica Sinica 48 (10), 894–901. Available at https://doi.org/10.1093/abbs/gmw082, http://abbs.oxfordjournals.org.

Sica, F., Pica, A., Merlino, A., Russo Krauss, I., Ercole, C., Picone, D., 2013. The multiple forms of bovine seminal ribonuclease: structure and stability of a C-terminal swapped dimer. FEBS Letters 587 (23), 3755–3762. Available at https://doi.org/10.1016/j.febslet.2013.10.003.

Sievers, K., Ficner, R., 2022. Structure of angiogenin dimer bound to double-stranded RNA. Acta Crystallographica. Section F, Structural Biology Communications. 78 (Pt 9), 330–337.

Simons, B.L., Kaplan, H., Fournier, S.M., Cyr, T., Hefford, M.A., 2007. A novel cross-linked RNase A dimer with enhanced enzymatic properties. Proteins: Structure, Function and Genetics 66 (1), 183–195. Available at https://doi.org/10.1002/prot.21144.

Siraj, S., Yameen, D., Bhati, S., Athar, T., Khan, S., Bhattacharya, J., et al., 2023. Sugar osmolyte inhibits and attenuates the fibrillogenesis in RNase A: an in vitro and in silico characterizations. International Journal of Biological Macromolecules 253 (Pt 8), 127378. Available at https://doi.org/10.1016/j.ijbiomac.2023.127378.

Skorupa, A., King, M.A., Aparicio, I.M., Dussmann, H., Coughlan, K., Breen, B., et al., 2012. Motoneurons secrete angiogenin to induce RNA cleavage in astroglia. Journal of Neuroscience 32 (15), 5024–5038. Available at https://doi.org/10.1523/JNEUROSCI.6366-11.2012Ireland, http://www.jneurosci.org/content/32/15/5024.full.pdf+html.

Smith, B.D., Raines, R.T., 2006. Genetic selection for critical residues in ribonucleases. Journal of Molecular Biology 362 (3), 459–478. Available at https://doi.org/10.1016/j.jmb.2006.07.020, https://www.journals.elsevier.com/journal-of-molecular-biology.

Smyth, D.G., Stein, W.H., Moore, S., 1963. The sequence of amino acid residues in bovine pancreatic ribonuclease: revisions and confirmations. The Journal of Biological Chemistry 238, 227–234. Available at https://doi.org/10.1016/s0021-9258(19)83984-4.

Solis, D., Bruix, M., Gonzalez, L., Diaz-Maurino, T., Rico, M., Jimenez-Barbero, J., et al., 2001. Carrier protein-modulated presentation and recognition of an N-glycan: observations on the interactions of Man8 glycoform of ribonuclease B with conglutinin. Glycobiology 11 (1), 31–36. Available at https://doi.org/10.1093/glycob/11.1.31.

Sorrentino, S., Libonati, M., 1994. Human pancreatic-type and nonpancreatic-type ribonucleases: a direct side-by-side comparison of their catalytic properties. Archives of Biochemistry and Biophysics 312 (2), 340–348. Available at https://doi.org/10.1006/abbi.1994.1318.

Sorrentino, S., 1998. Human extracellular ribonucleases: multiplicity, molecular diversity and catalytic properties of the major RNase types. Cellular and Molecular Life Sciences (CMLS) 54 (8), 785–794. Available at https://doi.org/10.1007/s000180050207.

Sorrentino, S., 2010. The eight human "canonical" ribonucleases: molecular diversity, catalytic properties, and special biological actions of the enzyme proteins. FEBS Letters 584 (11), 2194–2200. Available at https://doi.org/10.1016/j.febslet.2010.04.018.

Strydom, D.J., 1998. The angiogenins. Cellular and Molecular Life Sciences (CMLS) 54 (8), 811–824. Available at https://doi.org/10.1007/s000180050210.

Suhasini, A.N., Sirdeshmukh, R., 2007. Onconase action on tRNALys3, the primer for HIV-1 reverse transcription. Biochemical and Biophysical Research Communications 363 (2), 304–309. Available at https://doi.org/10.1016/j.bbrc.2007.08.157.

Sundlass, N.K., Eller, C.H., Cui, Q., Raines, R.T., 2013. Contribution of electrostatics to the binding of pancreatic-type ribonucleases to membranes. Biochemistry 52 (37), 6304–6312. Available at https://doi.org/10.1021/bi400619m.

SUZUKI, H., PARENTE, A., FARINA, B., GRECO, L., LA MONTAGNA, R., LEONE, E., 1987. Complete amino-acid sequence of bovine seminal ribonuclease, a dimeric protein from seminal plasma. Biological Chemistry Hoppe-Seyler 368 (2), 1305–1312. Available at https://doi.org/10.1515/bchm3.1987.368.2.1305.

Tarnowski, G.S., Kassel, R.L., Mountain, I.M., Blackburn, P., Wilson, G., Dalton, D., et al., 1976. Comparison of antitumor activities of pancreatic ribonuclease and its cross-linked dimer. Cancer Research 36, 4074–4078.

Teng, P.K., Anderson, N.J., Goldschmidt, L., Sawaya, M.R., Sambashivan, S., Eisenberg, D., 2012. Ribonuclease A suggests how proteins self-chaperone against amyloid fiber formation. Protein Science 21 (1), 26–37. Available at https://doi.org/10.1002/pro.754, http://onlinelibrary.wiley.com/doi/10.1002/pro.754/pdf.

Thomas, S.P., Hoang, T.T., Ressler, V.T., Raines, R.T., 2018. Human angiogenin is a potent cytotoxin in the absence of ribonuclease inhibitor. RNA (New York, N.Y.) 24 (8), 1018–1027. Available at https://doi.org/10.1261/rna.065516.117, http://rnajournal.cshlp.org/content/24/8/1005.full.pdf+html.

Thompson, J.E., Raines, R.T., 1994. Value of general acid-base catalysis to ribonuclease A. Journal of the American Chemical Society 116 (12), 5467–5468. Available at https://doi.org/10.1021/ja00091a060.

Titani, K., Takio, K., Kuwada, M., Nitta, K., Sakakibara, F., Kawauchi, H., et al., 1987. Amino acid sequence of sialic acid binding lectin from frog (Rana catesbeiana) eggs. Biochemistry 26 (8), 2189–2194. Available at https://doi.org/10.1021/bi00382a018.

Torrent, M., Odorizzi, F., Nogués, M.V., Boix, E., 2010. Eosinophil cationic protein aggregation: identification of an N-terminus amyloid prone region. Biomacromolecules 11 (8), 1983–1990. Available at https://doi.org/10.1021/bm100334u.

Turcotte, R.F., Raines, R.T., 2008. Interaction of onconase with the human ribonuclease inhibitor protein. Biochemical and Biophysical Research Communications 377 (2), 512–514. Available at https://doi.org/10.1016/j.bbrc.2008.10.032.

Ui, N., 1971. Isoelectric points and conformation of proteins. I. Effect of Urea on the Behavior of Some Proteins in Isoelectric Focusing. BBA - Protein Structure. 229 (3), 567–581. Available at https://doi.org/10.1016/0005-2795(71)90272-8.

Van Es, M.A., Schelhaas, H.J., Van Vught, P.W.J., Ticozzi, N., Andersen, P.M., Groen, E.J.N., et al., 2011. Angiogenin variants in Parkinson disease and amyotrophic lateral sclerosis. Annals of Neurology 70 (6), 964–973. Available at https://doi.org/10.1002/ana.22611.

Vasandani, V.M., Burris, J.A., Sung, C., 1999. Reversible nephrotoxicity of onconase and effect of lysine pH on renal onconase uptake. Cancer Chemotherapy and Pharmacology 44 (2), 164–169. Available at https://doi.org/10.1007/s002800050962.

Venge, P., Bystrom, J., 1998. Eosinophil cationic protein (ECP). The International Journal of Biochemistry & Cell Biology 30 (4), 433–437.

Vescia, D., Tramontano, G., Augusti-Tocco, G.D., 1980. 'Alessio, In vitro studies on selective inhibition of tumor cell growth by seminal ribonuclease. Cancer Research 40 (10), 3740–3744.

Vogelzang, N.J., Aklilu, M., Stadler, W.M., Dumas, M.C., Mikulski, S.M., 2001. A phase II trial of weekly intravenous ranpirnase (Onconase®), a novel ribonuclease in patients with metastatic kidney cancer. Investigational New Drugs 19 (3), 255–260. Available at https://doi.org/10.1023/A:1010633004157.

Vottariello, F., Costanzo, C., Gotte, G., Libonati, M., 2010. Zero-Length dimers of ribonuclease a: further characterization and no evidence of cytotoxicity. Bioconjugate Chemistry 21 (4), 635–645. Available at https://doi.org/10.1021/bc900407v, http://pubs.acs.org/journal/bcches.

Vottariello, F., Giacomelli, E., Frasson, R., Pozzi, N., De Filippis, V., Gotte, G., 2011. RNase A oligomerization through 3D domain swapping is favoured by a residue located far from the swapping domains. Biochimie 93 (10), 1846–1857. Available at https://doi.org/10.1016/j.biochi.2011.07.005.

Wang, D., Moore, S., 1977. Polyspermine-ribonuclease prepared by cross-linkage with dimethyl suberimidate. Biochemistry 16 (13), 2937–2942. Available at https://doi.org/10.1021/bi00632a021.

Wang, D., Wilson, G., Moore, S., 1976. Preparation of cross-linked dimers of pancreatic ribonuclease. Biochemistry 15 (3), 660–665. Available at https://doi.org/10.1021/bi00648a033.

Wegiel, J., Wisniewski, H.M., 1994. Rosenthal fibers, eosinophilic inclusions, and anchorage densities with desmosome-like structures in astrocytes in Alzheimer's disease. Acta Neuropathologica 87 (4), 355–361. Available at https://doi.org/10.1007/bf00313604.

Yasuda, T., Nadano, D., Takeshita, H., Kishi, K., 1993. Two distinct secretory ribonucleases from human cerebrum: purification, characterization and relationships to other ribonucleases. Biochemical Journal 296 (3), 617–625. Available at https://doi.org/10.1042/bj2960617.

CHAPTER 7

Ribonucleases targeting noncoding RNAs

Jiarui Li
Department of Biochemistry and Molecular Biology, Autonomous University of Barcelona, Cerdanyola del Vallès, Spain

7.1 Introduction

Noncoding RNAs (ncRNAs) refer to RNAs that do not encode proteins. It is estimated that only 2% of the human genome encodes proteins, which conversely manifests that ncRNAs occupy a vast number of transcripts from the genome (Knowling and Morris, 2011). Since the 1950s, when the ribosomal RNAs (rRNAs) and transfer RNAs (tRNAs) were discovered (Hoagland et al., 1958), and thanks to the rapid development and widespread applications of RNA sequencing methods, a large number of functional ncRNAs have been identified (Hombach and Kretz, 2016). Apart from the well-known roles of rRNAs and tRNAs in protein translation, novel ncRNAs, including microRNAs (miRNAs), small nuclear RNAs, small nucleolus RNAs, as well as long noncoding RNAs (lncRNAs) with specific biological functions, are reported nearly every day (Fu, 2014). ncRNAs involved in gene expression, translation, RNA splicing, and modification play important roles in innate immune regulation, cell proliferation, and apoptosis (Curtale and Citarella, 2013; Takeiwa et al., 2021; Kiełbowski et al., 2023). The studies on ncRNAs have gained great attention in current scientific research, aiming to unravel their key mechanisms in human metabolism and disease development and, in turn, would be beneficial for the discovery of new targets for fighting against diseases.

Pancreatic-type ribonucleases (ptRNases) are pyrimidine-specific endonucleases that cleave RNA substrates by producing 2'3' cyclic NMP intermediates (Raines, 1998). These ribonucleases (RNases) are critical for ncRNA metabolism, such as the degradation of extracellular RNAs (exRNAs), the cleavage of tRNAs, and transcription regulation by rRNA targeting (Lee et al., 2019). It has been demonstrated that RNase5 can target tRNAs to promote the formation of tRNA fragments, and RNase1 contributes to the production of extracellular tRFs under stress conditions (Fu et al., 2009; Nechooshtan et al., 2020). RNase5 also plays a role in rRNA biogenesis, which subsequently promotes endothelial cell proliferation (Kishimoto et al., 2005). Recently, RNase2 was found to selectively cleave tRNA substrates at U/C (B1) and A (B2) sites of anticodon and D-loops (Lu et al., 2022; Li et al., 2024a). In addition to RNases that

RNases in Health and Diseases
DOI: https://doi.org/10.1016/B978-0-443-27312-4.00001-8

act directly on ncRNAs, there are still ncRNAs with specific modifications naturally resistant to RNase degradation, exerting their biological structure and function for receptor sensing (Anderson et al., 2011).

Considering the increasing discoveries of the physiological roles of ncRNAs and the corresponding powerful functions of RNases, it is pertinent to summarize the current processes associated with RNases action on ncRNAs. Overall, in this chapter, the various ncRNAs, including their general functions and possible modification sites, are illustrated, followed by the exhaustive descriptions of reported ncRNAs that are targets or off-targets of RNases. Additionally, the techniques to study ncRNAs are also provided. Finally, a perspective is included at the end to emphasize the need for more thorough research between RNases and ncRNAs and their significance for future applications in the treatment of human diseases.

7.2 Noncoding RNA

7.2.1 Classification, biogenesis, and biological function

ncRNAs, considered first as nonfunctional "junk RNAs," occupy 98% of all transcriptional outputs in humans. However, the scientific community has reconceptualized the exact role of ncRNAs in the human body since the first discovery of regulatory RNAs in the 1980s (Mattick, 2001; Delihas, 2015). Currently, new ncRNAs with diverse biological roles are reported widely. The latest Nobel Prize in biology in 2024 is awarded "for the discovery of microRNA and its role in posttranscriptional gene regulation," indicating the studies on ncRNAs have gained prime attention all over the world.

In general, ncRNAs are classified into lncRNAs (>200 nt) and small ncRNAs (<200 nt) based on their size. And regarding the function of small ncRNAs, there are housekeeping ncRNAs such as tRNA and rRNA and regulatory ncRNAs, which include miRNA, small Interfering RNA (siRNA), and piRNA, among others (Fig. 7.1A). In this section, the characteristics of most studies of regulatory ncRNAs and examples of their known modes of action in gene regulation will be elucidated.

7.2.1.1 Long noncoding RNAs

lncRNAs normally refer to the long RNA transcripts (≥200 nts) that share many common features with mRNAs but do not encode proteins. Similar to mRNAs, lncRNAs are transcribed from their genes via the RNA polymerase II (Pol II), and most generated lncRNAs are decorated with a 5' methyl-cytosine cap and a polyadenylation (poly(A)) tail, although there are also lncRNAs without a poly(A) tail. Depending on their genomic origins, lncRNAs can be classified into intergenic, intronic, bidirectional, sense, and antisense ones (Ma et al., 2013). Intergenic lncRNAs are transcribed from

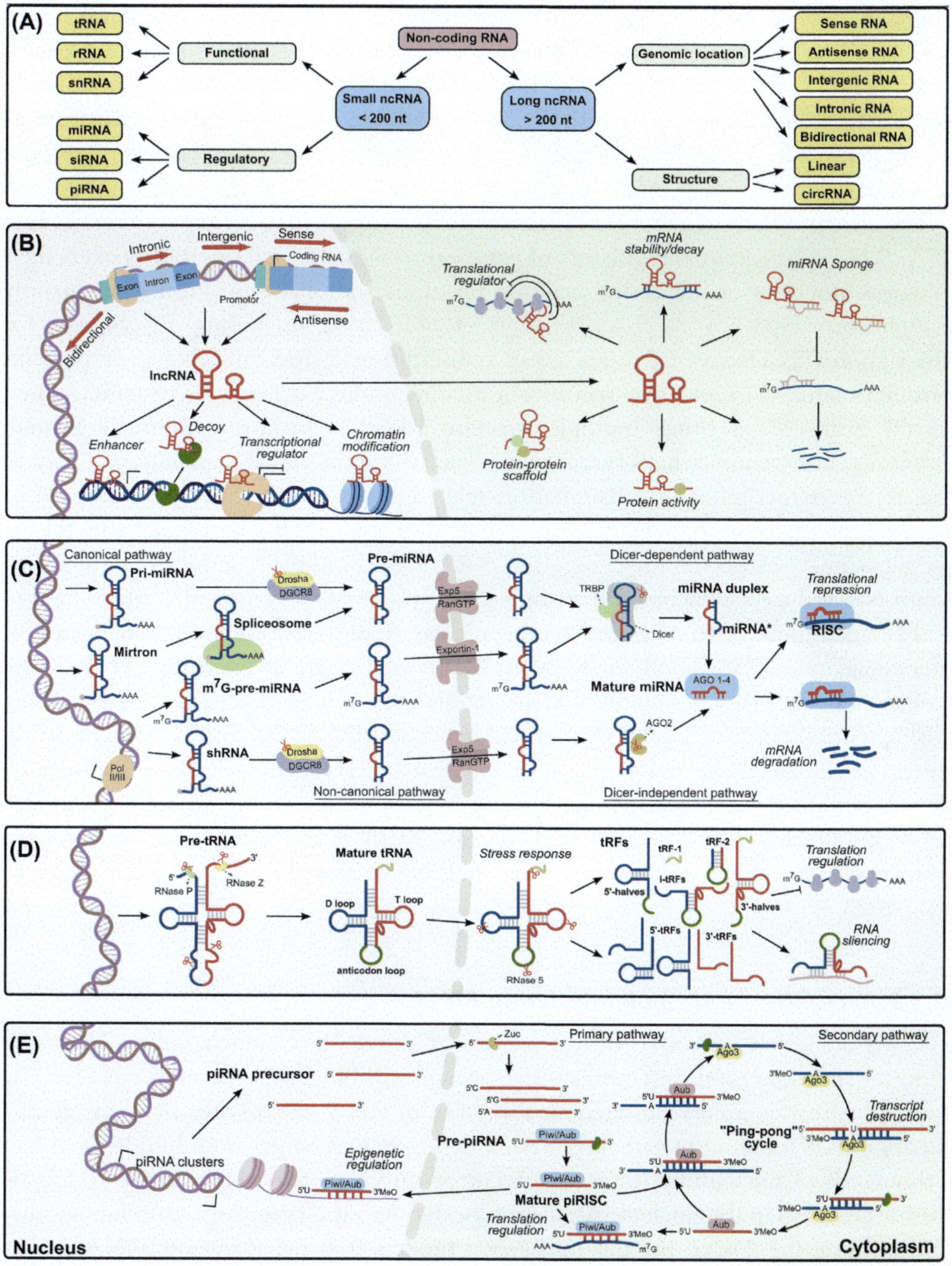

Figure 7.1 Biogenesis and biological functions of ncRNAs. (A) Classification of ncRNAs. Biogenesis and representative biological functions of (B) lncRNAs, (C) miRNAs, (D) tRNAs and tRFs, and (E) piRNAs. The functions of ncRNA are shown in italics.

intergenic regions between genes, whereas intronic lncRNAs are from introns of protein-coding genes. Sense lncRNAs are the ones from the sense strand of protein-coding genes, containing entire genes or parts of genes, and, on the contrary, antisense ones are from the antisense strand. The ones with the same promoter as a protein-coding gene but transcribed in the opposite direction are bidirectional lncRNAs (Fig. 7.1B).

The core mechanism of action of lncRNAs is based on their ability to interact with DNA, RNA, or protein as regulators of gene expression. Overall, lncRNAs exert their biological functions on transcriptional and posttranscriptional regulation levels mainly through four types of mechanisms (He et al., 2019), including (1) Signals for transcription; (2) Decoys to titrate away transcription factors, miRNAs, or proteins from chromatin; (3) Guides to recruit chromatin-modifying enzymes to target genes; and (4) Scaffolds to bring multiple proteins together to form ribonucleoprotein complexes. For example, lncRNAs can mediate the intracellular signaling pathway to regulate the transcription of inflammation-related genes under either physiological or pathological conditions (Ghahramani Almanghadim et al., 2025), and specific lncRNAs may act as miRNA decoys for competing mRNA for binding, leading to the blockage of miRNA-induced silencing and an increase in mRNA translation (Paraskevopoulou and Hatzigeorgiou, 2016). There have been many reports on inflammation-related or cancer epigenetics (Horsburgh et al., 2015; Mathy and Chen, 2017; Kumar et al., 2020; Nadhan et al., 2024), where lncRNAs are involved in epigenetic changes by recruiting epigenetic modifiers, such as DNA methyltransferases (DNMTs), for DNA methylation (Huang et al., 2022). It is also demonstrated that lncRNAs may serve as scaffolds by offering binding surfaces for specific histone modification enzyme assembly to induce histone modifications on target genes (Tsai et al., 2010). A large number of reviews have summarized the biological function of lncRNAs (Statello et al., 2021; Mattick et al., 2023).

7.2.1.2 MicroRNAs and small interfering RNAs

miRNAs are small ncRNAs with a length of about 18–22 nucleotides. There are canonical and noncanonical pathways generating miRNA (O'Brien et al., 2018). The canonical pathway, a major biogenesis source of miRNA, starts with the cleavage of a primary miRNA transcript (pri-miRNA), a stemloop that ranges from hundreds to tens of thousands of nucleotides, into the precursor miRNA (pre-miRNA, 60–70 bp) by Drosha/DGCR8 in the nucleus. After translocated into the cytoplasm, through another RNaseIII enzyme, Dicer, the pre-miRNA is processed to the mature miRNA duplex that contains mature miRNA and its complementary strand called miRNA*. Argonaute (AGO) proteins have been involved in forming an active RNA-induced silencing complex (RISC) with one miRNA strand. For the noncanonical pathway, there are either dicer-dependent or dicer-independent pathways. In the dicer-dependent pathway, mirtrons and 7-methylguanine-capped (m^7G)-pre-miRNA are exported into the cytoplasm to generate RISC depending on Dicer, as described before, while in the dicer-independent pathway, a small hairpin RNA from the nucleus to the cytoplasm is first cleaved by AGO2 and then trimmed by the poly(A)-specific ribonuclease (PARN) to form the mature miRNA (Fig. 7.1C).

After integration into RISC, mature miRNAs recognize the targeted mRNAs through base-pairing rules, and the translation is repressed, exerting the predominant mechanism of action of miRNAs, which plays a main role in gene expression regulation. One miRNA can regulate multiple genes, and conversely, one gene can be regulated by multiple miRNAs. The degree of miRNA-mRNA complementarity could determine how miRNAs silence the mRNA translation (Hudder and Novak, 2008). The targeted mRNAs will be degraded by AGO2 endonuclease activity if the miRNA and the mRNA are fully complementary, whereas nonperfect complementarity deprives the enzymatic ability of AGO2, and the mRNA translation is just suppressed (Jonas and Izaurralde, 2015). However, recent studies demonstrate that miRNAs could also induce gene expression (Vasudevan and Steitz, 2007; Place et al., 2008; Truesdell et al., 2012) and play their roles in the nucleus besides the cytoplasm. The nuclear-localized RISC could regulate mRNAs at both transcription and posttranscriptional levels. For example, miRNAs can interact with the promoter by preventing the recruitment of RNA polymerase and transcription factors to targeted genes (Ren et al., 2025), and the nuclear AGO1 can also directly interact with RNA polymerase II to regulate gene expression (Huang et al., 2013). More details about the biological functions of miRNAs can be found elsewhere (O'Brien et al., 2018).

siRNAs are ncRNAs with 20−30 nt in length. The biogenesis of siRNAs is similar to that of miRNAs, which depends on Dicer processing and then RISC loading, but the precursors to produce siRNAs or miRNAs are slightly different. In siRNAs, the initial cleavage occurs on either single-stranded RNAs (ssRNAs) or long biomolecular RNA duplexes to generate the long-extended hairpins, while in miRNAs, only ssRNAs are cleaved to form short hairpins (pri-miRNAs). Another key difference between siRNAs and miRNAs in gene silencing is that a miRNA could have multiple mRNA targets, but an siRNA only targets a specific mRNA. In addition, when mRNAs are targeted by siRNAs, they get degraded by AGO2 through its endonucleolytic activity (Lam et al., 2015).

7.2.1.3 Transfer RNAs and tRNA-derived small RNAs

tRNAs range between 70 and 100 nt in length. The typical tRNA has an L-shaped tertiary structure and a cloverleaf secondary structure with three loops (anticodon loop, D-loop, and TψC loop), four arms (D-stem, Anticodon stem, T-stem, and Acceptor arm), and a variable region. The biogenesis of tRNAs includes the transcription of precursor tRNAs (pre-tRNAs), followed by 5' leader and 3' trailer trimming by RNaseP and RNaseZ, respectively, CCA-terminus adding, introns splicing, and nucleoside modifications if there are in the nucleus (Simos and Hurt, 1999). The correctly folded and mature tRNAs are then translocated into the cytoplasm to exert their roles (Kim et al., 2020). Conceptually, tRNA isotypes refer to ones intended for the same amino acid and can be further classified as isoacceptors and isodecoders. tRNA isoacceptors mean tRNAs carry the same amino acid but with different anticodons, while tRNA isodecoders are tRNAs with the same anticodon but different sequences in the rest of the tRNA body (Hughes et al., 2023). Although tRNAs are one of the housekeeping ncRNAs, tRNA-derived small RNAs (tsRNAs) with 15–32 nts, including RNA-derived or stress-induced RNAs (tiRNAs) and tRNA-derived fragments (tRFs), have been discovered in particular cell types and play important roles in gene regulation and stress response. tsRNAs are a heterogeneous class of small ncRNAs generated from mature tRNAs or pre-tRNAs and termed as 5'-tiRNA (5' tRNA halves), 3'-tiRNA (3' tRNA halves), tRF-5, tRF-3, tRF-1, tRF-2, and i-tRF based on the cleavage site (Fig. 7.1D). Several RNases, including some of the ribonuclease A family members RNase1 and RNase5, DICER1, and RNaseL, have been identified to cleave specific tRNAs, and the produced tsRNAs may have modifications at the 5' and 3' ends that increase their stability.

Besides the canonical function of tRNAs in protein translation, tRNAs have the noncanonical functions to interact with specific elements through their structural features. Host tRNAs can behave as primers for viral reverse transcription (Kelly et al., 2003). To note, various biological functions of tsRNAs have been explosively reported nowadays with multiple mechanisms but are still based on the interaction with the target RNAs or the association with proteins ranging from transcriptional to posttranscriptional regulation (Zong et al., 2021). For example, tsRNAs can regulate translation by binding small ribosomal subunits to compete with mRNAs (Gebetsberger et al., 2017) or displacing the initiation factor from mRNAs to repress target mRNA translation (Ivanov et al., 2011). Interestingly, it has been considered that tsRNAs are similar to miRNAs in gene silencing, but there is no evidence about their interactions with AGOs, whereas the involvement of PIWI proteins is reported. In addition, tsRNAs also act as decoys to prevent transcripts from binding to proteins, leading to the decrease of transcript stability (Goodarzi et al., 2015). More roles of tsRNAs can be found in the following reviews (Pandey et al., 2021; Muthukumar et al., 2024; Jia and Zhang, 2025).

7.2.1.4 PIWI-interacting RNAs

Typical PIWI-interacting RNAs (piRNAs) are 26–30 nt in length, starting with a 5' uracil and having 2'-O-methylation at the 3' end. The biogenesis of piRNAs has been studied in *Drosophila melanogaster, Caenorhabditis elegans*, and *Mus musculus and* is currently being unveiled in *Homo sapiens* (Jiang et al., 2024). Almost all cellular transcripts from sequence clusters, individual genes, transposons, and intergenic regions in the genome could be converted into piRNA theoretically, but nearly 90% of piRNAs are from specific gene loci known as piRNA clusters (Czech et al., 2018). In the primary piRNA biogenesis, after being translocated into the cytoplasm, the piRNA precursor transcript is first cleaved by the mitochondria-associated endonuclease (Zucchini/PLD6) to produce a 5' phosphate and 3' hydroxyl piRNA (pre-piRNA). The base at the mature 5' end of pre-piRNAs is normally a uracil due to the subsequent preferential binding by PIWI proteins (Kawaoka et al., 2011), one subfamily of the AGO family. Together with the PIWI proteins, the 3' end of pre-piRNAs is trimmed by a 3'-to-5' exonuclease in the PARN family, followed by 2' oxygen methylation through the 3' methyltransferase. The generated mature piRNAs in this biogenesis are named primary piRNAs. There is also a secondary piRNA biogenesis, called the ping-pong cycle (Wu et al., 2020), producing secondary piRNAs. In this cycle in *Drosophila*, the primary piRNA acts as a guide, assisting Aub to recognize and cleave its complementary RNA-derived from the other strand of the same piRNA cluster or transposon. Following, the newly produced 5' end piRNA is loaded into Ago3 for trimming and maturation, which in turn induces the cleavage of its complementary RNA and thereby generates more piRNAs identical to the original piRNA that initiates the cycle (Czech and Hannon, 2016) (Fig. 7.1E).

piRNAs assemble in PIWI proteins to form the piRNA-induced silencing complex (pi-RISC) to silence the transposons' transcription at either transcriptional or posttranscriptional levels, called transcriptional gene silencing (TGS) and posttranscriptional gene silencing (PTGS), respectively. The former uses catalytically inactive PIWI proteins, while the latter relies highly on cleavage-competent PIWI proteins through the ping-pong cycle. In detail, in the TGS, the pi-RISC relocates to the nucleus, and the piRNA in the pi-RISC first binds to the nascent transposon transcripts by base-pairing. Then the PIWI proteins recruit cofactors and cooperate with histone-modifying enzymes or DNMTs to modify chromatin and silence transcription (Czech et al., 2018). In the PTGS, the catalytic PIWI proteins can degrade the actively targeted transposon transcripts directly and, in turn, the silencing triggers targeting elements that are amplified by the ping-pong cycle (Czech and Hannon, 2016). Although the main functions of piRNAs, firstly discovered in germ cells, are to repress transposon transcription in germ cells and to maintain the integrity of the gamete genome by protecting it, there is increasing evidence demonstrating that piRNAs play important roles in various diseases such as cancer and immune-related diseases (Jiang et al., 2024). For example, piRNAs have been identified in patients with HIV/HCV coinfection, and another study implies that piRNAs can be involved in antiviral actions (Martínez-González et al., 2020; Yu et al., 2020).

7.2.2 Modifications of noncoding RNAs

RNA posttranscriptional modifications are natural biological processes occurring on specific nucleotides within both coding RNA and ncRNAs. ncRNA modifications can be divided into reversible and nonreversible ones. The reversible modifications are the traditional chemical modifications, where over 100 types have been identified (Roundtree et al., 2017). The nonreversible modifications include adenosine deaminase (ADAR)-mediated RNA editing (Torsin et al., 2021). Here, the main identified natural modifications of regulatory ncRNAs and examples of their related biological roles are introduced. Fig. 7.2 lists the general modifications found in ncRNAs (Cui et al., 2022).

A large number of publications outline the modifications on mRNAs (McCown et al., 2020; Gilbert and Nachtergaele, 2023; Hou et al., 2024), most of which are also present in ncRNA modifications (Fig. 7.2A). Chemical modifications usually include reversible methylations, N^6-methyladenosine (m^6A), N^1-methyladenosine (m^1A), N^5-methylcytosine (m^5C), N^7-methylguanosine (m^7G), and 2'-O-methylation (Nm), as well as the nonreversible pseudouridylation (Ψ). There are three groups of RNA-modifying proteins, called "Writers," "Readers," and "Erasers," respectively, involved in depositing, recognizing and removing the RNA modifications (Esteve-Puig et al., 2020). Methyltransferases and demethylases are common "Writers" and "Erasers" of m^6A, m^1A, m^5C, m^7G, and Nm modifications, where a methyl group is added to or deleted from specific nucleotides, while "Writers" of Ψ are pseudouridine synthases. Moreover, there is also another nonreversible modification called RNA editing that refers to nucleotide substitution in an RNA sequence. Adenosine-to-inosine (A-to-I) and cytidine-to-uridine (C-to-U) are two types of RNA editing, and the former one is

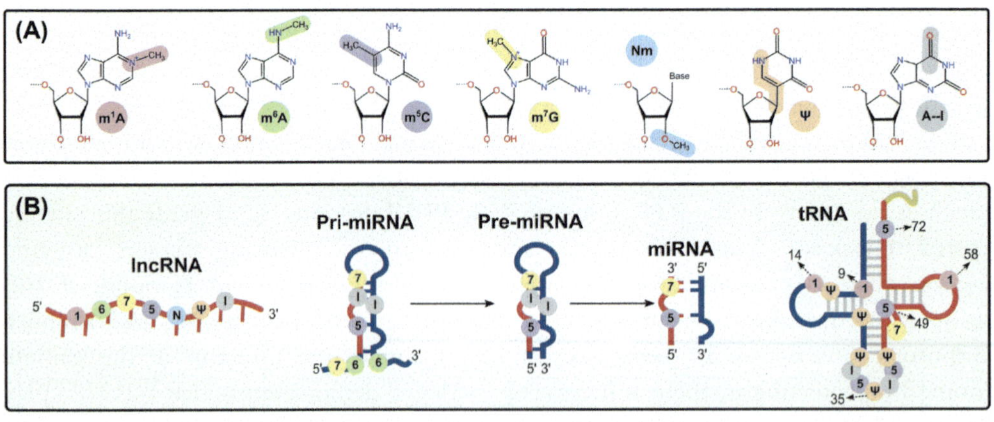

Figure 7.2 ncRNA modifications. (A) Chemical structures of common RNA modifications. (B) Modifications on different types of ncRNAs. m^1A: N^1-methyladenosine, m^6A: N^6-methyladenosine (m6A), m^5C: N^5-methylcytidine, m^7G: N^7-methylguanosine, Nm: 2'-O-methylation, Ψ: pseudouridylation, A–I: adenosine-to-inosine RNA editing.

dominant in ncRNAs. The A-to-I conversion is a deamination reaction on the adenosine C^6 catalyzed by RNA-specific ADAR (Nishikura, 2016; Christofi and Zaravinos, 2019).

To begin with, studies have shown that lncRNAs contain modified nucleotides such as Ψ, m^6A, m^1A, m^5C, m^7G, Nm, and A-to-I editing (Fig. 7.2B). The roles these modified nucleosides play in lncRNAs are not fully understood. The modifications could impair lncRNA stability, subellular distribution, or structure and thereby affect their biological functions (Li et al., 2024b). There is growing evidence that m^6A- and m^5C-related lncRNAs may be potential biomarkers in some types of tumors (Kazimierczyk and Wrzesinski, 2021). Moreover, an alteration of m^5C modification in host lncRNAs was noticed in A549 cells upon influenza A virus infection (Jiang et al., 2023), indicating that differentially modified lncRNAs are associated with host–pathogen recognition pathways. Subsequently, methylation modifications like m^6A, m^5C, m^7G, and A-to-I editing are also observed in miRNAs (Carissimi et al., 2021; Su et al., 2025). m^6A modification regulates the biosynthesis of miRNAs through recruiting DGCR8 to the target pri-miRNAs (Alarcón et al., 2015). The stability and target recognition (Konno et al., 2019) of a mature miRNA may also be altered because of the methyl marks. For example, an internal m^7G modification, mediated by the methyltransferase METTL1, in pri-let-7e, can block the formation of G-quadruplex structures, promote m^7G-modified let-7e-5p miRNA maturation, and suppress cancer cell migration (Pandolfini et al., 2019). There are many miRNAs found with A-to-I editing, resulting in suppression of miRNA biogenesis and changes of miRNA targets (Nishikura, 2016).

Notably, tRNAs and tsRNAs harbor the most diverse chemical modifications, including but not limited to A-to-I editing, dihydrouridine (D), Ψ, m^6A, m^1A, m^5C, m^7G, and Nm, at each region in their secondary structures (Fig. 7.2B). The numerous chemical modifications within tRNAs are involved in the processing, structure, and cellular localization of pre-tRNAs or mature tRNAs, playing key roles in controlling and monitoring the onset and development of biological processes and human diseases (Jackman and Alfonzo, 2013). It is demonstrated that tRNA modifications are highly conserved across species, suggesting that these modifications will not cause drastic changes in tRNA structures and have universal roles. For example, the nucleotides between positions 34 and 37 at the anticodon loop of each RNA are always modified to maintain the accuracy and stability of the tRNA–mRNA interaction (Smith et al., 2024). Modifications at either the D-loop or the TψC loop contribute to the stability of tRNA secondary and tertiary structures (Nobles et al., 2002). Besides, the modifications on specific tRNAs and tsRNAs have influences on cell cycle, stress response, and immune regulation and are related to diseases such as cancer, diabetes, and neurological diseases (Wu et al., 2024). Last but not least, tRNA modifications can regulate the tsRNAs' biogenesis. It is reported that tRNAs with Queuosine (Q), m^5C, and Nm modifications were resistant to the cleavage of angiogenin (ANG, RNase5) (Liang et al., 2025) and in turn affect downstream signaling. More details about multiple tRNA modifications and their related diseases can be found in review works (Jackman and Alfonzo, 2013; Cui et al., 2023).

For piRNA, the biogenesis concludes with the 2'-O-methylation at the 3' end to protect the piRNAs from target-directed degradation (Gainetdinov et al., 2021), but the reports on the internal modifications of piRNAs are still pending.

7.3 RNases-related noncoding RNAs

RNases are pivotal factors in the regulation of the ncRNA population. In previous sections, not only are the origins of ncRNAs and their potential functions in organisms explained, but also the roles of some ubiquitous RNases, such as RNaseIII, RNaseP, and RNaseZ, in ncRNA processing are introduced. Above all, the RNaseA superfamily, also called ptRNases, is a paradigm for RNA-binding studies. The family currently has 13 members in humans and cleaves RNA in two steps by the formation of a 2'3' cyclic phosphate (cp) intermediate (Fisher et al., 1998). The biological roles of ptRNases, such as antimicrobial activity and immune regulation, have been discovered and reported widely (Dyer and Rosenberg, 2006; Boix and Nogués, 2007; Lu et al., 2018). In addition, increasing studies have also shown that ncRNAs derived from cleavage of pri/pre-ncRNAs or mature ncRNAs and specifically regulated by ptRNases would further mediate the body's biological processes under physiological or pathological conditions. To get straight to the theme of this book, therefore, the roles of each ptRNase in RNA metabolism as well as ncRNA processing are summarized, while other key RNases found in humans are also outlined retrospectively to help clarify their similarities and differences (Table 7.1).

7.3.1 Pancreatic-type RNases

ptRNases, RNase A-like RNases, are vertebrate-specific secretory endoribonucleases that share structures and functions with bovine pancreatic RNaseA, a prime model of RNase studies since the 1960s. The typical tertiary structure of ptRNases consists of three α-helices and six β-sheets that fold a cavity to recognize RNA and cleave its phosphodiester bonds by a catalytic triad (two histidines and one lysine). The binding sites to bases of ptRNases are named B_0−B_n, where B_0 means the previous two bases, B_1 is the base before the cleaved phosphodiester bond, and B_2 is the base after the cleaved phosphodiester bond (del Cardayré and Raines, 1994). It is already demonstrated that the identification of RNA by ptRNases is base preference, where B_1 is pyrimidine-specific, and B_2 prefers purines, indicating ptRNases can target some ncRNAs and make specific ncRNA products.

Table 7.1 RNase-targeting ncRNAs and pathogen RNAs.

Human RNases	Target ncRNA	Outputs	Study models	Possible biological function	References
RNase1	• exRNA[a]	• RNA clearance	• In mouse models	• Block the adverse effects of exRNAs	Cabrera-Fuentes et al. (2014), Garnett and Raines (2022)
	• tRNA	• tRNA halves ($tRNA^{AspGTC}$-3' halves and $tRNA^{GlyGCC}$-5' halves)	• In human leukemia cell line K562		Nechooshtan et al. (2020)
	• Y RNA	• Y RNA fragments			
RNase2	• tRNA	• tiRNAs (from $tRNA^{AspGTC}$, $tRNA^{GlnCTG}$, $tRNA^{LysTTT}$, $tRNA^{LysCTT}$, $tRNA^{MetCAT}$)	• In THP-1 induced macrophages		Lu et al. (2022)
		• 5'-tRNA halves (from $tRNA^{ValCAC}$, $tRNA^{HisGUG}$)	• In A549 cells	• Activate TLR7	Shigematsu et al. (2025)
	• Synthetic, bacterial, and protozoan RNA	• Release uridines synergistically with RNaseT2	• In PMA-differentiated THP-1	• Activate TLR8	Ostendorf et al. (2020)
RNase5	• miRNA	• Decay of mature miR-141	• In human umbilical endothelial cells	• Promote proliferation, migration, tube formation, and angiogenesis	Weng et al. (2022)

(*Continued*)

Table 7.1 (Continued)

Human RNases	Target ncRNA	Outputs	Study models	Possible biological function	References
	• pRNA[b]	• Cleave pRNA within a TIP5·pRNA complex	• In HeLa cells.	• Derepress rDNA transcription	Hoang and Raines (2017)
	• rRNA	• Degrade 28S and 18S rRNAs	• In vitro degradation		Shapiro et al. (1986)
	• tRNA	• tRNA halves (from $tRNA^{Glu}$, $tRNA^{Gly}$, $tRNA^{Lys}$, $tRNA^{Val}$, $tRNA^{His}$, $tRNA^{Asp}$, and $tRNA^{SeC}$)	• In HEK293T cells		Su et al. (2019)
		• $tRNA^{HisGTG}$-5' halves, and $tRNA^{AspGTC}$-3' halves	• In U2OS cells	• Oxidative stress response	
		• $tRNA^{ValAAC}$-5' halves	• In HepG2 cells	• Stress response	Fu et al. (2009)
		• 5'-tiRNAs	• In human U2OS cells	• Inhibit protein synthesis	Ivanov et al. (2011), Yamasaki et al. (2009)
		• tRF5-Glu^{CTC}	• In RSV-infected A549 cells	• Control RSV replication	Wang et al. (2013)
		• SHOT-RNAs[c]	• In cancer cell lines	• Enhance cell proliferation	Honda et al. (2015)
RNase6	• Bacterial RNA	• Release uridines-terminated products	• In BLaER1 cells	• Activate TLR8	Nunes et al. (2024)
Onconase	• dsRNA	• Generate different-sized dsRNAs	• In vitro degradation	• RNA interference mechanism involving siRNA and miRNA	Saxena et al. (2009)

	• pre-miRNA	• Less production of mature miRNA	• In vitro degradation		Qiao et al. (2012)
	• siRNA	• Cleave siRNA within RISC	• In A549 cells	• Reverse gene silencing	Zhao et al. (2008)
	• tRNA	•tRFs (from $tRNA^{Lys}$)	• In vitro degradation		Suhasini and Sirdeshmukh (2006, 2007)
RNaseL	• cellular RNA	• Small RNAs	• In C57BL/6 mice	• RIG-I, MDA5 and IPS-1 signaling	Malathi et al. (2007)
	• ssRNA[d]	• Cleave poliovirus RNA	• In HeLa cells		Cooper et al. (2014)
	• noncoding mRNA	• Translation of mRNA decay fragments	• In A549 cells	• Prolonged activation of RNaseL	Karasik et al. (2021)
	• rRNA	• rRNA fragments (28S and 18S rRNAs)	• In A549 cells		Cooper et al., 2015
	• tRNA	• Cleavage of $tRNA^{His}$, $tRNA^{Glu}$, and $tRNA^{Pro}$ at anticodon loop	• In HeLa/T47D cells	• Essential for translation	Donovan et al. (2017)
		• noncanonical tDRs (from $tRNA^{Leu}$, $tRNA^{SeC}$, and $tRNA^{Ser}$)			Takenaka et al. (2025)
	• viral RNA	• Cleavage within large IAV[e] RNA segments	• In A549 cells		Cooper et al., 2015
	• Y RNA	• Y RNA fragments	• In HeLa cells		Donovan et al. (2017)

(*Continued*)

Table 7.1 (Continued)

Human RNases	Target ncRNA	Outputs	Study models	Possible biological function	References
RNaseT2	• rRNA	• Clear rRNAs	• In a RNase T2-mutated zebrafish model		Haud et al. (2011)
	• ssRNA	• Release uridines	• In BLaER1 cells	• Activate TLR8	Greulich et al. (2019)
	• mtRNA[f]	• Degradation	• In HEK293 cells		Liu et al. (2017)
RNaseP	• pre-tRNA	• Remove the 5' leader from pre-tRNA			Lai et al. (2010)
RNaseZ	• pre-tRNA	• Remove 3' trailer from pre-tRNA			Rossmanith (2011)
DIS3	• lncRNA	• Degradation	• In DIS3 auxin-degron cell model (DIS3-AID)		Foretek et al. (2024)

[a] exRNA: extracellular RNA (including mRNA, miRNA, tRNA, etc.).
[b] pRNA: promoter-associated RNA.
[c] SHOT-RNAs: Sex Hormone-dependent TRNA-derived RNAs.
[d] ssRNA: single-stranded RNA.
[e] IAV: Influenza A virus.
[f] mtRNA: mitochondrial RNA.

7.3.1.1 RNase1

RNase1 presents the highest catalytic activity within the family and is active against Poly(U), Poly(C), Poly(A), 2',3'-cyclic nucleotides, double-stranded RNAs (dsRNAs), and RNA–DNA hybrid chains. It prefers Poly(C) over Poly(U) (Sorrentino, 2010). RNase1 has been considered for a long time as an RNA scavenger in the body's circulation. A study showed that the plasma of mice with *RNASE1* knock-out (KO) had more RNAs than that of normal mice, suggesting one of the physiological roles of RNase1 would be to degrade exRNAs in plasma (Garnett et al., 2019). exRNAs, even though usually protected from degradation by loading into membranous vesicles or associating with lipids and/or proteins, are RNAs present outside of the cells and consist of mRNA, miRNA, tRNA, and other ncRNAs. The accumulation of exRNAs could be a sign of disease progression or exacerbation, and the addition of RNase1 significantly blocked the adverse effects of exRNAs in inflammation response in mouse models (Cabrera-Fuentes et al., 2014; Garnett and Raines, 2022). However, the exact targeted exRNAs of RNase1 have not been fully understood.

Recently, RNase1 was found to contribute to the production of tRNA halves and distinct Y RNA fragments in the extracellular space (Nechooshtan et al., 2020). The authors sequenced the nonextracellular vesicle exRNAs by comparing *RNASE1* wild-type (WT) and mutant cells, demonstrating that the tRNA halves population in *RNASE1* WT cells was significantly higher than that in *RNASE1*-KO cells (17.3% vs 1.6%). Among the tRNA halves products of RNase1, tRNA-Asp-GTC 3' halves (25.8%) and tRNA-Gly-GCC 5' halves (24.9%) occupied over 50% of the population. The 3' ends of tRNAs were also different between these two cell groups. In *RNASE1*-WT cells, the 3' end was a single cytosine after the end of the mature tRNA, whereas in KO cells, most tRNAs ended with CCA tails. This result was consistent with the substrate specificity of RNase1 that targets cytidine at the B1 site. Moreover, several Y RNA genes were also mapped in WT cells but not in cells lacking RNase1. Interestingly, an in vitro assay has shown that naked tRNA-Lys-UUU was relatively resistant to the action of RNase1 compared with tRNA-Gly-GCC, but once cleaved by RNase1, it cannot survive as a nicked tRNA. However, in the case of tRNA-Gly-GCC, the situation was reversed, where the naked full-length tRNA-Gly-GCC was sensitive to RNase1 cleavage, but the produced nicked tRNA was stable (Costa et al., 2023). This suggests that RNase1 has different effects on tRNAs with different sequences, thus resulting in different regulatory factors. Additionally, the bovine analogs of RNase1 (70% sequence identity), RNaseA, could recognize and cleave RNA containing 8-oxo-7,8-dihydroguanosine, a nucleotide generated under oxidative stress, with a similar preference to uracil but lower than a cytosine at B_1 in vitro (Herbert et al., 2018).

7.3.1.2 RNase2

RNase2, also known as eosinophil-derived neurotoxin, is one of the four main proteins secreted by human eosinophil granules. The catalytic substrate preference of RNase2 is opposite to that of RNase1, preferring poly(U) to poly(C) and being inactive against poly(A), dsRNAs, and 2',3'-cyclic nucleotides (Sorrentino, 2010).

RNase2 was first pointed out to have antiviral function on respiratory syncytial virus relying on its ribonuclease activity by directly targeting viral ssRNAs (Domachowske et al., 1998). Later research focused on the activation of pattern recognition receptors such as Toll-like receptors (TLRs) by RNA fragments obtained from selective cleavage of RNase2. Ostendorf and coworkers have reported that RNase2 worked synergistically with RNaseT2 to process RNA into TLR8 ligands by releasing uridine from oligoribonucleotides (Ostendorf et al., 2020). Later, a clinical case report noticed that the subject who received two doses of COVID-19 mRNA vaccines had elevated expression levels of RNase2 and suffered from cardiac complications after the booster (the third dose). The authors speculated that the increased RNase2 level may induce the release of more TLR ligands with 3' uridine, although the original mRNA vaccines contain Ψ to avoid overstimulation of RNA sensors that would lead to acute/chronic inflammation (Ong et al., 2023).

By comparing the tRNA population between RNase 2-WT and RNase 2-KO THP-1-induced macrophages through cP-RNA sequencing, Lu et al. have identified five top tRNAs ($tRNA^{AspGTC}$, $tRNA^{GlnCTG}$, $tRNA^{LysTTT}$, $tRNA^{LysCTT}$, and $tRNA^{MetCAT}$) related to RNase2 cleavage with base specificity at B_1 (U/C) and B_2 (A) sites (Lu et al., 2022), and the following study from the same group confirmed the results by sequencing the in vitro degradation products of these tRNAs (Li et al., 2024a). It is also observed that there were more in vitro cleavage sites of the synthesized $tRNA^{GlnCTG}$ than those of $tRNA^{GlnCTG}$ in cells. Interestingly, the additional cleaved sites of $tRNA^{GlnCTG}$ in vitro are mainly located at the nucleotide modification sites that were reported in vivo, indicating natural modifications on tRNAs can protect tRNAs from degradation by RNases. RNase2 is also reported to target tRNAs to produce tiRNAs. The generated RNase 2-cleaved $tRNA^{ValCAC}$ and $tRNA^{HisGUG}$ would activate the TLR7 receptor, thereby enhancing the levels of pro-inflammatory cytokines in asthma (Shigematsu et al., 2025).

7.3.1.3 RNase5

RNase5, also called angiogenin, can promote angiogenesis. RNase5 shows relatively weak ribonuclease activity on dinucleotides, with the order being CpA > CpG > UpA > UpG (Sheng and Xu, 2016), and its roles in RNA metabolism are the most noteworthy features within the family.

Earlier studies have shown that RNase5 can cleave rRNAs in vitro with base preference (Shapiro et al., 1986; Shapiro and Vallee, 1987; Rybak and Vallee, 1988). Later, it was discovered that RNase5 in the nucleus can promote rRNA transcription upon binding to the rDNA promoter ex vivo (Sheng et al., 2014). Nuclear RNase5 was found afterward to directly cleave pRNAs (promoter-associated RNAs), thereby inhibiting the formation of TIP5-pRNA complexes and leading to rRNA transcription derepression (Hoang and Raines, 2017).

Notably, RNase5 can also specifically cleave tRNA (Saxena et al., 1992). In the cytoplasm, RNase5 is a stress-activated ribonuclease that can cleave tRNAs within the anticodon loop to generate tiRNAs that trigger stress granule assembly and inhibit protein synthesis (Yamasaki et al., 2009). Recently, more studies have revealed that multiple tRNA isotypes were targets of this protein, and the cleaved sites of tRNAs may not be limited to the anticodon loop. For example, the overexpression of RNase5 in HEK293T cells without stress conditions could promote the production of tiRNAs derived from $tRNA^{Glu}$, $tRNA^{Gly}$, $tRNA^{Lys}$, $tRNA^{Val}$, $tRNA^{His}$, $tRNA^{Asp}$, and $tRNA^{SeC}$, whereas the levels of specific miRNAs did not decrease. However, specific tRNA fragments, mainly $tRNA^{HisGTG}$-5' halves and $tRNA^{AspGTC}$-3' halves, were significantly induced in WT but not in RNase 5-KO cells under oxidative stress (Su et al., 2019). RNase5 could also target the TψC loop of tRNA to generate the tRF-3 fragments and may remove 3'-CCA termini of tRNA during stress, both of which are controversial, as the study was based on in vitro studies (Czech et al., 2013). Another study showed that RNase5 mainly targeted anticodon rather than the 3'-CCA termini (Akiyama et al., 2022). Furthermore, it is pointed out that m^5C modifications by Dnmt2 proteins would diminish ANG-mediated tRNA cleavage (Schaefer et al., 2010). Nevertheless, accumulating evidence suggests that RNase5 is not the only RNase in cells that target tRNAs to generate tiRNAs.

7.3.1.4 Onconase

Onconase, also called ranpirnase, is a homolog of RNaseA from the leopard frog (*Rana pipiens*) that exhibits antitumor activity through cytotoxicity and cell-cycle regulation. The ribonuclease activity of Onconase is about 1% that of RNaseA. rRNAs, miRNAs, tRNAs, and dsRNAs have been proposed to be its substrates. Onconase has already advanced to clinical trials for the treatment of cancers (Iordanov et al., 2000; Menegazzi and Gotte, 2022).

The cytotoxic mechanism of Onconase was postulated to be dependent on its ribonucleolytic activity and to start with an initial binding to the cell surface receptor. After internalization into the cytoplasm, Onconase degrades rRNA, resulting in the inhibition of protein synthesis (Wu et al., 1993). However, the study from Iordanov et al. pointed out that at least 28S and 18S rRNAs were almost not degraded by

Onconase. Instead, the tRNA population decreased significantly in the total RNA extracted from Onconase-treated HeLa cells. The authors also proposed that Onconase triggered apoptosis (Iordanov et al., 2000). Coincidentally, further studies have shown that Onconase preferentially cleaves tRNAs instead of rRNAs in cells (Saxena et al., 2002; Suhasini and Sirdeshmukh, 2006), although rRNA can be degraded in vitro by Onconase at much higher concentrations. Moreover, Onconase cleaves tRNAs with base preference on G–G couples that were located at the variable or D-arm region (Suhasini and Sirdeshmukh, 2006). The authors examined the activity of Onconase on in vitro transcribed tRNA (Lys3) and found that the GG site from a UGG triplet in the variable loop was the predominant target of this RNase. Interestingly, nucleotide mutations of this cleaved site did not deprive the cleavage ability of Onconase (Suhasini and Sirdeshmukh, 2007), implying that the selective cleavage of $tRNA^{Lys3}$ by Onconase is determined by the tRNA secondary structure rather than its nucleotide type content, as confirmed later by Raines' lab (Lee et al., 2008).

In fact, while studying the biological mechanisms of Onconase, researchers have noticed the activation of signaling pathways and the regulation of gene expression in cell cycles, implying RNA interference (RNAi)-involved ncRNAs that could be potential targets of Onconase (Darzynkiewicz, 2017). In human lung adenocarcinoma A549 cells, added Onconase can target siRNA, likely within the RISC complex; thus, the GAPDH siRNA-induced gene silencing was counteracted (Zhao et al., 2008). Furthermore, recombinant Onconase added to mesothelioma cells could downregulate the levels of mature miRNAs (miR-155 and miR-21) resulting from the breakage of pre-miRNAs. Onconase, preferring the UG or UU nucleotides, was demonstrated to cleave pre-miRNAs at both loop and stem because the hairpin structures of pre-miRNAs, like parts of tRNAs, were appropriate for recognition by Onconase (Qiao et al., 2012). Interestingly, a recent study observed upregulation of some miRNAs, miR-34a-5p, miR-20a-3p, and miR-29a-3p mediated by Onconase in A375 melanoma cells. The authors speculated that the upstream lncRNAs or circRNAs corresponding to these miRNAs may be ribonucleolytic targets of Onconase; however, this hypothesis still needs further exploration (De Tomi et al., 2022). Again, the extraordinary activity of Onconase against dsRNA in vitro suggested an alternative mechanism for inducing apoptosis via RNAi involving siRNA and miRNA (Saxena et al., 2009).

In summary, the roles of the RNaseA superfamily on ncRNAs are highly dependent on their ribonucleolytic activities. Among the ncRNAs already reported, tRNAs are the predominant substrates of these family members. The tiRNAs and tRFs produced by RNase cleavage have played key roles in stress response and antiviral infection, as well as inflammation regulation of the body. It is also noteworthy that RNases not only mediate the local ncRNA population but also target foreign ncRNAs. RNase2 and Onconase can exert antiviral activity by directly degrading viral RNA (Li and Boix, 2021). Most recently, RNase6 has been found to target bacterial RNA to release uridine-terminated products that activate TLR8, and this response can be suspended by 2'O-RNA methyl modification of the first uridine in the UA dinucleotides (Nunes et al., 2024). However, most of the identified ncRNA substrates of RNases are based on in vitro gel electrophoresis assays, which may not be consistent with the actual situations in the body, and more validations in vivo are pending.

7.3.2 Other RNase

RNA decay regulated by RNases has been studied for a long time by using *E. coli* as a model; thus, most RNases with straightforward names such as RNaseD, E, G, H, J, P, R, T1, I, II, and III in the bibliography were initially reported in prokaryotes. As relevant research deepens, the human RNase homologs to *E. coli* are revealed gradually. For example, DIS3, homologous to bacterial RNaseII, plays an important role in lncRNA regulation (Foretek et al., 2024). More information about human RNase-mediated ncRNAs can be found in Table 7.1.

As mentioned earlier, there are various RNases with specialized functions involved in the regular biogenesis of ncRNAs in mammalian cells. Again, the human RNaseIII family, a group of dsRNA-specific endoribonucleases, participates in the dsRNA metabolism where Drosha processes pri-miRNAs to pre-miRNAs by cleaving the 3' and 5' strands of the stem-loop in pri-miRNAs, and Dicer cleaves pre-miRNAs or long dsRNAs to mature miRNAs or siRNAs with 21 nt to 23 nt length. Another two endonucleases, nuclear RNaseP and RNaseZ (ELAC2), are responsible for removing the 5' leader and 3' trailer from a pre-tRNA (Jarrous et al., 2022). We must remind here that the nuclear RNaseP is a ribozyme, a protein–RNA complex, whereas mitochondrial RNaseP is a protein-only RNase consisting of multiple subunits and processing mitochondrial pre-tRNAs. Notably, there are two other RNases in humans, RNaseL and RNaseT2, worth highlighting for their major ability to regulate ncRNAs under pathological conditions.

7.3.2.1 RNaseL

RNaseL is a 2'5'-linked oligoadenylates (2–5A)-dependent endonuclease that is normally inhibited at the basal expression level and can be activated by the 2–5A, an oligonucleotide specifically synthesized by the 2–5A synthetase (OAS). RNaseL targets ssRNAs with a preference for UU and UA sites in vitro (Brennan-Laun et al., 2014; Schwartz and Conn, 2019).

For decades, the 2–5A/RNaseL system has been reported to fight against viral infection through the combined mechanisms of directly cleaving ssRNAs from the viral genome, inducing antiviral gene expression by dsRNA-sensing pathways, and apoptosis. When pathogens, bacteria, or viruses invade the host, a series of genes is upregulated by the interferon (IFN)-induced immune response, including OAS. OAS is a synthetase that is activated by a pathogen recognition system that senses dsRNA and converts ATP into 2–5A. The 2–5A then binds to RNaseL and makes it shift from inactive monomer to active dimer, which in turn cleaves viral RNAs and stimulates the production of IFN (Schwartz and Conn, 2019).

RNaseL also targets other cellular RNAs to exert antiviral activity, and these processes are regarded as unusual roles of this protein (Karasik and Guydosh, 2024). To some extent, RNaseL is believed to cleave the RNAs globally in infected cells, causing cell apoptosis and thus clearing the virus in the host. However, growing evidence demonstrates that RNaseL can select its substrates. Upon activation by 2–5As, mRNA turnover mediated by RNaseL could strongly inhibit protein synthesis. The levels of gene expression related to mRNA stability and translation, differentiation, and IFN were decreased in RNaseL overexpression cells, revealing that specific mRNAs were recognized as substrates of RNaseL (Brennan-Laun et al., 2014). Interestingly, short peptides translated from noncoding mRNA decay fragments that were cleaved by RNaseL were suggested to contribute to antiviral response (Karasik et al., 2021).

Besides, RNaseL also exerts its ribonucleolytic activity on ncRNAs, including rRNAs, Y-RNAs, and tRNAs (Donovan et al., 2017). Previously, it was thought that RNase L-induced rRNA degradation in host cells directly limited the synthesis of viral proteins. Recent studies showed that in some viral infections, the activation of RNaseL did not suspend the translation of viral protein through rRNA cleavage, as viral protein products were detected in the infected cells. However, the specific underlying mechanisms are still to be unraveled (Burke et al., 2021). It is also proposed that the blockage of the translation machinery by RNaseL activation is not merely mediated by RNA decay (Donovan et al., 2017). In the same study, the authors also noticed a strong upregulation of tRNA and Y-RNA populations. The RNase L-mediated cleavage on tRNA-His, tRNA-Glu, and tRNA-Pro was identified at the anticodon loop and, on Y-RNA, at the upper region between nucleotides 24 and 32 at UN*N sites.

7.3.2.2 RNaseT2

RNaseT2 is the only member of the RNaseT2 family in humans, but its structure and function are conserved across diverse organisms, including bacteria and plants (MacIntosh, 2011). Like RNaseA, RNaseT2 hydrolyze RNA to 3' mononucleotides via a 2', 3'-cp intermediate, and like RNaseL, RNaseT2 is also associated with the host antiviral response.

It is reported that RNase T2-deficient/mutated leukoencephalopathy mimics congenital cytomegalovirus (CMV) infection in infancy (Henneke et al., 2009; Kameli et al., 2019). Using an RNase T2-mutated zebrafish model, Haud et al. noted that undigested rRNAs accumulated in lysosomes within brain neurons, and white matter lesions in these animals were comparable to those in infants with RNaseT2 deficiency, suggesting that one possible role of RNaseT2 in CMV infection is to digest rRNAs (Haud et al., 2011). A later study also noticed that the mitochondrial RNA (mtRNA) degradation rates varied in RNase T2-overexpressed or knocked-down cells and demonstrated that the RNase T2-mediated mtRNA degradation occurred in the mitochondrial intermembrane space (Liu et al., 2017).

Additionally, together with RNase2, RNaseT2 was found to cleave ssRNAs between purines and uridine, which in turn activated TLR8 to induce the innate immune response (Greulich et al., 2019). Interestingly, the enrichment in uridine products cleaved by these two RNases is achieved by the complementary cleavage preferences: RNaseT2 at N*U but RNase2 at U*N (N is a purine) (Li and Boix, 2021).

7.4 Identification of noncoding RNAs

For decades, identification of ncRNAs has been challenging not only because of their diversity in length and spatial structures but also for the various modified nucleotides embedded. Fortunately, with the advancement of contemporary science, a variety of technologies have been applied to study ncRNAs as well as their potential posttranscriptional modifications (Hüttenhofer and Vogel, 2006). The ongoing methods to identify the ncRNA sequences include:

1 Direct RNA sequencing. In this method, the desired ncRNAs are labeled and extracted by cutting the corresponding band from a denaturing gel. Then the labeled ncRNAs are sequenced through 2D RNA fingerprinting or by enzymatic or chemical sequencing with base-specific RNases (RNaseA, RNaseT2, et al.) or reagents. The results are finally visualized via electrophoresis according to the size of the proposed fragments. This method would cost a large amount of biological samples in order to obtain sufficient ncRNAs and was initially used for the study of tRNA and rRNA. However, other novel RNAs with modifications can also be identified.

2 Specialized cDNA libraries. This method also starts with the selection by a denaturing PAGE, but ncRNAs are extracted based on size range or specific functions. After the addition of the poly(A/C) tails by polymerase or ligation with linkers by RNA ligase, the ncRNAs are reverse transcribed to cDNA with corresponding primers, which are later cloned into a vector for cycle sequencing. As the different structures and modifications of ncRNAs would cause different reverse transcription efficiencies, not all selected ncRNAs can be cloned.

3 Microarray analysis. Microarray, which consists of DNA probes attached to a slide, has enabled the detection of ncRNAs at the whole-genome level. The profiles of either miRNAs or lncRNAs have already been well established.

4 Next Generation Sequencing (NGS). The advent of NGS has provided a more rapid and precise method to map the micro-scale ncRNAs on the human genome. The specific sequencing technology and the abundant bioinformatic workflows make it possible to identify and quantify different types of ncRNAs as well as the related modifications. Reviews (Micheel et al., 2021; Satam et al., 2023) about different NGS methods can be easily found in the database.

The analysis of modified nucleotides and sites of ncRNAs is also vital to explain the resistance mechanism of ncRNAs to RNases, which can assist in designing RNase-resistant RNAs and also be beneficial for the development of RNA-based drugs or vaccines. The review from Zhang et al. (2022) has summarized in detail how to detect RNA modifications. Briefly, in the article, the techniques are classified into four categories.

1 Quantification methods, which are based on chemical analysis and include (1) Two-dimensional thin-layer chromatography (2D-TLC) where the RNase A/T1/T2-cleaved RNAs are labeled with ^{32}P at the 5' terminal and the corresponding spots can be detected on the TLC plate; (2) Dot blot by using a specific antibody; (3) Liquid chromatography–mass spectrometry (LC–MS) to determine the different modifications depending on retention time and molecular mass. The main drawback of this method is that the RNAs used should be pure enough and in large amounts, and that it does not provide sequence information.

2 Locus-specific detection methods that can locate various RNA modifications. This category further contains (1) a primer extension method where the synthesis of cDNA suspends at the modified sites, and the reverse transcribed fragments are then separated and visualized by denaturing PAGE; (2) an RNase H-based approach (site-specific cleavage and radioactive labeling followed by ligation-assisted extraction and thin-layer chromatography, SCARLET). A complementary DNA probe is first used to guide RNaseH to the cleavage site of interest. After a splint ligation with an 116-nucleotide single-stranded DNA oligonucleotide and subsequent RNase A/T1 and Nuclease P1 treatment, the ^{32}P-labeled modified or nonmodified nucleotide can be checked and quantified by TLC. (3) Semiquantitative PCR or qPCR-based approach that is similar to the primer extension method, depending on the blocking of reverse transcription at the modified nucleotide. (4) ESI-MS-based approach. Different from the simple LC–MS, in this method RNAs are digested by selective endoribonucleases, such as RNaseA, RNaseT1, and RNaseU2, which can give both site and stoichiometry information for the modifications of interest. This method is improved as the first three locus-specific detection methods (1–3) require knowledge of the RNA sequence.

3 Next-generation sequencing-based detection technologies. Under the assistance of chemicals, antibodies, or enzyme/protein treatments, the modified sites and nucleotides of RNAs can be identified through the transcriptome coverage. Many types of NGS-based technologies are developed to test various types of RNA modifications.

4 Nanopore direct RNA sequencing-based technologies are a recently developed method that detects the nucleic acid modifications in unamplified RNA samples. The molecule passing through the "nanopore" can be directly decoded by computer algorithms. This protocol, also referred to as third-generation sequencing, has the advantage of not being biased by the presence of RNA modification.

Clearly, for recent research on RNase-related ncRNAs, the most applied approaches normally start with the construction of biological samples with overexpression or KO of targeted RNases by CRISPR-based genome editing. Following these procedures, with/without additional treatments, ncRNAs from cells in different groups are extracted and purified for NGS to compare. Among the various innovative NGS technologies, 2',3'-cyclic phosphate (cP) RNA sequencing (**cP-RNA-Seq**) is highlighted, as cP is a unique signature of cleavage products by endoribonucleases, such as RNaseA, RNaseL, orRNaseT2 . This technique also helps to identify the specific recognition sites and cleavage products of RNases (Honda et al., 2016). The workflow of cP-RNA-Seq is shown in Fig. 7.3.

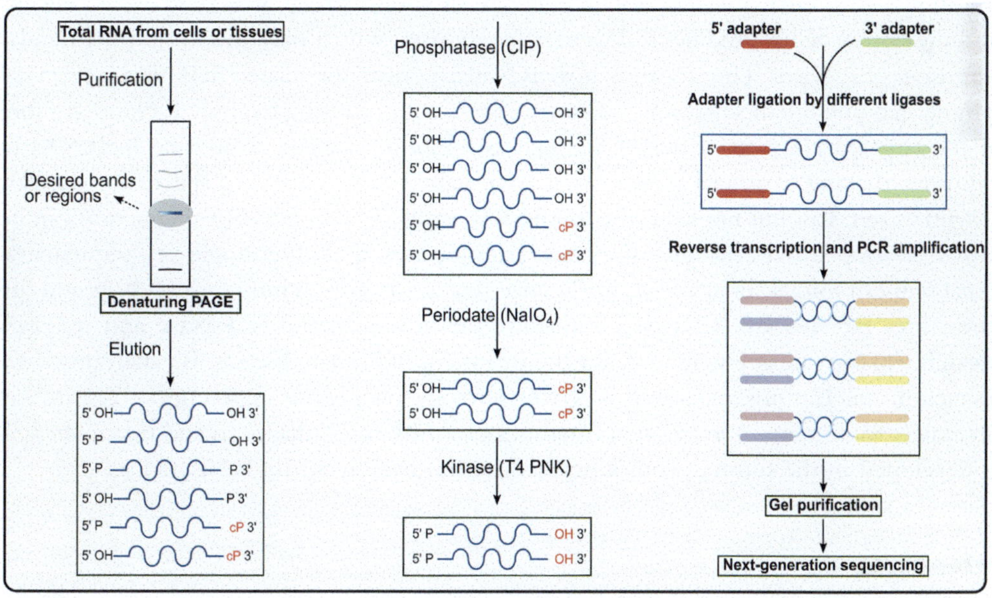

Figure 7.3 Flow diagram of cP-RNA-Seq methodology. The extracted RNAs are purified, and after a series of enzyme treatments, only the RNA fragments containing cP at the 3' end can be ligated with the adapters by ligases.

On the other hand, the introduction of modified nucleotides in an RNA at sites susceptible to RNases can also affect RNase-dependent biological processes. It was demonstrated that the 2–5A system activity was reduced when RNA contains modified nucleosides; thereby, the activation of OAS and RNaseL cleavage was decreased (Anderson et al., 2011). Besides, it was found that the 3'-methylenephosphonate RNA modification is highly resistant to the action of RNaseA (Ghidini et al., 2013), and human RNase4 was able to cleave oligonucleotides containing 1-methylpseudouridine ($m^1\Psi$), dihydrouridine (D), 4-thiouridine (s^4U), and 5-methyluridine (m^5U) but not oligonucleotides containing 2′-O-methyluridine (Um) and 5,2′-O-dimethyluridine (m^5Um) (Wolf et al., 2022).

The rapid development of structural biology and bioinformatics has greatly promoted the research on the function of RNases and RNase-targeting ncRNAs. Summarily, RNases are involved in the processing and maturation of ncRNAs, thereby regulating gene expression. Targeted degradation of ncRNAs by RNases helps to maintain ncRNA homeostasis. In turn, when the body is under stress, the expression of RNases can precisely regulate the population of ncRNAs and affect their metabolism, which afterward contributes to the occurrence and development of diseases. Therefore research on RNases and ncRNAs should still focus on the following aspects. Firstly, it is important to continually reveal the mechanism of action of RNases on ncRNA regulation, such as exploring the new roles of different types of RNases on their targeting ncRNAs or the dynamic regulation of RNases on different ncRNAs under different physiological conditions. Abnormal metabolism of ncRNAs is found to be associated with cancers, immune system failure, and neurodegenerative diseases, which may arise from the irregular expression of RNases. Thus to study how RNases regulate the related ncRNAs is also vital to understanding the mechanism of diseases. Next, it is promising to apply RNases as a tool for the targeted cleavage of ncRNAs, such as inhibiting cancer-promoting lncRNAs in antitumor therapy. Artificial RNases that have stability and improved specificity can also be synthesized. Last but not least, the methods to study RNase-ncRNA interaction can be further developed with the advent of new technologies, including single-cell sequencing, spatial transcriptomics, or artificial intelligence that assists in searching for ncRNAs and the action sites targeted by RNases. Ultimately, the advancement of RNase and ncRNA research will provide a new strategy for the treatment of human diseases. By analyzing their interaction mechanisms and using RNA technology for precise intervention, researchers can expect to develop more efficient methods of RNA-based diagnosis and treatment will be developed in the future, promoting medical innovation in the RNA fields.

References

Akiyama, Y., Lyons, S.M., Fay, M.M., Tomioka, Y., Abe, T., Anderson, P.J., Ivanov, P., 2022. Selective cCleavage at CCA eEnds and aAnticodon lLoops of tRNAs by sStress-iInduced RNases. Front Mol Biosci 9, 791094. Available at https://doi.org/10.3389/fmolb.2022.791094.

Alarcón, C.R., Lee, H., Goodarzi, H., Halberg, N., Tavazoie, S.F., 2015. N^6-methyladenosine marks primary microRNAs for processing. Nature 519 (7544), 482–485. Available at https://doi.org/10.1038/nature14281.

Anderson, B.R., Muramatsu, H., Jha, B.K., Silverman, R.H., Weissman, D., Karikó, K., 2011. Nucleoside modifications in RNA limit activation of 2'-5'-oligoadenylate synthetase and increase resistance to cleavage by RNase L. Nucleic Acids Res 39 (21), 9329–9338. Available at https://doi.org/10.1093/nar/gkr586.

Boix, E., Nogués, M.V., 2007. Mammalian antimicrobial proteins and peptides: Overview on the RNase A superfamily members involved in innate host defence. Mol Biosyst 3 (5), 317–335. Available at https://doi.org/10.1039/b617527a.

Brennan-Laun, S.E., Ezelle, H.J., Li, X.L., Hassel, B.A., et al., 2014. RNase-L control of cellular mRNAs: roles in biologic functions and mechanisms of substrate targeting. J Interferon Cytokine Res 34 (4), 275–288. Available at https://doi.org/10.1089/jir.2013.0147.

Burke, J.M., Gilchrist, A.R., Sawyer, S.L., Parker, R., 2021. RNase L limits host and viral protein synthesis via inhibition of mRNA export. Sci Adv 7 (23), eabh2479. Available at https://doi.org/10.1126/sciadv.abh2479.

Cabrera-Fuentes, H.A., Ruiz-Meana, M., Simsekyilmaz, S., Kostin, S., Inserte, J., Saffarzadeh, M., et al., 2014. RNase1 prevents the damaging interplay between extracellular RNA and tumour necrosis factor-α in cardiac ischaemia/reperfusion injury. Thromb Haemost 112 (6), 1110–1119. Available at https://doi.org/10.1160/TH14-08-0703.

Carissimi, C., Laudadio, I., Lorefice, E., Azzalin, G., De Paolis, V., Fulci, V., 2021. Bisulphite miRNA-seq reveals widespread CpG and non-CpG 5-(hydroxy)methyl-Cytosine in human microRNAs. RNA Biol 18 (12), 2226–2235. Available at https://doi.org/10.1080/15476286.2021.1927423.

Christofi, T., Zaravinos, A., 2019. RNA editing in the forefront of epitranscriptomics and human health. J Transl Med 17 (1), 319. Available at https://doi.org/10.1186/s12967-019-2071-4.

Cooper, D.A., Jha, B.K., Silverman, R.H., Hesselberth, J.R., Barton, D.J., 2014. Ribonuclease L and metal-ion-independent endoribonuclease cleavage sites in host and viral RNAs. Nucleic Acids Res 42 (8), 5202–5216. Available at https://doi.org/10.1093/nar/gku118.

Cooper, D.A., Banerjee, S., Chakrabarti, A., García-Sastre, A., Hesselberth, J.R., Silverman, R.H., Barton, D.J., 2015. RNase L targets distinct sites in influenza A virus RNAs. J Virol 89 (5), 2764–2776. Available at https://doi.org/10.1128/JVI.02953-14.

Costa, B., Li Calzi, M., Castellano, M., Blanco, V., Cuevasanta, E., Litvan, I., et al., 2023. Nicked tRNAs are stable reservoirs of tRNA halves in cells and biofluids. Proc Natl Acad Sci U S A 120 (4), e2216330120. Available at https://doi.org/10.1073/pnas.2216330120.

Cui, L., Ma, R., Cai, J., Guo, C., Chen, Z., Yao, L., et al., 2022. RNA modifications: Importance in immune cell biology and related diseases. Sig Transduct Target Ther 7 (1), 334. Available at https://doi.org/10.1038/s41392-022-01175-9.

Cui, W., Zhao, D., Jiang, J., Tang, F., Zhang, C., Duan, C., 2023. tRNA modifications and modifying enzymes in disease, the potential therapeutic targets. Int J Biol Sci 19 (4), 1146–1162. Available at https://doi.org/10.7150/ijbs.80233.

Curtale, G., Citarella, F., 2013. Dynamic nature of noncoding RNA regulation of adaptive immune response. Int J Mol Sci 14 (9), 17347–17377. Available at https://doi.org/10.3390/ijms140917347.

Czech, A., Wende, S., Mörl, M., Pan, T., Ignatova, Z., 2013. Reversible and rapid transfer-RNA deactivation as a mechanism of translational repression in stress. PLoS Genet 9 (8), e1003767. Available at https://doi.org/10.1371/journal.pgen.1003767.

Czech, B., Hannon, G.J., 2016. One loop to rule them all: The ping-pong cycle and piRNA-guided silencing. Trends Biochem Sci 41 (4), 324–337. Available at https://doi.org/10.1016/j.tibs.2015.12.008.

Czech, B., Munafò, M., Ciabrelli, F., Eastwood, E.L., Fabry, M.H., Kneuss, E., Hannon, G.J., et al., 2018. piRNA-guided genome defense: From biogenesis to silencing. Annu Rev Genet 52, 131–157. Available at https://doi.org/10.1146/annurev-genet-120417-031441.

Darzynkiewicz, Z., 2017. microRNA-494 and ATF3 the targets of onconase(?). Oncotarget 8 (7), 10769–10770. Available at https://doi.org/10.18632/oncotarget.13799.

De Tomi, E., Campagnari, R., Orlandi, E., Cardile, A., Zanrè, V., Menegazzi, M., Gomez-Lira, M., Gotte, G., 2022. Upregulation of miR-34a-5p, miR-20a-3p and miR-29a-3p by Onconase in A375 melanoma cells correlates with the downregulation of specific Onco-Proteins. Int J Mol Sci 23 (3), 1647. Available at https://doi.org/10.3390/ijms23031647.

delCardayré, S.B., Raines, R.T., 1994. Structural determinants of enzymatic processivity. Biochemistry 33 (20), 6031–6037. Available at https://doi.org/10.1021/bi00186a001.

Delihas, N., 2015. Discovery and characterization of the first non-coding RNA that regulates gene expression, micF RNA: A historical perspective. World J Biol Chem 6 (4), 272–280. Available at https://doi.org/10.4331/wjbc.v6.i4.272.

Domachowske, J.B., Bonville, C.A., Dyer, K.D., Rosenberg, H.F., 1998. Evolution of antiviral activity in the ribonuclease A gene superfamily: Evidence for a specific interaction between eosinophil-derived neurotoxin (EDN/RNase 2) and respiratory syncytial virus. Nucleic Acids Res 26 (23), 5327–5332. Available at https://doi.org/10.1093/nar/26.23.5327.

Donovan, J, Rath, S, Kolet-Mandrikov, D, Korennykh, A, 2017. Rapid RNase L–driven arrest of protein synthesis in the dsRNA response without degradation of translation machinery. RNA 23 (11), 1660–1671. Available at https://doi.org/10.1261/rna.062000.117.

Dyer, K.D., Rosenberg, H.F., 2006. The RNase a superfamily: Generation of diversity and innate host defense. Mol Divers 10 (4), 585–597. Available at https://doi.org/10.1007/s11030-006-9028-2.

Esteve-Puig, R., Bueno-Costa, A., Esteller, M., 2020. Writers, readers and erasers of RNA modifications in cancer. Cancer Lett 474, 127–137. Available at https://doi.org/10.1016/j.canlet.2020.01.021.

Fisher, B.M., Schultz, L.W., Raines, R.T., 1998. Coulombic effects of remote subsites on the active site of ribonuclease A. Biochemistry 37 (50), 17386–17401. Available at https://doi.org/10.1021/bi981369s.

Foretek, D., Gabriel, M., Hatin, I., Jarroux, J., Pinskaya, M., Pepermans, E., et al., 2024. Accumulation of lncRNAs in cytoplasm upon DIS3 depletion leads to production of cryptic peptides detected in Multiple Myeloma. Preprint: Research Square. Available at https://doi.org/10.21203/rs.3.rs-3006132/v3

Fu, H., Feng, J., Liu, Q., Sun, F., Tie, Y., Zhu, J., et al., 2009. Stress induces tRNA cleavage by angiogenin in mammalian cells. FEBS Lett 583 (2), 437–442. Available at https://doi.org/10.1016/j.febslet.2008.12.043.

Fu, X.D., 2014. Non-coding RNA: A new frontier in regulatory biology. Natl Sci Rev 1 (2), 190–204. Available at https://doi.org/10.1093/nsr/nwu008.

Gainetdinov, I., Colpan, C., Cecchini, K., Arif, A., Jouravleva, K., Albosta, P., et al., 2021. Terminal modification, sequence, length, and PIWI-protein identity determine piRNA stability. Mol Cell 81 (23), 4826–4842.e8. Available at https://doi.org/10.1016/j.molcel.2021.09.012.

Garnett, E.R., Lomax, J.E., Mohammed, B.M., Gailani, D., Sheehan, J.P., Raines, R.T., 2019. Phenotype of ribonuclease 1 deficiency in mice. RNA 25 (8), 921–934. Available at https://doi.org/10.1261/rna.070433.119.

Garnett, E.R., Raines, R.T., 2022. Emerging biological functions of ribonuclease 1 and angiogenin. Crit Rev Biochem Mol Biol 57 (3), 244–260. Available at https://doi.org/10.1080/10409238.2021.2004577.

Gebetsberger, J., Wyss, L., Mleczko, A.M., Reuther, J., Polacek, N., 2017. A tRNA-derived fragment competes with mRNA for ribosome binding and regulates translation during stress. RNA Biol 14 (10), 1364–1373. Available at https://doi.org/10.1080/15476286.2016.1257470.

Ghahramani Almanghadim, H., Karimi, B., Valizadeh, S., Ghaedi, K., 2025. Biological functions and affected signaling pathways by Long Non-Coding RNAs in the immune system. Noncoding RNA Res 10, 70–90. Available at https://doi.org/10.1016/j.ncrna.2024.09.001.

Ghidini, A., Ander, C., Winqvist, A., Strömberg, R., 2013. An RNA modification with remarkable resistance to RNase A. Chem Commun 49 (79), 9036–9038. Available at https://doi.org/10.1039/c3cc44621b.

Gilbert, W.V., Nachtergaele, S., 2023. mRNA regulation by RNA modifications. Annu Rev Biochem 92, 175–198. Available at https://doi.org/10.1146/annurev-biochem-052521-035949.

Goodarzi, H., Liu, X., Nguyen, H.C., Zhang, S., Fish, L., Tavazoie, S.F., 2015. Endogenous tRNA-derived fragments suppress breast cancer progression via YBX1 displacement. Cell 161 (4), 790–802. Available at https://doi.org/10.1016/j.cell.2015.02.053.

Greulich, W., Wagner, M., Gaidt, M.M., Stafford, C., Cheng, Y., Linder, A., et al., 2019. TLR8 is a sensor of RNase T2 degradation products. Cell 179 (6), 1264–1275.e13. Available at https://doi.org/10.1016/j.cell.2019.11.001.

Haud, N., Kara, F., Diekmann, S., Henneke, M., Willer, J.R., Hillwig, M.S., et al., 2011. rnaset2 mutant zebrafish model familial cystic leukoencephalopathy and reveal a role for RNase T2 in degrading ribosomal RNA. Proc Natl Acad Sci U S A 108 (3), 1099–1103. Available at https://doi.org/10.1073/pnas.1009811107.

He, R.Z., Luo, D.X., Mo, Y.Y., 2019. Emerging roles of lncRNAs in the post-transcriptional regulation in cancer. Genes Dis 6 (1), 6–15. Available at https://doi.org/10.1016/j.gendis.2019.01.003.

Henneke, M., Diekmann, S., Ohlenbusch, A., Kaiser, J., Engelbrecht, V., Kohlschütter, A., et al., 2009. RNASET2-deficient cystic leukoencephalopathy resembles congenital cytomegalovirus brain infection. Nat Genet 41 (7), 773–775. Available at https://doi.org/10.1038/ng.398.

Herbert, C., Dzowo, Y.K., Urban, A., Kiggins, C.N., Resendiz, M.J.E., 2018. Reactivity and specificity of RNase T_1, RNase A, and RNase H toward oligonucleotides of RNA containing 8-Oxo-7,8-dihydroguanosine. Biochemistry 57 (20), 2971–2983. Available at https://doi.org/10.1021/acs.biochem.8b00277.

Hoagland, M.B., Stephenson, M.L., Scott, J.F., Hecht, L.I., Zamecnik, P.C., 1958. A soluble ribonucleic acid intermediate in protein synthesis. J Biol Chem 231 (1), 241–257. Available at https://www.jbc.org/article/S0021-9258(19)77302-5/pdf.

Hoang, T.T., Raines, R.T., 2017. Molecular basis for the autonomous promotion of cell proliferation by angiogenin. Nucleic Acids Res 45 (2), 818–831. Available at https://doi.org/10.1093/nar/gkw1192.

Hombach, S., Kretz, M., 2016. Non-coding RNAs: Classification, biology and functioning. In: Slaby, O., Calin, G.A. (Eds.), Non-coding RNAs in Colorectal Cancer. Advances in Experimental Medicine and Biology 937. Springer, Switzerland, pp. 3–17. Available at https://doi.org/10.1007/978-3-319-42059-2_1.

Honda, S., Loher, P., Shigematsu, M., Palazzo, J.P., Suzuki, R., Imoto, I., et al., 2015. Sex hormone-dependent tRNA halves enhance cell proliferation in breast and prostate cancers. Proc Natl Acad Sci U S A 112 (29), E3816–E3825. Available at https://doi.org/10.1073/pnas.1510077112.

Honda, S., Morichika, K., Kirino, Y., 2016. Selective amplification and sequencing of cyclic phosphate-containing RNAs by the cP-RNA-seq method. Nat Protoc 11 (3), 476–489. Available at https://doi.org/10.1038/nprot.2016.025.

Horsburgh, S., Robson-Ansley, P., Adams, R., Smith, C., 2015. Exercise and inflammation-related epigenetic modifications: focus on DNA methylation. Exerc Immunol Rev 21, 26–41. Available at https://journals.ub.uni-giessen.de/public/eir/2015/eir-2015-026-article.pdf.

Hou, X., Shi, J., Xiao, Y., 2024. mRNA medicine: Recent progresses in chemical modification, design, and engineering. Nano Research 17 (10), 9015–9030. Available at https://doi.org/10.1007/s12274-024-6978-6.

Huang, V., Zheng, J., Qi, Z., Wang, J., Place, R.F., Yu, J., et al., 2013. Ago1 interacts with RNA polymerase II and binds to the promoters of actively transcribed genes in human cancer cells. PLoS Genet 9 (9), e1003821. Available at https://doi.org/10.1371/journal.pgen.1003821.

Huang, W., Li, H., Yu, Q., Xiao, W., Wang, D.O., 2022. LncRNA-mediated DNA methylation: An emerging mechanism in cancer and beyond. J Exp Clin Cancer Res 41 (1), 100. Available at https://doi.org/10.1186/s13046-022-02319-z.

Hudder, A., Novak, R.F., 2008. miRNAs: Effectors of environmental influences on gene expression and disease. Toxicol Sci 103 (2), 228–240. Available at https://doi.org/10.1093/toxsci/kfn033.

Hughes, L.A., Rudler, D.L., Siira, S.J., McCubbin, T., Raven, S.A., Browne, J.M., et al., 2023. Copy number variation in tRNA isodecoder genes impairs mammalian development and balanced translation. Nat Commun 14 (1), 2210. Available at https://doi.org/10.1038/s41467-023-37843-9.

Hüttenhofer, A., Vogel, J., 2006. Experimental approaches to identify non-coding RNAs. Nucleic Acids Res 34 (2), 635–646. Available at https://doi.org/10.1093/nar/gkj469.

Iordanov, M.S., Ryabinina, O.P., Wong, J., Dinh, T.H., Newton, D.L., Rybak, S.M., et al., 2000. Molecular determinants of apoptosis induced by the cytotoxic ribonuclease onconase: evidence for cytotoxic mechanisms different from inhibition of protein synthesis. Cancer Res 60 (7), 1983–1994. Available at https://aacrjournals.org/cancerres/article/60/7/1983/507158/Molecular-Determinants-of-Apoptosis-Induced-by-the.

Ivanov, P., Emara, M.M., Villen, J., Gygi, S.P., Anderson, P., 2011. Angiogenin-induced tRNA fragments inhibit translation initiation. Mol Cell 43 (4), 613–623. Available at https://doi.org/10.1016/j.molcel.2011.06.022.

Jackman, J.E., Alfonzo, J.D., 2013. Transfer RNA modifications: Nature's combinatorial chemistry playground. Wiley Interdiscip Rev RNA 4 (1), 35–48. Available at https://doi.org/10.1002/wrna.1144.

Jarrous, N., Mani, D., Ramanathan, A., 2022. Coordination of transcription and processing of tRNA. FEBS J 289 (13), 3630–3641. Available at https://doi.org/10.1111/febs.15904.

Jiang, S., Hu, J., Bai, Y., Hao, R., Liu, L., Chen, H., 2023. Transcriptome-wide 5-methylcytosine modification profiling of long non-coding RNAs in A549 cells infected with H1N1 influenza A virus. BMC Genomics 24 (1), 316. Available at https://doi.org/10.1186/s12864-023-09432-z.

Jiang, M., Hong, X., Gao, Y., Kho, A.T., Tantisira, K.G., Li, J., 2024. piRNA associates with immune diseases. Cell Commun Signal 22 (1), 347. Available at https://doi.org/10.1186/s12964-024-01724-5.

Jia, H., Zhang, L., 2025. tRNA-derived small RNAs in disease immunity. Theranostics 15 (1), 245–257. Available at https://doi.org/10.7150/THNO.102650.

Jonas, S., Izaurralde, E., 2015. Towards a molecular understanding of microRNA-mediated gene silencing. Nat Rev Genet 16 (7), 421–433. Available at https://doi.org/10.1038/nrg3965.

Kameli, R., Amanat, M., Rezaei, Z., Hosseionpour, S., Nikbakht, S., Alizadeh, H., et al., 2019. RNASET2-deficient leukoencephalopathy mimicking congenital CMV infection and aicardi-goutieres syndrome: A case report with a novel pathogenic variant. Orphanet J Rare Dis 14 (1), 184. Available at https://doi.org/10.1186/s13023-019-1155-9.

Karasik, A., Jones, G.D., Depass, A.V., Guydosh, N.R., 2021. Activation of the antiviral factor RNase L triggers translation of non-coding mRNA sequences. Nucleic Acids Res 49 (11), 6007–6026. Available at https://doi.org/10.1093/nar/gkab036.

Karasik, A., Guydosh, N.R., 2024. The unusual role of ribonuclease L in innate immunity. Wiley Interdiscip Rev RNA 15 (6), e1878. Available at https://doi.org/10.1002/wrna.1878.

Kawaoka, S., Izumi, N., Katsuma, S., Tomari, Y., 2011. 3' end formation of PIWI-interacting RNAs in vitro. Mol Cell 43 (6), 1015–1022. Available at https://doi.org/10.1016/j.molcel.2011.07.029.

Kazimierczyk, M., Wrzesinski, J., 2021. Long non-coding RNA epigenetics. Int J Mol Sci 22 (11), 6166. Available at https://doi.org/10.3390/ijms22116166.

Kelly, N.J., Palmer, M.T., Morrow, C.D., 2003. Selection of retroviral reverse transcription primer is coordinated with tRNA biogenesis. J Virol 77 (16), 8695–8701. Available at https://doi.org/10.1128/JVI.77.16.8695-8701.2003.

Kiełbowski, K., Ptaszyński, K., Wójcik, J., Wojtyś, M.E., 2023. The role of selected non-coding RNAs in the biology of non-small cell lung cancer. Adv Med Sci 68 (1), 121–137. Available at https://doi.org/10.1016/j.advms.2023.02.004.

Kim, H.K., Yeom, J.H., Kay, M.A., 2020. Transfer RNA-derived small RNAs: Another layer of gene regulation and novel targets for disease therapeutics. Mol Ther 28 (11), 2340–2357. Available at https://doi.org/10.1016/j.ymthe.2020.09.013.

Kishimoto, K., Liu, S., Tsuji, T., Olson, K.A., Hu, G.F., 2005. Endogenous angiogenin in endothelial cells is a general requirement for cell proliferation and angiogenesis. Oncogene 24 (3), 445–456. Available at https://doi.org/10.1038/sj.onc.1208223.

Knowling, S., Morris, K.V., 2011. Non-coding RNA and antisense RNA. Nature's trash or treasure? Biochimie 93 (11), 1922–1927. Available at https://doi.org/10.1016/J.BIOCHI.2011.07.031.

Konno, M., Koseki, J., Asai, A., Yamagata, A., Shimamura, T., Motooka, D., et al., 2019. Distinct methylation levels of mature microRNAs in gastrointestinal cancers. Nat Commun 10 (1), 3888. Available at https://doi.org/10.1038/s41467-019-11826-1.

Kumar, S., Gonzalez, E.A., Rameshwar, P., Etchegaray, J.P., 2020. Non-coding RNAs as mediators of epigenetic changes in malignancies. Cancers 12 (12), 3657. Available at https://doi.org/10.3390/cancers12123657.

Lai, L.B., Vioque, A., Kirsebom, L.A., Gopalan, V., 2010. Unexpected diversity of RNase P, an ancient tRNA processing enzyme: Challenges and prospects. FEBS Lett 584 (2), 287–296. Available at https://doi.org/10.1016/j.febslet.2009.11.048.

Lam, J.K.W., Chow, M.Y.T., Zhang, Y., Leung, S.W.S., 2015. siRNA versus miRNA as therapeutics for gene silencing. Mol Ther Nucleic Acids 4 (9), e252. Available at https://doi.org/10.1038/mtna.2015.23.

Lee, J.E., Bae, E., Bingman, C.A., Phillips Jr, G.N., Raines, R.T., 2008. Structural basis for catalysis by Onconase. J Mol Biol 375 (1), 165–177. Available at https://doi.org/10.1016/j.jmb.2007.09.089.

Lee, H.H., Wang, Y.N., Hung, M.C., 2019. Functional roles of the human ribonuclease A superfamily in RNA metabolism and membrane receptor biology. Mol Aspects Med 70, 106–116. Available at https://doi.org/10.1016/j.mam.2019.03.003.

Li, J., Boix, E., 2021. Host defence RNases as antiviral agents against enveloped single stranded RNA viruses. Virulence 12 (1), 444–469. Available at https://doi.org/10.1080/21505594.2021.1871823.

Li, J., Kang, X., Guidi, I., Lu, L., Fernández-Millán, P., Prats-Ejarque, G., et al., 2024. Structural determinants for tRNA selective cleavage by RNase 2/EDN. Structure 32 (3), 328–341.e4. Available at https://doi.org/10.1016/j.str.2023.12.012.

Li, J., Wang, X., Wang, H., 2024. RNA modifications in long non-coding RNAs and their implications in cancer biology. Bioorg Med Chem 113, 117922. Available at https://doi.org/10.1016/j.bmc.2024.117922.

Liang, Y., Ji, D., Ying, X., Ma, R., Ji, W., 2025. tsRNA modifications: An emerging layer of biological regulation in disease. J Adv Res 74, 403–414. Available at https://doi.org/10.1016/j.jare.2024.09.010.

Liu, P., Huang, J., Zheng, Q., Xie, L., Lu, X., Jin, J., et al., 2017. Mammalian mitochondrial RNAs are degraded in the mitochondrial intermembrane space by RNASET2. Protein Cell 8 (10), 735–749. Available at https://doi.org/10.1007/s13238-017-0448-9.

Lu, L., Li, J., Moussaoui, M., Boix, E., 2018. Immune modulation by human secreted RNases at the extracellular space. Front Immunol 9, 1012. Available at https://doi.org/10.3389/fimmu.2018.01012.

Lu, L., Li, J., Wei, R., Guidi, I., Cozzuto, L., Ponomarenko, J., et al., 2022. Selective cleavage of ncRNA and antiviral activity by RNase2/EDN in THP1-induced macrophages. Cell Mol Life Sci 79 (4), 209. Available at https://doi.org/10.1007/s00018-022-04229-x.

Ma, L., Bajic, V.B., Zhang, Z., 2013. On the classification of long non-coding RNAs. RNA Biol 10 (6), 924–933. Available at https://doi.org/10.4161/rna.24604.

MacIntosh, G.C., 2011. RNase T2 family: Enzymatic properties, functional diversity, and evolution of ancient ribonucleases. In: Nicholson, Allen W. (Ed.), Nucleic Acids and Molecular Biology. Springer, Berlin, Heidelberg, pp. 0933–1891. Available at http://doi.org/10.1007/978-3-642-21078-5_4.

Malathi, K., Dong, B., Gale Jr., M., Silverman, R.H., 2007. Small self-RNA generated by RNase L amplifies antiviral innate immunity. Nature 448 (7155), 816–819. Available at https://doi.org/10.1038/nature06042.

Martínez-González, E., Brochado-Kith, Ó., Gómez-Sanz, A., Martín-Carbonero, L., Jimenez-Sousa, M.Á., Martínez-Román, P., et al., 2020. Comparison of methods and characterization of small RNAs from plasma extracellular vesicles of HIV/HCV coinfected patients. Sci Rep 10 (1), 11140. Available at https://doi.org/10.1038/s41598-020-67935-1.

Mathy, N.W., Chen, X.M., 2017. Long non-coding RNAs (lncRNAs) and their transcriptional control of inflammatory responses. J Biol Chem 292 (30), 12375–12382. Available at https://doi.org/10.1074/jbc.R116.760884.

Mattick, J.S., Amaral, P.P., Carninci, P., Carpenter, S., Chang, H.Y., Chen, L.L., et al., 2023. Long non-coding RNAs: definitions, functions, challenges and recommendations. Nat Rev Mol Cell Biol 24 (6), 430–447. Available at https://doi.org/10.1038/s41580-022-00566-8.

Mattick, J.S., 2001. Non-coding RNAs: The architects of eukaryotic complexity. EMBO Rep 2 (11), 986–991. Available at https://doi.org/10.1093/embo-reports/kve230.

McCown, P.J., Ruszkowska, A., Kunkler, C.N., Breger, K., Hulewicz, J.P., Wang, M.C., et al., 2020. Naturally occurring modified ribonucleosides. Wiley Interdiscip Rev RNA 11 (5), e1595. Available at https://doi.org/10.1002/wrna.1595.

Menegazzi, M., Gotte, G., 2022. Role of the ribonuclease ONCONASE in miRNA biogenesis and tRNA processing: focus on cancer and viral infections. Int J Mol Sci 23 (12), 6556. Available at https://doi.org/10.3390/ijms23126556.

Micheel, J., Safrastyan, A., Wollny, D., 2021. Advances in non-coding RNA sequencing. Noncoding RNA 7 (4), 70. Available at https://doi.org/10.3390/ncrna7040070.

Muthukumar, S., Li, C.T., Liu, R.J., Bellodi, C., 2024. Roles and regulation of tRNA-derived small RNAs in animals. Nat Rev Mol Cell Biol 25 (5), 359–378. Available at https://doi.org/10.1038/s41580-023-00690-z.

Nadhan, R., Isidoro, C., Song, Y.S., Dhanasekaran, D.N., 2024. LncRNAs and the cancer epigenome: mechanisms and therapeutic potential. Cancer Lett 605, 217297. Available at https://doi.org/10.1016/j.canlet.2024.217297.

Nechooshtan, G., Yunusov, D., Chang, K., Gingeras, T.R., 2020. Processing by RNase 1 forms tRNA halves and distinct y RNA fragments in the extracellular environment. Nucleic Acids Res 48 (14), 8035–8049. Available at https://doi.org/10.1093/nar/gkaa526.

Nishikura, K., 2016. A-to-I editing of coding and non-coding RNAs by ADARs. Nat Rev Mol Cell Biol 17 (2), 83–96. Available at https://doi.org/10.1038/nrm.2015.4.

Nobles, K.N., Yarian, C.S., Liu, G., Guenther, R.H., Agris, P.F., 2002. Highly conserved modified nucleosides influence Mg^{2+}-dependent tRNA folding. Nucleic Acids Res 30 (21), 4751–4760. Available at https://doi.org/10.1093/nar/gkf595.

Nunes, I.V., Breitenbach, L., Pawusch, S., Eigenbrod, T., Ananth, S., Schad, P., et al., 2024. Bacterial RNA sensing by TLR8 requires RNase 6 processing and is inhibited by RNA 2'O-methylation. EMBO Rep 25 (11), 4674–4692. Available at https://doi.org/10.1038/s44319-024-00281-9.

O'Brien, J., Hayder, H., Zayed, Y., Peng, C., 2018. Overview of microRNA biogenesis, mechanisms of actions, and circulation. Front Endocrinol (Lausanne) 9, 402. Available at https://doi.org/10.3389/fendo.2018.00402.

Ong, E.Z., Koh, C.W.T., Tng, D.J.H., Ooi, J.S.G., Yee, J.X., Chew, V.S.Y., et al., 2023. RNase2 is a possible trigger of acute-on-chronic inflammation leading to mRNA vaccine-associated cardiac complication. Med 4 (6), 353–360.e2. Available at https://doi.org/10.1016/j.medj.2023.04.001.

Ostendorf, T., Zillinger, T., Andryka, K., Schlee-Guimaraes, T.M., Schmitz, S., Marx, S., et al., 2020. Immune sensing of synthetic, bacterial, and protozoan RNA by Toll-like receptor 8 requires coordinated processing by RNase T2 and RNase 2. Immunity 52 (4), 591–605.e6. Available at https://doi.org/10.1016/j.immuni.2020.03.009.

Pandey, K.K., Madhry, D., Ravi Kumar, Y.S., Malvankar, S., Sapra, L., Srivastava, R.K., et al., 2021. Regulatory roles of tRNA-derived RNA fragments in human pathophysiology. Mol Ther Nucleic Acids 26, 161–173. Available at https://doi.org/10.1016/j.omtn.2021.06.023.

Pandolfini, L., Barbieri, I., Bannister, A.J., Hendrick, A., Andrews, B., Webster, N., et al., 2019. METTL1 promotes let-7 microRNA processing via m7G methylation. Mol Cell 74 (6), 1278–1290.e9. Available at https://doi.org/10.1016/j.molcel.2019.03.040.

Paraskevopoulou, M.D., Hatzigeorgiou, A.G., 2016. Analyzing miRNA–lncRNA interactions. In: Feng, Yi, Zhang, Lin (Eds.), Long Non-Coding RNAs. Methods in Molecular Biology 1402. Humana Press, New York, pp. 271–286. Available at https://doi.org/10.1007/978-1-4939-3378-5_21.

Place, R.F., Li, L.C., Pookot, D., Noonan, E.J., Dahiya, R., 2008. MicroRNA-373 induces expression of genes with complementary promoter sequences. Proc Natl Acad Sci U S A 105 (5), 1608–1613. Available at https://doi.org/10.1073/pnas.0707594105.

Qiao, M., Zu, L.D., He, X.H., Shen, R.L., Wang, Q.C., Liu, M.F., 2012. Onconase downregulates microRNA expression through targeting microRNA precursors. Cell Res 22 (7), 1199–1202. Available at https://doi.org/10.1038/cr.2012.67.

Raines, R.T., 1998. Ribonuclease A. Chem Rev 98 (3), 1045–1065. Available at https://doi.org/10.1021/cr960427h.

Ren, X., Liu, G., Zhou, J., 2025. Nuclear-activating miRNAs: Unveiling the intricacies of subcellular miRNA function and regulation in cancer and immunity disease. Cancer Cell Int 25 (1), 147. Available at https://doi.org/10.1186/s12935-025-03760-8.

Rossmanith, W., 2011. Localization of human RNase Z isoforms: Dual nuclear/mitochondrial targeting of the ELAC2 gene product by alternative translation initiation. PLoS One 6 (4), e19152. Available at https://doi.org/10.1371/journal.pone.0019152.

Roundtree, I.A., Evans, M.E., Pan, T., He, C., 2017. Dynamic RNA modifications in gene expression regulation. Cell 169 (7), 1187–1200. Available at https://doi.org/10.1016/j.cell.2017.05.045.

Rybak, S.M., Vallee, B.L., 1988. Base cleavage specificity of angiogenin with *Saccharomyces cerevisiae* and *Escherichia coli* 5S RNAs. Biochemistry 27 (7), 2288–2294. Available at https://doi.org/10.1021/bi00407a007.

Satam, H., Joshi, K., Mangrolia, U., Waghoo, S., Zaidi, G., Rawool, S., et al., 2023. Next-generation sequencing technology: current trends and advancements. Biology 12 (7), 997. Available at https://doi.org/10.3390/biology12070997.

Saxena, S.K., Rybak, S.M., Davey Jr, R.T., Youle, R.J., Ackerman, E.J., 1992. Angiogenin is a cytotoxic, tRNA-specific ribonuclease in the RNase A superfamily. J Biol Chem 267 (30), 21982–21986. Available at https://www.jbc.org/article/S0021-9258(19)36710-9/pdf.

Saxena, S.K., Sirdeshmukh, R., Ardelt, W., Mikulski, S.M., Shogen, K., Youle, R.J., 2002. Entry into cells and selective degradation of tRNAs by a cytotoxic member of the RNase A family. J Biol Chem 277 (17), 15142–15146. Available at https://doi.org/10.1074/jbc.M108115200.

Saxena, A., Shailendra, K., Saxena, K., 2009. Shogen, Effect of Onconase on double-stranded RNA in vitro. Anticancer Res 29 (4), 1067–1071. Available at https://ar.iiarjournals.org/content/29/4/1067.long.

Schaefer, M., Pollex, T., Hanna, K., Tuorto, F., Meusburger, M., Helm, M., et al., 2010. RNA methylation by Dnmt2 protects transfer RNAs against stress-induced cleavage. Genes Dev 24 (15), 1590–1595. Available at https://doi.org/10.1101/gad.586710.

Schwartz, S.L., Conn, G.L., 2019. RNA regulation of the antiviral protein 2′-5′-oligoadenylate synthetase. Wiley Interdiscip Rev RNA 10 (4), e1534. Available at https://doi.org/10.1002/wrna.1534.

Shapiro, R., Riordan, J.F., Vallee, B.L., 1986. Characteristic ribonucleolytic activity of human angiogenin. Biochemistry 25 (12), 3527–3532. Available at https://doi.org/10.1021/bi00360a008.

Shapiro, R., Vallee, B.L., 1987. Human placental ribonuclease inhibitor abolishes both angiogenic and ribonucleolytic activities of angiogenin. Proc Natl Acad Sci U S A 84 (8), 2238–2241. Available at https://doi.org/10.1073/pnas.84.8.2238.

Sheng, J., Yu, W., Gao, X., Xu, Z., Hu, G., 2014. Angiogenin stimulates ribosomal RNA transcription by epigenetic activation of the ribosomal DNA promoter. J Cell Physiol 229 (4), 521–529. Available at https://doi.org/10.1002/jcp.24477.

Sheng, J., Xu, Z., 2016. Three decades of research on angiogenin: a review and perspective. Acta Biochim Biophys Sin 48 (5), 399–410. Available at https://doi.org/10.1093/abbs/gmv131.

Shigematsu, M., Minami, A., Kawamura, T., Shah, S.D., Ogawa, T., Deshpande, D.A., et al., 2025. RNase 2/EDN cleaves the anticodon loops of tRNAs to produce immunostimulatory tRNA halves in asthma. Preprint: bioRxiv. Available from: https://doi.org/10.1101/815803.

Simos, G., Hurt, E., 1999. Transfer RNA biogenesis: A visa to leave the nucleus? Curr Biol 9 (7), R238–R241. Available at https://doi.org/10.1016/s0960-9822(99)80152-3.

Smith, T.J., Giles, R.N., Koutmou, K.S., 2024. Anticodon stem-loop tRNA modifications influence codon decoding and frame maintenance during translation. Semin Cell Dev Biol 154, 105–113. Available at https://doi.org/10.1016/j.semcdb.2023.06.003.

Sorrentino, S., 2010. The eight human "canonical" ribonucleases: Molecular diversity, catalytic properties, and special biological actions of the enzyme proteins. FEBS Lett 584 (11), 2194–2200. Available at https://doi.org/10.1016/j.febslet.2010.04.018.

Statello, L., Guo, C.-J., Chen, L.-L., Huarte, M., 2021. Gene regulation by long non-coding RNAs and its biological functions. Nat Rev Mol Cell Biol 22 (2), 96–118. Available at https://doi.org/10.1038/s41580-020-00315-9.

Su, Z., Kuscu, C., Malik, A., Shibata, E., Dutta, A., 2019. Angiogenin generates specific stress-induced tRNA halves and is not involved in tRF-3-mediated gene silencing. J Biol Chem 294 (45), 16930–16941. Available at https://doi.org/10.1074/jbc.ra119.009272.

Su, N., Yu, X., Duan, M., Shi, N., 2025. Recent advances in methylation modifications of microRNA. Genes Dis 12 (1), 101201. Available at https://doi.org/10.1016/j.gendis.2023.101201.

Suhasini, A.N., Sirdeshmukh, R., 2006. Transfer RNA cleavages by onconase reveal unusual cleavage sites. J Biol Chem 281 (18), 12201–12209. Available at https://doi.org/10.1074/jbc.M504488200.

Suhasini, A.N., Sirdeshmukh, R., 2007. Onconase action on $tRNA^{Lys3}$, the primer for HIV-1 reverse transcription. Biochem Biophys Res Commun 363 (2), 304–309. Available at https://doi.org/10.1016/j.bbrc.2007.08.157.

Takeiwa, T., Ikeda, K., Horie-Inoue, K., Inoue, S., 2021. Mechanisms of apoptosis-related long non-coding RNAs in ovarian cancer. Front Cell Dev Biol 9, 641963. Available at https://doi.org/10.3389/fcell.2021.641963.

Takenaka, Y., Yamada, A., Tomioka, Y., Akiyama, Y., Ivanov, P., 2025. RNase L produces tRNA-derived RNAs that contribute to translation inhibition. RNA 31 (7), 961–972. Available at https://doi.org/10.1261/rna.080419.125.

Torsin, L.I., Petrescu, G.E.D., Sabo, A.A., Chen, B., Brehar, F.M., Dragomir, M.P., et al., 2021. Editing and chemical modifications on non-coding rnas in cancer: A new tale with clinical significance. Int J Mol Sci 22 (2), 581. Available at https://doi.org/10.3390/ijms22020581.

Truesdell, S.S., Mortensen, R.D., Seo, M., Schroeder, J.C., Lee, J.H., LeTonqueze, O., et al., 2012. MicroRNA-mediated mRNA translation activation in quiescent cells and oocytes involves recruitment of a nuclear microRNP. Sci Rep 2, 842. Available at https://doi.org/10.1038/srep00842.

Tsai, M.C., Manor, O., Wan, Y., Mosammaparast, N., Wang, J.K., Lan, F., et al., 2010. Long noncoding RNA as modular scaffold of histone modification complexes. Science 329 (5992), 689–693. Available at https://doi.org/10.1126/science.1192002.

Vasudevan, S., Steitz, J.A., 2007. AU-rich-element-mediated upregulation of translation by FXR1 and Argonaute 2. Cell 128 (6), 1105–1118. Available at https://doi.org/10.1016/j.cell.2007.01.038.

Wang, Q., Lee, I., Ren, J., Ajay, S.S., Lee, Y.S., Bao, X., 2013. Identification and functional characterization of tRNA-derived RNA fragments (tRFs) in respiratory syncytial virus infection. Mol Ther 21 (2), 368–379. Available at https://doi.org/10.1038/mt.2012.237.

Weng, C., Dong, H., Bai, R., Sheng, J., Chen, G., Ding, K., et al., 2022. Angiogenin promotes angiogenesis via the endonucleolytic decay of miR-141 in colorectal cancer. Mol Ther Nucleic Acids 27, 1010–1022. Available at https://doi.org/10.1016/j.omtn.2022.01.017.

Wolf, E.J., Grünberg, S., Dai, N., Chen, T.H., Roy, B., Yigit, E., et al., 2022. Human RNase 4 improves mRNA sequence characterization by LC-MS/MS. Nucleic Acids Res 50 (18), e106. Available at https://doi.org/10.1093/nar/gkac632.

Wu, X., Pan, Y., Fang, Y., Zhang, J., Xie, M., Yang, F., et al., 2020. The biogenesis and functions of piRNAs in human diseases. Mol Ther Nucleic Acids 21, 108–210. Available at https://doi.org/10.1016/j.omtn.2020.05.023.

Wu, Y.N., Mikulski, S.M., Ardelt, W., Rybak, S.M., Youle, R.J., 1993. A cytotoxic ribonuclease. Study of the mechanism of onconase cytotoxicity. J Biol Chem 268 (14), 10686–10693. Available at https://www.jbc.org/article/S0021-9258(18)82252-9/pdf.

Wu, Z., Zhou, R., Li, B., Cao, M., Wang, W., Li, X., 2024. Methylation modifications in tRNA and associated disorders: Current research and potential therapeutic targets. Cell Prolif 57 (9), e13692. Available at https://doi.org/10.1111/cpr.13692.

Yamasaki, S., Ivanov, P., Hu, G.-F., Anderson, P., 2009. Angiogenin cleaves tRNA and promotes stress-induced translational repression. J Cell Biol 185 (1), 35–42. Available at https://doi.org/10.1083/jcb.200811106.

Yu, Bin, Ikhlas, S., Ruan, C., Zhong, X., Cai, D., 2020. Innate and adaptive immunity of murine neural stem cell-derived piRNA exosomes/microvesicles against pseudotyped SARS-CoV-2 and HIV-based lentivirus. iScience 23 (12), 101806. Available at https://doi.org/10.1016/j.isci.2020.101806.

Zhang, Y., Lu, L., Li, X., 2022. Detection technologies for RNA modifications. Exp Mol Med 54 (10), 1601–1616. Available at https://doi.org/10.1038/s12276-022-00821-0.

Zhao, H., Ardelt, B., Ardelt, W., Shogen, K., Darzynkiewicz, Z., 2008. The cytotoxic ribonuclease onconase targets RNA interference (siRNA). Cell Cycle 7 (20), 3258–3261. Available at https://doi.org/10.4161/cc.7.20.6855.

Zong, T., Yang, Y., Zhao, H., Li, L., Liu, M., Fu, X., et al., 2021. tsRNAs: Novel small molecules from cell function and regulatory mechanism to therapeutic targets. Cell Prolif 54 (3), e12977. Available at https://doi.org/10.1111/cpr.12977.

SECTION II

Ribonucleases in diseases

CHAPTER 8

Ribonucleases in infectious diseases

Raúl Anguita Contreras
Department of Biochemistry and Molecular Biology, Faculty of Biosciences, Universitat Autònoma de Barcelona, Cerdanyola del Vallès, Spain

8.1 Introduction

In the 1970s pioneering studies revealed the critical role of granulocyte function in combating infections. Elevated levels of cationic proteins secreted by granulocytes were described in response to infection, particularly those produced by eosinophils (Mahmoud et al., 1975; Venge et al., 1978). Eosinophils were shown to possess the ability to kill pathogens in vitro, and subsequent research on eosinophil granule proteins demonstrated that purified RNase3 exhibits potent antimicrobial activity against both bacteria (Lehrer et al., 1989) and parasites (McLaren et al., 1981; Ackerman et al., 1985). Further studies with RNaseA family members also revealed activities against viral (Paganelli et al., 1991; Domachowske et al., 1998a; Lee-Huang et al., 1999) and fungal (Harder and Schröder, 2002; Huang et al., 2007) pathogens. Today, this family of proteins is well known for its contribution to innate immunity (Boix and Nogués, 2007; Lu et al., 2018). Notably, antibacterial activity is also found in distantly related RNases from lower and higher order vertebrates, suggesting that their role in host defense appeared early on in RNaseA superfamily evolution (Cho and Zhang, 2007; Pizzo and D'Alessio, 2007; Rosenberg, 2008). The evolutionary significance of this family of proteins in host protection becomes apparent when observing the high sequence divergence and rapid rate of evolution of some of its members, such as the eosinophil RNases, which is a hallmark of immune system proteins that must rapidly adapt to combat emerging pathogens (Zhang et al., 1998).

The RNaseA family comprises small secretory proteins (13–15 kDa) that comply with the characteristic properties associated with antimicrobial peptides (AMPs). They present a highly positive net charge with the presence of hydrophobic regions and secondary structures with amphipathic nature capable of interacting with microbial membranes and inducing their disruption (Boix et al., 2012). In contrast, whether their ribonuclease activity contributes or not to their antimicrobial function, particularly against bacterial pathogens, remains a subject of debate. Rosenberg and colleagues first determined that the catalytic activity of RNase3 is independent of its capacity to kill bacteria (Rosenberg, 1995). Nevertheless, more recent research revealed that ribonuclease activity plays an important role in host defense, not only through direct action against single-stranded RNA (ssRNA) viruses (Sikriwal et al., 2012) and other

RNases in Health and Diseases
DOI: https://doi.org/10.1016/B978-0-443-27312-4.00005-5

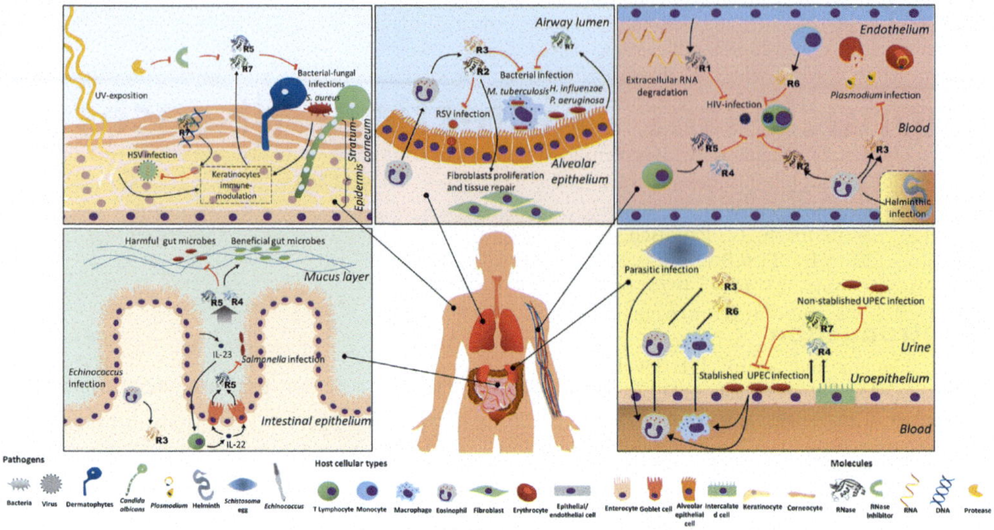

Figure 8.1 Host defense functions described for RNaseA superfamily members. Activation, secretion, and extravasation-related processes are indicated by black lines; inhibition and microbicidal-related processes are indicated by red lines. *This figure is based on most of the references cited in this section, which include studies performed in humans, in mouse models, and in vitro.*

pathogens (Salazar et al., 2019; Abengózar et al., 2022), but also by contributing to the activation of immune responses (Greulich et al., 2019).

The following sections summarize current knowledge about the RNaseA family proteins that contribute to host defense against infectious diseases. We also include a brief overview of the human noncanonical RNases, a subgroup characterized by lower sequence identity (15%–30%) and the absence of catalytic activity (Cho et al., 2005). In addition, other human RNases with key roles in host defense, such as RNaseT2 and RNaseL (Lu et al., 2018), are discussed. Fig. 8.1 illustrates the tissue-specific expression of RNases in the human body, providing a systemic overview of their potential to protect the host against infectious agents.

8.2 Bacterial infections

Human RNaseA family members possess antimicrobial activity against a broad number of bacterial species. The primary mechanism for bacterial eradication is shared with other AMPs and relies on membrane leakage (Torrent and Sánchez, 2009a). These proteins present a highly cationic surface charge, which facilitates the formation of electrostatic bonds with negatively charged bacterial membranes, ultimately leading to destablization through a carpet-like mechanism (Boix et al., 2012). The key bacterial membrane components responsible for RNase high-affinity binding are lipid A or complete lipopolysaccharides (LPS) and peptidoglycans (Torrent et al., 2008). Other bacterial membrane elements, such as *Pseudomonas aeruginosa* Outer Membrane Protein I, can also mediate RNase binding (Lin et al., 2010). Besides, the presence of aggregation-prone regions in the RNase structure can cause self-aggregation and induce bacterial agglutination (Torrent et al., 2012). These antimicrobial properties are attributed to the presence of highly cationic clusters, mainly constituted by spatially exposed arginines, and of hydrophobic patches that conform amphipathic secondary structures upon interacting with bacterial membranes (Carreras et al., 2003). Notably, the antibacterial activity of RNases is mostly confined to the N-terminal region, and synthetic peptides derived from these sequences retain membrane-disruptive, cell-agglutinating, and LPS-binding capabilities (Pulido et al., 2013; Pulido et al., 2016a; Torrent and de la Torre, 2009b).

8.2.1 RNase3, RNase6, and RNase7 as key antibacterial effectors

Among the RNaseA family members, RNase3, RNase6, and RNase7 are the best characterized in terms of their structural properties and mechanisms underlying their bactericidal activity against a broad range of Gram-negative and Gram-positive bacteria (Torrent and Sánchez, 2009a; Torrent et al., 2010; Pulido et al., 2016b). RNase3, also named as eosinophil cationic protein, is primarily secreted by eosinophils in response to infection and inflammation, although neutrophils and macrophages can also contribute to its release (Venge et al., 1999; Bystrom et al., 2011). In addition, RNase3 can activate mast cells, leading to prostaglandin D2 synthesis, which in turn induces further recruitment of neutrophils (Bystrom et al., 2011).

Both RNase6 and RNase7 play important roles in combating bacteria in the urinary tract, though they differ in their cellular origins and expression patterns (Becknell et al., 2015; Eichler et al., 2019; Ruiz-Rosado et al., 2023). RNase6 is secreted by myeloid cells, including resident leukocytes, infiltrating granulocytes, and macrophages. In contrast, RNase7 is expressed by the bladder urothelium and the kidney's intercalated cells. Besides, RNase6 secretion is induced in response to urinary tract infection (UTI) and pyelonephritis, while RNase7 is constitutively expressed. This suggests that RNase7 functions as a sentinel AMP to prevent bacterial colonization, while RNase6 is more involved in combating established infections (Becknell et al., 2015). RNase3 is also secreted in the urinary tract by infiltrating eosinophils and neutrophils during parasitic infections, although its specific contribution in the clearance of uropathogenic *Escherichia coli* (UPEC) requires further exploration (Leutscher et al., 2000; Stecher et al., 2017; Becknell et al., 2019). A potential protective role for RNase4 against UTI should not be overlooked. Like RNase7, RNase4 is constitutively expressed in the uroepithelium and is effective in killing UPEC strains, including multidrug resistant isolates (Murtha et al., 2018). The expression of both RNase4 and RNase7 in uroepithelial and intercalated cells is regulated by insulin receptor activation via the phosphatidylinositol 3-kinase/AKT (AKT) signaling pathway, providing a mechanistic explanation for the increased susceptibility to UTIs and poorer outcomes observed in patients with diabetes mellitus (Murtha et al., 2018; Eichler et al., 2016).

The epithelial expression of RNase7 is not limited to the urinary tract; it is also constitutively secreted by keratinocytes and plays a key role in skin defense against a wide range of pathogens, including bacteria, viruses, and fungi (Rademacher et al., 2019). RNase7 expression on skin is not only constitutive but can also be induced by interleukins (IL), ultraviolet radiation, and bacterial challenge (Harder and Schröder, 2002; Gläser et al., 2009; Mohammed et al., 2011). Stimulation of human primary keratinocytes with both IL-17 and interferon gamma (IFN-γ) synergistically potentiates the expression of RNase7 more than any other tested cytokine alone or in combination (Simanski et al., 2013). RNase7's antibacterial activity against *P. aeruginosa*, *Enterococcus faecium,* and *Propionibacterium acnes* has been demonstrated, but its protective role in skin has been especially well characterized in *Staphylococcus aureus* and methicillin-resistant *S. aureus* infections (Harder and Schröder, 2002; Rademacher et al., 2019; Haisma et al., 2013). It has been described that lower constitutive expression of RNase7 is associated with a higher risk of *S. aureus* infection of the skin (Zanger et al., 2009). Additionally, RNase inhibitor (RI) can block both the catalytic and antimicrobial activities of RNase7 in skin, promoting *S. aureus* growth and RNA-mediated inflammation in atopic dermatitis patients (Abtin et al., 2009; Rademacher et al., 2024). It has also been proposed that RNase7 could contribute to respiratory tract defense against *Haemophilus influenzae* and *Mycobacterium tuberculosis* infections, as its expression is induced after infection of airway epithelial cells (Amatngalim et al., 2015; Torres-Juarez et al., 2018). Indeed, several human RNases, including RNase7, have demonstrated both extracellular and intracellular antimycobacterial activity *in vitro* (Pulido et al., 2013; Lu et al., 2019).

8.2.2 RNase4 and RNase5: emerging regulators of gut microbiota and host–microbe balance

RNases4 and 5, the latter being also known as angiogenin, have been associated with antimicrobial activity and the regulation of bacterial populations in mouse models. Both proteins are expressed by intestinal epithelial cells, such as Paneth and goblet cells, and play essential roles in regulating gut microbiota and preventing inflammatory bowel disease (IBD) (Sun et al., 2021, 2024).

RNase4 protects mice from colitis by limiting the growth of *Parasutterella* and preventing the suppression of tryptophan metabolism that this microorganism induces upon colonization. An increase in *Parasutterella* population was detected in both mouse and human stool samples from patients with IBD and correlated with decreased levels of RNase4, suggesting a conserved function of RNase4 in microbiota regulation across species (Sun et al., 2024). The expression of murine angiogenin 4 (Ang 4) is mediated by bacterial exposure. For instance, its levels in germ-free mice guts were increased when colonized by *Bacteroides thetaiotaomicron*, a predominant member of gut microbiota (Hooper et al., 2003). A cytokine-modulated lymphocyte–enterocyte crosstalk was proposed to increase Ang 4 levels in response to *Salmonella*. Briefly, exposure to *Salmonella* induces IL-23 secretion by enterocytes, which stimulates IL-22 secretion by intestinal intraepithelial lymphocytes, thereby inducing Ang 4 production by goblet cells to counteract the infection. This suggests that RNase expression in the gut is dynamically regulated through host–microbe interactions (Walker et al., 2013). As for murine Ang 1, it promotes colonization by beneficial Lachnospiraceae in the mouse gut by suppressing α-Proteobacteria growth. Similarly, IBD in both mice and humans has been associated with decreased Lachnospiraceae abundance and lower RNase5 levels (Sun et al., 2021).

These findings, along with sequence homology, similar tissue expression patterns, and experimental data, indicate that human RNase4 and RNase5 likely perform gut microbiota regulatory functions comparable to their murine counterparts. The antimicrobial activity of RNases4 and 5 is attributed to their ability to bind bacterial membranes and cause membrane leakage. In the murine Ang 4, key cationic residues located in the N-terminal region and the cell-binding segment are responsible for membrane permeabilization and bactericidal activity against *S. typhimurium* (Sultana et al., 2022).

The ability of RNase4 and RNase5 to selectively inhibit certain bacterial populations while promoting the growth of others has been preliminarily attributed to the increased susceptibility of Gram-negative bacteria to RNase-induced killing. In mouse models, expression of RNase4 and Ang1 reduced the abundance of Parasutterella and α-Proteobacteria, both Gram-negative groups associated with IBD, while increasing the population of beneficial Gram-positive Lachnospiraceae (Sun et al., 2021, 2024). Supporting this selectivity, *in vitro* incubation with Ang 1 demonstrated potent activity against *Brevundimonas diminuta* and *Sphingomonas paucimobilis*, Gram-negative species, but not against the Gram-positive *Anaerostipes* sp. and *Blautia* sp. Other studies also described Ang1's limited antibacterial activity against the Gram-positive species *E. faecalis* and *Listeria monocitogenes*. Similar preferential activity toward Gram-negative bacteria has been observed for other RNases as well (Pulido et al., 2016a). This pattern may be explained by the absence of LPS in Gram-positive bacteria, LPS being a key component mediating electrostatic interactions with RNases, and the presence of a thicker peptidoglycan layer compared to Gram-negative strains. However, administration of Ang 4 to mice resulted in an increase in the Gram-negative gut resident beneficial bacteria *Akkermansia*, while a reduction in the Gram-positive populations *Roseburia*, *Enterorhabdus,* and *Eubacterium brachy*. Interestingly, Ang 4 demonstrated high activity *in vitro* against *E. faecalis* and *L. monocitogenes*, but little activity was detected against *L. innocua* and the Gram-negative species *E. coli K12* and *B. thetaiomicron* (Hooper et al., 2003; Walker et al., 2013). These findings suggest that Ang 4 may favor growth of beneficial bacterial populations independently of their Gram classification. Interestingly, RNase5 and the murine Ang 1 and Ang 4 demonstrated different killing selectivity in vitro for the Gram-positive species *E. faecalis*, *L. monocitogenes*, and *Streptococcus pneumoniae*, indicating possible differences in target specificity. Further research is needed to clarify the molecular mechanisms underlying the selective targeting of bacterial populations by RNase4 and RNase5 and to better understand their roles in maintaining gut microbiota homeostasis. More information on RNases associated with gut diseases can be found in Chapter 10.

8.2.3 RNase1 and RNase2: nonantibacterial effectors

RNase1 and RNase2, the second also known as eosinophil-derived neurotoxin, are not primarily associated with antibacterial defense, as both exhibit poor membrane binding and limited membrane-disruptive capacity (Lu et al., 2018, 2019). RNase1 is mainly involved in the clearance of extracellular RNA from body fluids (Lomax et al., 2017) and may help mitigate inflammation associated with tissue damage (Cabrera-Fuentes et al., 2014; Zernecke and Preissner, 2016). Interestingly, antibacterial activity was identified for RNase1 against *E. coli* strains 25922 and SYY89 but not DR115, indicating strain-specific effects. However, the mechanisms underlying this strain-specificity remain unclear (Kosgey et al., 2020).

RNase2 exhibits only minimal bactericidal activity (Kosgey et al., 2020). Notably, RNase 2's antimicrobial potency can be significantly enhanced, approaching that of RNase3, through just four amino acid substitutions in its N-terminal region. However, this enhancement comes at the cost of nearly a 40-fold decrease in catalytic activity (Tran et al., 2024). These observations support the hypothesis that following gene duplication from a common eosinophil RNase ancestor, RNase3 underwent rapid evolutionary adaptation in primates to enhance its antimicrobial function, sacrificing catalytic efficiency. In contrast, RNase2 retained its ribonucleolytic activity to exert antiviral functions (Tran et al., 2024; Rosenberg et al., 1995).

One should not discard the possibility that RNases act in combination with other AMPs in a catalytically dependent mechanism. For instance, RNase1 has been shown to act synergistically with the membrane-permeabilizing AMP LL-37. In this interaction, LL-37 facilitates membrane disruption, allowing RNase1 to enter the bacterial cytoplasm and degrade intracellular RNA, resulting in cell death (Eller and Raines, 2020). Furthermore, a chimeric construct combining RNase1 and RNase3 demonstrated synergistic effects when used in combination with colistin and was able to delay the development of colistin resistance in *Acinetobacter baumannii* (Prats-Ejarque et al., 2019).

8.2.4 RNase8 and the noncanonical RNases

RNase8 is the closest homolog to RNase7 in the human RNaseA superfamily, sharing 78% sequence identity. Even though this protein also possesses a wide spectrum of antibacterial activity, its expression is restricted to the placenta, and its biological relevance still needs to be fully elucidated (Lu et al., 2018; Rudolph et al., 2006; Zhang et al., 2002). The four noncanonical RNases, also named as RNases9–13, are another example of RNaseA family members with poorly described functions. These proteins share only 15%–30% identity with the canonical RNases and lack the catalytic triad required for RNA hydrolysis (Goo and Cho, 2013). Noncanonical RNase expression is restricted to the testis, and their functions are mainly associated with the male reproductive system (Shaffer et al., 2024; Liu et al., 2008). Antibacterial activity against *E. coli* was discovered for RNase9 at the low micromolar range, suggesting that it may contribute to maintaining the sterility of male reproductive tissues (Cheng et al., 2009; Krutskikh et al., 2012). Unlike the canonical RNases, the noncanonical members lack the highly conserved N-terminal region typically associated with antimicrobial function (Lu et al., 2018), suggesting that RNase9's bactericidal activity is mediated through alternative secondary structures within its protein sequence.

8.2.5 RNase-mediated detection of bacterial pathogens

The role of RNases in dealing with bacterial infections is not only based on direct action against the microorganisms. Members of the RNaseA superfamily collaborate with RNaseT2 to degrade pathogenic RNA and release uridine, which, in turn, activates Toll-like receptor (TLR) 8, triggering cytokine release and initiating innate immune response (Greulich et al., 2019; Ostendorf et al., 2020; Nunes et al., 2024). For example, TLR 8 activation in monocytes and macrophages initiates a strong T-helper-1 (Th1) response by the release of IL-18 and IL-12p70, which subsequently activates natural killer (cells and promotes IFN-γ release (Coch et al., 2019)).

While RNaseA family members preferentially cleave RNA at UX sequences, RNaseT2 cleaves at XU sites, where X is a purine, facilitating the release of uridine. These uridines, along with ssRNA fragments, function as TLR8 ligands. RNase2 and RNase6 are both highly expressed in monocytes and are essential for TLR8 pathway activation in response to bacterial RNAs from *E. coli*, *S. aureus*, *S. pyogenes*, and other species (Greulich et al., 2019; Ostendorf et al., 2020; Nunes et al., 2024). The presence of 2′O-Me-modified uridines in the bacterial RNA can impair RNase6-mediated cleavage and subsequent TLR8 activation (Nunes et al., 2024). Interestingly, tRNA from *E. coli* tRNA methyltransferase H (trmH) knockout induced higher IFN-γ levels than tRNA from the wild-type strain, suggesting that bacteria may utilize enzymes such as TrmH to modify their RNA and evade immune detection (Freund et al., 2019).

8.3 Viral infections

Numerous antiviral activities have been identified among human members of the RNaseA superfamily, particularly against ssRNA viruses (Li and Boix, 2021). The protective effects of these proteins can be partially attributed to the hydrolysis of viral RNA, both intracellularly and extracellularly, which leads to inhibition of viral replication (Bedoya et al., 2006). In addition to their direct antiviral activity, RNases also contribute to host defense by activating immune responses and triggering antiviral signaling pathways within infected cells (Greulich et al., 2019; Ostendorf et al., 2020; Yang et al., 2008; Kopfnagel et al., 2018, 2020). Among the RNaseA family members, RNase2 is likely the most extensively studied in terms of antiviral function. It has been described to inhibit infections caused by respiratory syncytial virus (RSV) (Domachowske et al., 1998a, b; Sikriwal et al., 2012; Lu et al., 2022), human immunodeficiency virus (HIV) (Lee-Huang et al., 1999; Bedoya et al., 2006; Rugeles et al., 2003), hepatitis B virus (HBV) (Li et al., 2010), and parainfluenza virus (PIV) (Domachowske et al., 1998a, b).

8.3.1 Eosinophil RNase protection from respiratory viruses

Both RNase2 and RNase3 are secreted by eosinophils in response to RSV infection (Harrison et al., 1999; Kim et al., 2013), and elevated levels of these proteins are considered markers of wheezing (Koller et al., 1997; Garofalo et al., 1992). The antiviral activity of eosinophils against RSV is blocked when adding RI, confirming that it is dependent on RNases (Domachowske et al., 1998a, b). The C-terminal loop L7 of RNase2 mediates its interaction with the virion's envelope, enabling RNase penetration and subsequent degradation of the viral genome, thereby reducing infectivity in epithelial cells (Sikriwal et al., 2012). Key substitutions at the position of this loop in RNase3 correlate with its lower activity against RSV particles (Domachowske et al., 1998a, b). Both eosinophil RNases are also highly expressed in macrophages. *RNASE2* knockout in THP-1 cells differentiated with phorbol 12-myristate 13-acetate leads to more susceptibility to RSV infection (Lu et al., 2022), while RNase3 overexpression is protective (Lu et al., 2021). Antiviral activity against PIV virions was also detected for RNase2, although to a lesser extent than against RSV (Domachowske et al., 1998a, b). Eosinophil RNases may also have a protective role against coronavirus (CoV), as RNase2 was found to be upregulated in response to CoV infections and proposed as a clinical biomarker (Lee et al., 2005; Dosanjh, 2020). Besides, increased eosinophil levels correlate with better disease outcomes (Dosanjh, 2020; Liu et al., 2020; Rodrigo-Muñoz et al., 2021), while eosinopenia is related to a higher risk of CoV infection (Lindsley et al., 2020; Lippi and Henry, 2020). RNase1 has also been linked to coronavirus disease 2019 patient outcomes, with increased levels during infection correlating with renal dysfunction and poorer recovery prognosis (Zechendorf et al., 2023).

Studies in mice provide valuable insights into how mouse eosinophil-associated ribonucleases (EARs) influence the progression of viral infections. For example, mEAR 2 reduces the infectivity of the pneumonia virus of mice in vitro, a virus closely related to human RSV. However, levels of mEAR 2 detected in the lungs of infected mice were diminished, which the authors interpret as a viral evasion mechanism to escape RNase-mediated antiviral effects (Moreau et al., 2003). Another example of how reduced RNase expression can increase susceptibility to viral respiratory infections is observed in neonates with bronchopulmonary dysplasia (BPD). BPD, often caused by oxygen exposure in premature infants, is linked to recurrent viral infections requiring hospitalization. Adult mice exposed to hyperoxia as neonates show decreased numbers of alveolar epithelial type II cells and consequently lower levels of mEAR11, resulting in heightened sensitivity to influenza A virus (IAV) infection. Heterologous expression of mouse Ear11 in the lungs of BPD-affected mice protected them from IAV infection and reduced associated leukocyte recruitment (O'Reilly et al., 2012).

8.3.2 Activity against human immunodeficiency virus and hepatitis B virus

Several studies have identified antiHIV activities for RNaseA superfamily members. The antiviral activity of RNase2 against HIV was discovered in supernatants from mixed lymphocyte reactions of peripheral blood mononuclear cells from healthy human leukocyte antigen-discordant individuals. The addition of specific antiRNase2 antibody, or RI, partially blocked antiviral activity of the supernatants (Rugeles et al., 2003). AntiHIV activity was also identified in human chorionic gonadotropin preparations, although the specific RNases involved were not identified (Lee-Huang et al., 1999). Interestingly, RNase3 levels are increased in HIV-infected patients, although, to our knowledge, no studies have confirmed a direct role in combating the infection (Paganelli et al., 1991; Onorato et al., 1999). Both RNases4 and 5 are produced by primary T-cells and, together with other soluble factors, are capable of inhibiting HIV replication (Cocchi et al., 2012). AntiHIV activity has also been reported for RNase6, which is downregulated in Th 17 polarized cells upon HIV infection (Christensen-Quick et al., 2016). Notably, recombinant RNase1, RNase2, and RNase5 demonstrated the ability to reduce HIV primary infection of phytohemagglutinin-stimulated T cell blasts. The inhibitory effects were observed in three different experimental assays: (1) cells infected prior to exposure to RNase, (2) HIV-1 incubated with RNase prior to cell infection, and (3) cells incubated with RNase prior to HIV-1 infection. These results suggest that these RNases are active not only against replicating HIV-1 within the cytosol but also provide protection when taken up by noninfected cells and may even act directly against free virions (Bedoya et al., 2006).

An RNase2 fusion protein with HBV core protein demonstrated to reduce HBV replication in the human hepatoblastoma HepG2.2.15 cell line when transfected (Li et al., 2010). The most probable target molecule for this ribonucleic fusion protein is the HBV pregenomic RNA, which serves as the template for viral protein and DNA synthesis. This suggests that RNA intermediates in the viral life cycle could potentially be targeted by RNases (Li et al., 2010; Wu et al., 2019).

8.3.3 Immunomodulatory and regulatory roles of RNases during viral infections

RNase2 is also an important contributor to immune system activation during viral infections. The uridine-rich ssRNA40 sequence from the HIV-1 long terminal repeat can activate strong RNase2-dependent TLR8-mediated responses (Greulich et al., 2019; Ostendorf et al., 2020). It is also described that TLR2 can be activated by RNase2 acting as an endogenous ligand, leading to dendritic cell activation and cytokine production (Yang et al., 2008). Thus RNase2 contributes to antiviral defense both through direct action on viral particles and by stimulating immune cells. After viral clearance, RNase2 may also participate in tissue remodeling by promoting fibroblast proliferation (Munitz and Levi-Schaffer, 2004). In mice, mEAR11 functions as a macrophage chemoattractant in a mechanism independent of RNA hydrolisis activity or TLR2 activation (Yamada et al., 2015).

RNase7 is also an important alarmin in skin that is capable of inducing antiviral responses. Although RNase7 has no direct activity against HSV-1, it can bind self-DNA released from skin lesions to induce antiviral interferon-stimulated genes in plasmacytoid DCs and keratinocytes, effectively inhibiting HSV-1 infection. While TLR9 was identified as the target for RNase7/DNA complex binding in pDCs, human keratinocytes are likely activated through an intracellular DNA receptor, such as the IFI16/cGAS-STING pathway (Kopfnagel et al., 2018, 2020).

In response to viral infections, human RNases can induce the production of tRNA fragments (tRFs) that may act as modulators to protect host cells. RNase5 and, more recently, RNase2 have been identified as contributors to tRF generation during RSV infections. Three RNase5-derived tRFs (tRF5-GluCTC, tRF5-GlyCCC, and tRF5-LysCTT) have been characterized as promoters of RSV replication (Wang et al., 2013; Zhou et al., 2017), and several RNase2-derived tRFs have also been identified (Lu et al., 2022). However, it is still unknown whether RNase-derived tRFs are contributing to the host's antiviral response against RSV.

Regarding HIV, it is well known that tRNAs can bind to primer binding sites and promote replication of the retroviral genome (Das et al., 1995; Das and Berkhout, 1995). Interestingly, 3' terminal tRFs were proposed as potential antiretroviral effectors as they were found to be strongly complementary to human endogenous retroviruses, capable of associating with Ago2 to degrade target RNAs (Li et al., 2012). RNase5 expression was also associated with dengue virus replication, though the implication of tRFs was not explored (Madhry et al., 2023). Cleavage of specific tRNAs necessary for viral replication is an additional mechanism by which RNases could mediate the host response (Suhasini and Sirdeshmukh, 2007). Further research is needed to clarify the cellular functions of RNase-derived tRFs and their contribution to antiviral responses.

8.3.4 Bovine seminal RNase and Onconase

Nonhuman RNaseA family members have also been studied for their capacity to protect human cells from HIV infection. Both bovine seminal (BS) RNase, a reference member of the RNaseA superfamily, and Onconase, an amphibian RNase extensively studied for its antitumoral properties, demonstrated the ability to inhibit HIV infection in H9 cells. The antiviral activity of BS-RNase and Onconase is not based on a direct effect against viral particles (Youle et al., 1994). Bovine and human IFN-γ bind to double-stranded RNA (dsRNA) through their C-terminal regions, increasing its susceptibility to degradation by BS-RNase. The authors hypothesize that since dsRNA is an intermediate in the replication of many viruses, its cleavage via an IFN-γ/RNase-dependent mechanism may contribute to infection clearance (Schein et al., 1990; Schein and Haugg, 1995).

Regarding Onconase, this amphibian RNase selectively degrades viral RNA within infected cells to exert its antiviral effects. Its full antiviral activity requires the N-terminal nine amino acids and a pyroglutamyl residue (Saxena et al., 1996). Onconase also demonstrated selective degradation of $tRNA^{Lys3}$, which is the primer necessary for HIV reverse transcription. Apparently, Onconase specificity for $tRNA^{Lys3}$ is based on structural determinants rather than the cleavage site sequence. However, the extent to which $tRNA^{Lys3}$ degradation contributes to Onconase's anti-HIV activity remains unclear (Suhasini and Sirdeshmukh, 2007). Additionally, activation of activating transcription factor 3 by Onconase may further explain its antiviral properties, as this pathway inhibits viral replication (Vert et al., 2017).

8.3.5 Other RNases with antiviral properties

Beyond the RNases of the RNaseA family, other human RNases also play important roles in antiviral defense. As noted above, RNaseT2 is an important contributor to TLR8 pathway activation thanks to its capacity, in combined action with RNase2 or RNase6, to release uridine from pathogenic RNA (Greulich et al., 2019; Ostendorf et al., 2020; Nunes et al., 2024). RNaseL and ISG-20 are two IFN response genes that play key roles in combating multiple viral infections (Silverman, 2007). In the presence of dsRNA, 2'5'-oligoadenylate synthetase is activated and starts to generate 2′5′-

oligoadenylates (2–5A), which act as activators of RNaseL (Behera et al., 2002). Antiviral activities associated with RNaseL have been detected against multiple viral pathogens, such as hepatitis C virus (HCV) (Malathi et al., 2007, 2010), HIV (Silverman, 2007), RSV (Behera et al., 2002), mouse hepatitis virus (Ireland et al., 2009), and IAV (Cooper et al., 2015). The mechanism of action of RNaseL relies on the degradation of both viral and host RNA. Apart from exerting a direct effect against virus replication, the RNA fragments generated by RNaseL enable the amplification of IFN signaling and stress granule formation thanks to the activation of multiple pattern recognition receptors (Manivannan et al., 2020).

Regarding ISG20, the mechanisms behind its broad-spectrum antiviral activity remain incompletely understood but are thought to include direct degradation of viral RNA, degradation of deaminated viral DNA, and inhibition of viral protein translation (Weiss et al., 2018). In any case, it is well known that ISG20 catalytic activity is essential for its antiviral performance (Deymier et al., 2022). Regnase-1, also known as MCPIP1, is a CCCH-type zinc-finger protein that is induced during viral infection. It exhibits unique antiviral activity compared to closely related human homologs by cleaving viral RNA and by degrading inflammatory gene transcripts to control excessive immune activation (Lin et al., 2013). The zinc-finger antiviral protein is another important host defense factor (Bick et al., 2003). Finally, RNaseP has been shown to specifically cleave HCV genomic regions containing RNA secondary structures (Nadal et al., 2002).

8.4 Fungal infections

Human RNases also exhibit antifungal activity, being effective not only against pathogenic yeasts but also against dermatophytes (Abtin et al., 2009; Fritz et al., 2012). The fungicidal activity of RNases relies on their capacity to interact with components of the fungal envelope and to induce its disruption (Salazar et al., 2019). Although the mechanisms involved in targeting fungal and bacterial pathogens are similar, certain RNases display selectivity (Hooper et al., 2003).

8.4.1 Activity against *Candida albicans*

Candida albicans is one of the most common fungal pathogens, which is especially threatening for hospitalized and immunocompromised patients. Among human RNases, RNase3, RNase5, and RNase7 have been described as the most effective in killing *C. albicans* (Abtin et al., 2009; Salazar et al., 2016). These RNases can bind to *C. albicans* β-D-glucans and induce membrane permeabilization, depolarization, and leakage (Salazar et al., 2019, 2016). Electrostatic interactions with anionic phospholipids present in fungal membranes, such as phosphatidylserine and phosphatidylinositol, or with wall components, such as mannoproteins, are also likely to contribute, as observed with LL-37 and other AMPs (Salazar et al., 2019; Swidergall and Ernst, 2014; Tsai et al., 2011).

RNase7's antifungal activity is primarily exerted in human skin, where it is expressed by keratinocytes alongside RNase1, RNase4, and RNase5 (Harder and Schröder, 2002; Abtin et al., 2009). RNase5 was also reported to be secreted by human reconstructed epidermis in vitro (Rendl et al., 2001). Anticandidal activity of RNase5 on skin is modulated by the presence of RI and stratum corneum proteases, as previously mentioned for RNase7 activity against *E. faecium*. Interestingly, although RNase5 possesses high anticandidal activity, the protein was demonstrated to be inefficient in killing bacterial species such as *E. coli, P. aeruginosa, S. aureus, E. faecalis*, or *L. monocitogenes* (Abtin et al., 2009; Hooper et al., 2003). This suggests that its antimicrobial spectrum may be specialized for targeting *C. albicans* and potentially other fungal pathogens. Notably, mAng 1 has also shown anticandidal activity, unlike mAng 4. The fact that these proteins also differ in their bactericidal specificities suggests that rapid evolutionary divergence and mutation following gene duplication may have led to functional specialization against distinct pathogens (Hooper et al., 2003). Anticandidal activity was also found for bovine milk RNase4 and RNase5. However, these proteins are found at concentrations in milk below those required for effective *C. albicans* killing, indicating they may not serve an intrinsic antimicrobial function in that context (Harris et al., 2010).

The role of catalytic activity in the antifungal function of RNases has been investigated. An initial study involving RNases5 and 7 showed that their potent in vitro activity against *C. albicans* was inhibited by RNH1 or by pretreatment with diethyl pyrocarbonate (DEPC) or benzopurpurin B. These inhibitory effects are likely due to RNH1 masking cationic surface patches or DEPC modifying key histidine residues, both of which interfere with the RNases' ability to interact with fungal membranes (Abtin et al., 2009). As such, these findings do not offer definitive conclusions regarding the role of catalytic activity in the antifungal mechanisms of RNases5 and 7. However, a subsequent in vitro study using recombinant RNases3 and 7 demonstrated that the catalytically active forms were more effective than their catalytically inactive H15A mutants in killing *C. albicans*. Both RNases were able to translocate into the cytosol of *C. albicans* and exert antifungal effects through nonspecific RNA degradation, supporting a role for catalytic activity in their antimicrobial action (Salazar et al., 2016).

Given that the membrane-leakage capabilities of RNases are primarily attributed to their N-terminal regions, one study investigated the antifungal activity of the N-terminal peptides from all eight canonical human RNaseA superfamily members against *C. albicans* (Salazar et al., 2019). This approach provides valuable insight into the antifungal mechanisms of RNases by comparing the effects of the isolated N-terminal peptides with those of the full-length proteins. While the N-terminal peptides of RNases3 and 7 retained strong anticandidal activity, the N-terminal peptide of RNase5 was completely inactive. This observation suggests two possibilities: either key residues required for membrane disruption in RNase5 are located outside its N-terminal region, or its catalytic activity is essential for its antimicrobial function. These findings, combined with RNase5's unusually species-specific antimicrobial spectrum, highlight its unique role as a host defense protein (Hooper et al., 2003). A similar pattern is observed with RNase8, whose N-terminal peptide possesses no activity against *C. albicans*, while the full protein is reported to effectively kill the yeast in vitro (Rudolph et al., 2006). Interestingly, N-terminal-derived peptides from RNase1 and RNase4 also demonstrated notable capacity to kill *C. albicans*, which aligns with evidence from other in vitro studies using the full-length proteins (Kosgey et al., 2020; Harris et al., 2010).

Another important mechanism through which RNases may help combat *C. albicans* infections is biofilm inhibition and eradication. N-terminal peptides from RNases3 and 7 were shown to suppress the expression of genes involved in hyphal growth, biofilm formation, and extracellular matrix production. In addition, mature biofilms exposed to these peptides exhibited a significant loss of viability and structural integrity (Salazar et al., 2019).

RNase activity against other yeast species has also been explored. A comparative in vitro study testing recombinant RNase7 against various pathogens found it to be even more potent against *Pichia pastoris* than against *C. albicans* (Huang et al., 2007).

8.4.2 Activity against dermatophytes

Dermatophytes are fungi capable of invading animal and human skin tissues, causing infections typically known as tineas (Brasch, 2009). Although the role of RNases in combating dermatophytes has not been extensively studied, emerging evidence suggests they may play a significant role in host defense.

RNase7 demonstrated the ability to inhibit *Trichophyton rubrum, T. mentagrophytes*, and *Epidermophyton floccosum* growth after 2 hours of incubation with their respective germinating conidia. While *T. rubrum* and *T. mentagrophytes* growth was inhibited for 3 days, *E. floccosum* was inhibited for at least 7 days, as no growth for the dermatophyte was detected in all the experimental timelines. *Microscoporum canis* was also included in the study, but no antimicrobial activity was found for RNase7 against this dermatophyte (Fritz et al., 2012). Overexpression of RNase7, along with other AMPs and TLRs, was found in sections of skin injured by tinea and pytiriasis versicolor (Brasch et al., 2014). Additionally, in vitro studies revealed that infection of primary keratinocytes with *T. rubrum* induces the expression of RNase7 and human beta-defensin-2. This induction is further enhanced in the presence of IL-17A and IFN-γ (Firat et al., 2014). The EGFR has been identified as a central signaling mediator for AMPs expression in skin. The exposure of keratinocytes to the EGFR inhibitor erlotinib diminishes *T. rubrum*-mediated induction of RNase7 expression. In fact, patients undergoing anti-EGFR treatment are more susceptible to suffering tineas, likely due to diminished levels of

RNase7 and other AMPs in the skin (Lichtenberger et al., 2013; Eilers et al., 2010). Collectively, these findings highlight RNase7 as a critical antimicrobial effector in protecting the skin against recurrent and chronic dermatophytic infections.

8.4.3 Parasitic infections

The role of human RNases in parasitic infections has been explored in numerous studies. Among all human RNases belonging to the RNaseA superfamily, RNase3 is the most extensively studied for its involvement in defending against parasitic infections caused by protozoa, nematodes, trematodes, and cestodes. Both peripheral and local eosinophil levels are commonly increased in response to parasitic infections together with their granulocyte content, which includes RNase3, RNase2, eosinophil peroxidase (EPO), and major basic protein (MBP) (Rajamanickam et al., 2018). Incubation of *Leishmania donovani* promastigotes with human eosinophils leads to fast killing of the protozoal pathogen (Pearson et al., 1987). Besides, mice treated with eosinophil antiserum before infection with *Schistosoma mansoni* cercariae develop reduced numbers of schistosomula (Mahmoud et al., 1975).

Research into the role of eosinophil-derived RNases during parasitic infections has primarily focused on two areas: (1) their cytotoxic effects against protozoan and helminth pathogens and (2) their potential use as clinical biomarkers for diagnosis.

8.4.4 Activity against protozoan parasites

Although the activity of RNase3 against human parasites has been studied for decades, the underlying mechanisms of parasite killing are still being elucidated. Similar to its action against bacteria, RNase3 kills *Leishmania donovani* through membrane permeabilization mediated by key cationic residues. Mutations in these residues, primarily arginines and histidines located in the N-terminal region, result in reduced antiprotozoal activity (Singh and Batra, 2011). Interestingly, mouse eosinophil RNases (mEARs) also exhibit activity against *L. donovani*, though their efficacy varies. While mEAR 2 displays minimal antileishmanial activity, mEAR 11 shows activity comparable to human RNase3, and mEAR 5 and mEAR 7 are even more potent. However, no correlation was initially found between catalytic activity and antileishmanial efficacy, suggesting that parasite eradication by eosinophil RNases was primarily due to membrane permeabilization (Attery and Batra, 2017). More recent findings, however, have demonstrated that RNase3's catalytic activity is indeed essential for effective *Leishmania* promastigotes killing. The authors described that RNase3 is internalized into the protozoan cytoplasm, mainly via endocytosis, and degrades pathogen RNA, complementing its membrane-disrupting function. The H15A catalytic mutant of RNase3 exhibited significantly lower activity than the wild-type protein and failed to degrade protozoal RNA (Abengózar et al., 2022).

Even though ribonuclease activity contributes to *Leishmania* killing, the capacity of RNase3 to destabilize protozoa membranes is likely to be essential, as neither RNase1 nor RNase2, neither of which possesses membrane-destabilizing activities, is active against *Leishmania* (Singh and Batra, 2011). RNase3 activity against protozoa is not exclusive to *Leishmania*. Eosinophil secretory products containing RNase3 also demonstrated high in vitro activity against *Plasmodium falciparum*, and the effect was inhibited when adding RNase3 antibody to the mixture (Waters et al., 1987). Interestingly, the *RNASE3* rs2073342 polymorphism, which results in an arginine-to-threonine substitution at position 97, is more frequent in malaria-endemic populations, as it is suggested to confer protection against severe and cerebral malaria (Adu et al., 2011; Diop et al., 2018). Apparently, RNase3-R97 is more effective against several pathogens but also more cytotoxic than the RNase3-T97 variant, which may have a survival advantage for populations exposed to parasitic infections (Adu et al., 2011). *P. falciparum* detection by the immune system is partly mediated by TLR8 activation in monocytes through degradation of pathogenic RNA by RNase2 (Ostendorf et al., 2020). RNase3 also demonstrated to effectively kill *Trypanosoma cruzi* trypomastigotes and amastigotes (Molina et al., 1988).

8.4.5 Activity against helminthic parasites

RNase3 also plays a significant role in the immune response against helminthic parasites and has demonstrated potent antiparasitic activity across a range of species. In vitro studies have shown that RNase3 is highly toxic to *Schistosoma mansoni* schistosomula, proving even more potent than MBP (Ackerman et al., 1985). Eosinophil granule extracts, apparently enriched in RNases, were capable of causing severe damage to the schistosomula's envelope, including tegumental membrane detachment, ultimately leading to parasite death (McLaren et al., 1981). In mouse models, expression of mEAR 6 increases in the spleen and liver following *S. mansoni* infection, suggesting a role in host defense (Nitto et al., 2004). Moreover, the *RNASE3* rs2073342 (T97) polymorphism has been associated with protection against *S. mansoni*, correlating with reduced infection severity and lower levels of liver fibrosis (Eriksson et al., 2007).

RNase3, similarly to the other main eosinophil proteins MBP and EPO, was also found to be toxic against *Brugia pahangi* and *B. malayi*. Notably, its antiparasitic activity remains effective even in the presence of the RI RNasin (Hamann et al., 1990). In the case of *Onchocerca volvulus*, immunohistochemistry and immunoelectron microscopy studies have demonstrated that eosinophils contribute to the clearance of microfilariae following antiparasitic chemotherapy, with RNase3 being degranulated and accumulating on the surface of the parasite (Wildenburg et al., 1994; Gutiérrez-Peña et al.,

1998). Further supporting its broad antihelminthic activity, RNase3 also damages the protoscoleces of *Echinococcus granulosus*, inducing significant surface damage. Both RNase3 and RNase2 have been detected in the laminated layer of hydatid cysts and within the hydatid fluid, suggesting that eosinophil degranulation near the cyst surface enables RNase penetration into the cyst wall (Ramos et al., 2006). *Trichinella spiralis* is also sensitive to both RNase3 treatment and MBP, confirming the widespread parasiticidal potential of RNase3 across different helminths (Hamann et al., 1987; Specht et al., 2006).

Unlike RNase3, RNase2 lacks consistent direct parasiticidal activity against helminths (Hamann et al., 1987; Specht et al., 2006). However, it should not be discarded that RNase2 could indirectly contribute to an antiparasitic response through activation of immune cells. The *RNASE2* rs2013109 polymorphism causes a reduction in RNase2 production by eosinophils, which is associated with a higher risk of visceral leishmaniosis. The authors suggested that RNase2 may contribute to host defense response against *Leishmania* by promoting dendritic cell maturation through the TLR2-MyD88 pathway (Yang et al., 2008; Blom et al., 2018). Surprisingly, RNase2 was described to be capable of destroying *T. cruzi* trypomastigotes and amastigotes at comparable levels to RNase3. However, unlike RNase3, RNase2's activity was inhibited by RNH1 and yeast RNA (Molina et al., 1988). While there is no direct evidence of RNase5 involvement in helminthic infections, its murine homolog, Ang 4, is upregulated during gastrointestinal infection with *Trichuris muris* in mice. This observation suggests a potential, though still unconfirmed, role in antihelminthic defense (D'Elia et al., 2009).

8.4.6 RNases as diagnostic markers of parasitic infections

The medical interest in RNase3 extends beyond its ability to destroy parasites. RNase3 has emerged as a promising biomarker in the diagnosis and monitoring of helminthic infections. Various studies highlight its potential to improve detection accuracy and assess treatment efficacy across a range of parasitic diseases.

The current standard methods for diagnosing helminthic infections rely on microscopic detection of parasite eggs in feces using the Kato–Katz technique or through urine filtration. However, these methods tend to underestimate infection intensity (Gyan et al., 2015; Amoani et al., 2019). Moreover, the predictive value of common clinical markers such as total immunoglobulin E levels and eosinophil counts remains controversial (Hotz et al., 2022). In this context, RNase3 has attracted attention as a potential biomarker, as its levels correlate with eosinophil degranulation triggered during infection by various parasites. Measuring RNase3 levels has been explored as a means of assessing infection intensity. For instance, in schistosomiasis, RNase3 concentrations positively correlate with egg detection in urine (Leutscher et al., 2000; Gyan et al., 2015) and are indicators of infection clearance after praziquantel treatment (Stecher et al., 2017). Individuals infected with *Necator americanus*, as well as those coinfected with *P. falciparum*, exhibit significantly elevated RNase3 levels compared to uninfected individuals. These increased levels also correlate with higher eosinophil counts, and RNase3 has shown high predictive value for classifying hookworm infections. In contrast, individuals infected solely with *P. falciparum* do not show significantly elevated RNase3 levels, suggesting that RNase3 is more specific to helminthic infections (Amoani et al., 2019). Patients with symptomatic cystic echinococcosis also present significantly higher levels of RNase3 than nonsymptomatic patients (Cicioğlu Aridoğan et al., 2009). In addition, RNase3 levels are reduced in both cystic echinococcosis and alveolar echinococcosis after surgical and benzimidazole treatment (Hotz et al., 2022). Toxocariasis also increases significantly RNase3 levels and has been associated with cough and rhinitis symptoms (Magnaval et al., 2006; Kaneva et al., 2020).

8.5 Conclusions

Despite the wide diversity of pathogens that threaten human health, RNases exhibit potent activity against most of them. Their broad antimicrobial spectrum arises in part from their ability to disrupt pathogen envelopes and to degrade RNA. Whereas RNase activity against bacteria is mainly associated with membrane targeting, antiviral activity depends largely on RNA cleavage, and antifungal and antiparasitic activities appear to require a combination of both properties. In addition to their direct antimicrobial effects, RNases modulate immune responses, a function that largely relies on their ability to interact with or hydrolyze nucleic acids. In the context of emerging resistance to conventional antipathogenic therapies, clinical interest in AMPs is increasing, and RNases may be particularly valuable. Several strategies have been explored to translate current knowledge of RNases into clinical applications, including the design of N-terminal-derived peptides, RNase chimeras, and fusion proteins with viral capsid subunits. Moreover, RNases represent promising novel biomarkers for the early detection of infections and for predicting disease outcomes. A deeper understanding of the mechanisms by which RNases protect the host from infection may contribute to the development of new therapeutic approaches against a broad range of pathogens.

References

Abengózar, M.Á., Fernández-Reyes, M., Salazar, V.A., Torrent, M., De La Torre, B.G., Andreu, D., et al., 2022. Essential role of enzymatic activity in the leishmanicidal mechanism of the eosinophil cationic protein (RNase 3). ACS Infectious Diseases 8 (7), 1207–1217. Available at https://doi.org/10.1021/acsinfecdis.1c00537, http://pubs.acs.org/journal/aidcbc.

Abtin, A., Eckhart, L., Mildner, M., Ghannadan, M., Harder, J., Schröder, J.M., et al., 2009. Degradation by stratum corneum proteases prevents endogenous RNase inhibitor from blocking antimicrobial activities of RNase 5 and RNase 7. Journal of Investigative Dermatology 129 (9), 2193–2201. Available at https://doi.org/10.1038/jid.2009.35, https://www.sciencedirect.com/science/journal/0022202X.

Ackerman, S.J., Gleich, G.J., Loegering, D.A., Richardson, B.A., Butterworth, A.E., 1985. Comparative toxicity of purified human eosinophil granule cationic proteins for schistosomula of *Schistosoma mansoni*. American Journal of Tropical Medicine and Hygiene 34 (4), 735–745. Available at https://doi.org/10.4269/ajtmh.1985.34.735.

Adu, B., Dodoo, D., Adukpo, S., Gyan, B.A., Hedley, P.L., Goka, B., et al., 2011. Polymorphisms in the RNASE3 gene are associated with susceptibility to cerebral malaria in Ghanaian children. PLoS One 6 (12). Available at https://doi.org/10.1371/journal.pone.0029465Denmark, http://www.plosone.org/article/fetchObjectAttachment.action?uri=info%3Adoi%2F10.1371%2Fjournal.pone.0029465&representation=PDF.

Amatngalim, G.D., Van Wijck, Y., De Mooij-Eijk, Y., Verhoosel, R.M., Harder, J., Lekkerkerker, A.N., et al., 2015. Basal cells contribute to innate immunity of the airway epithelium through production of the antimicrobial protein RNase 7. Journal of Immunology 194 (7), 3340–3350. Available at https://doi.org/10.4049/jimmunol.1402169, http://www.jimmunol.org/content/194/7/3340.full.pdf+html.

Amoani, B., Adu, B., Frempong, M.T., Sarkodie-Addo, T., Nuvor, S.V., Wilson, M.D., et al., 2019. Levels of serum eosinophil cationic protein are associated with hookworm infection and intensity in endemic communities in Ghana. PLoS One 14 (9). Available at https://doi.org/10.1371/journal.pone.0222382, https://journals.plos.org/plosone/article/file?id=10.1371/journal.pone.0222382&type=printable.

Attery, A., Batra, J.K., 2017. Mouse eosinophil associated ribonucleases: mechanism of cytotoxic, antibacterial and antiparasitic activities. International Journal of Biological Macromolecules 94, 445–450. Available at https://doi.org/10.1016/j.ijbiomac.2016.10.041, www.elsevier.com/locate/ijbiomac.

Becknell, B., Eichler, T.E., Beceiro, S., Li, B., Easterling, R.S., Carpenter, A.R., et al., 2015. Ribonucleases 6 and 7 have antimicrobial function in the human and murine urinary tract. Kidney International 87 (1), 151–161. Available at https://doi.org/10.1038/ki.2014.268, https://www.journals.elsevier.com/kidney-international.

Becknell, B., Ching, C., Spencer, J.D., 2019. The responses of the ribonuclease a superfamily to urinary tract infection. Frontiers in Immunology 10. Available at https://doi.org/10.3389/fimmu.2019.02786, https://www.frontiersin.org/journals/immunology#.

Bedoya, V.I., Boasso, A., Hardy, A.W., Rybak, S., Shearer, G.M., Rugeles, M.T., 2006. Ribonucleases in HIV type 1 inhibition: effect of recombinant RNases on infection of primary T cells and immune activation-induced RNase gene and protein expression. AIDS Research and Human Retroviruses 22 (9), 897–907. Available at https://doi.org/10.1089/aid.2006.22.897.

Behera, A.K., Kumar, M., Lockey, R.F., Mohapatra, S.S., 2002. 2′-5′ oligoadenylate synthetase plays a critical role in interferon-γ inhibition of respiratory syncytial virus infection of human epithelial cells. Journal of Biological Chemistry 277 (28), 25601–25608. Available at https://doi.org/10.1074/jbc.M200211200.

Bick, M.J., Carroll, J.W.N., Gao, G., Goff, S.P., Rice, C.M., MacDonald, M.R., 2003. Expression of the zinc-finger antiviral protein inhibits alphavirus replication. Journal of Virology 77 (21), 11555–11562. Available at https://doi.org/10.1128/JVI.77.21.11555-11562.2003.

Blom, K., Elshafie, A.I., Jönsson, U.B., Rönnelid, J., Håkansson, L.D., Venge, P., 2018. The genetically determined production of the alarmin eosinophil-derived neurotoxin is reduced in visceral leishmaniasis. APMIS: Acta Pathologica, Microbiologica, et Immunologica Scandinavica 126 (1), 85–91. Available at https://doi.org/10.1111/apm.12780, http://onlinelibrary.wiley.com/journal/10.1111/(ISSN)1600-0463.

Boix, E., Nogués, M.V., 2007. Mammalian antimicrobial proteins and peptides: overview on the RNase A superfamily members involved in innate host defence. Molecular Biosystems 3 (5), 317–335. Available at https://doi.org/10.1039/b617527a.

Boix, E., Salazar, V.A., Torrent, M., Pulido, D., Nogués, M.V., Moussaoui, M., 2012. Structural determinants of the eosinophil cationic protein antimicrobial activity. Biological Chemistry 393 (8), 801–815. Available at https://doi.org/10.1515/hsz-2012-0160.

Brasch, J., 2009. Current knowledge of host response in human tinea. Mycoses 52 (4), 304–312. Available at https://doi.org/10.1111/j.1439-0507.2008.01667.x.

Brasch, J., Mörig, A., Neumann, B., Proksch, E., 2014. Expression of antimicrobial peptides and toll-like receptors is increased in tinea and pityriasis versicolor. Mycoses 57 (3), 147–152. Available at https://doi.org/10.1111/myc.12118.

Bystrom, J., Amin, K., Bishop-Bailey, D., 2011. Analysing the eosinophil cationic protein - a clue to the function of the eosinophil granulocyte. Respiratory Research 12 (1). Available at https://doi.org/10.1186/1465-9921-12-10.

Cabrera-Fuentes, H.A., Ruiz-Meana, M., Simsekyilmaz, S., Kostin, S., Inserte, J., Saffarzadeh, M., et al., 2014. RNase1 prevents the damaging interplay between extracellular RNA and tumour necrosis factor-α in cardiac ischaemia/reperfusion injury. Thrombosis and Haemostasis 112 (6), 1110–1119. Available at https://doi.org/10.1160/TH14-08-0703, http://dx.doi.Org/10.1160/TH14-08-0703.

Carreras, E., Boix, E., Rosenberg, H.F., Cuchillo, C.M., Nogués, M.V., 2003. Both aromatic and cationic residues contribute to the membrane-lytic and bactericidal activity of eosinophil cationic protein. Biochemistry 42 (22), 6636–6644. Available at https://doi.org/10.1021/bi0273011.

Cheng, G.Z., Li, J.Y., Li, F., Wang, H.Y., Shi, G.X., 2009. Human ribonuclease 9, a member of ribonuclease A superfamily, specifically expressed in epididymis, is a novel sperm-binding protein. Asian Journal of Andrology 11 (2), 240–251. Available at https://doi.org/10.1038/aja.2008.30.

Cho, S., Zhang, J., 2007. Zebrafish ribonucleases are bactericidal: implications for the origin of the vertebrate RNase a superfamily. Molecular Biology and Evolution 24 (5), 1259–1268. Available at https://doi.org/10.1093/molbev/msm047.

Cho, S., Beintema, J.J., Zhang, J., 2005. The ribonuclease a superfamily of mammals and birds: identifying new members and tracing evolutionary histories. Genomics 85 (2), 208–220. Available at https://doi.org/10.1016/j.ygeno.2004.10.008.

Christensen-Quick, A., Lafferty, M., Sun, L., Marchionni, L., DeVico, A., Garzino-Demo, A., et al., 2016. Human T h 17 cells lack HIV-inhibitory RNases and are highly permissive to productive HIV infection. Journal of Virology 90 (17), 7833–7847. Available at https://doi.org/10.1128/jvi.02869-15.

Cicioğlu Aridoğan, Kaya, Cetin, S., Taş, Demirci, 2009. Evaluation of eosinophil cationic protein levels with clinical symptoms and laboratory findings of patients with cystic*Echinococcosis*. Mikrobiyoloji Bulteni 43 (2), 285–292.

Cocchi, F., DeVico, A.L., Lu, W., Popovic, M., Latinovic, O., Sajadi, M.M., et al., 2012. Soluble factors from T cells inhibiting X4 strains of HIV are a mixture of β chemokines and RNases. Proceedings of the National Academy of Sciences of the United States of America 109 (14), 5411–5416. Available at https://doi.org/10.1073/pnas.1202240109, http://www.pnas.org/content/109/14/5411.full.pdf+html.

Coch, C., Hommertgen, B., Zillinger, T., Daßler-Plenker, J., Putschli, B., Nastaly, M., et al., 2019. Human TLR8 senses RNA from plasmodium falciparum-infected red blood cells which is uniquely required for the IFN-γ response in NK cells. Frontiers in Immunology 10. Available at https://doi.org/10.3389/fimmu.2019.00371, https://www.frontiersin.org/journals/immunology#.

Cooper, D.A., Banerjee, S., Chakrabarti, A., García-Sastre, A., Hesselberth, J.R., Silverman, R.H., et al., 2015. RNase L targets distinct sites in influenza a virus RNAs. Journal of Virology 89 (5), 2764–2776. Available at https://doi.org/10.1128/JVI.02953-14, http://jvi.asm.org/content/89/5/2764.full.pdf.

Das, A.T., Berkhout, B., 1995. Efficient extension of a misaligned tRNA-primer during replication of the HIV-1 retrovirus. Nucleic Acids Research 23 (8), 1319–1326. Available at https://doi.org/10.1093/nar/23.8.1319.

Das, A.T., Klaver, B., Berkhout, B., 1995. Reduced replication of human immunodeficiency virus type 1 mutants that use reverse transcription primers other than the natural tRNA(3Lys). Journal of Virology 69 (5), 3090–3097. Available at https://doi.org/10.1128/jvi.69.5.3090-3097.1995.

D'Elia, R., de Schoolmeester, M.L., Zeef, L.A.H., Wright, S.H., Pemberton, A.D., Else, K.J., 2009. Expulsion of *Trichuris muris* is associated with increased expression of angiogenin 4 in the gut and increased acidity of mucins within the goblet cell. BMC Genomics 10, 492. Available at https://doi.org/10.1186/1471-2164-10-492.

Deymier, S., Louvat, C., Fiorini, F., Cimarelli, A., 2022. ISG20: an enigmatic antiviral RNase targeting multiple viruses. FEBS Open Biology 12 (6), 1096–1111. Available at https://doi.org/10.1002/2211-5463.13382.

Diop, G., Derbois, C., Loucoubar, C., Mbengue, B., Ndao, B.N., Thiam, F., et al., 2018. Genetic variants of RNASE3 (ECP) and susceptibility to severe malaria in Senegalese population. Malaria Journal 17 (1). Available at https://doi.org/10.1186/s12936-018-2205-9, http://www.malariajournal.com/home/.

Domachowske, J.B., Dyer, K.D., Adams, A.G., Leto, T.L., Rosenberg, H.F., 1998a. Eosinophil cationic protein/RNase 3 is another RNase A-family ribonuclease with direct antiviral activity. Nucleic Acids Research 26 (14), 3358–3363. Available at https://doi.org/10.1093/nar/26.14.3358.

Domachowske, J.B., Dyer, K.D., Bonville, C.A., Rosenberg, H.F., 1998b. Recombinant human eosinophii-derived neurotoxin/RNase 2 functions as an effective antiviral agent against respiratory syncytial virus. Journal of Infectious Diseases 177 (6), 1458–1464. Available at https://doi.org/10.1086/515322, http://jid.oxfordjournals.org/content/current.

Dosanjh, A., 2020. Eosinophil-derived neurotoxin and respiratory tract infection and inflammation: implications for COVID-19 management. Journal of Interferon & Cytokine Research 40 (9), 443–445. Available at https://doi.org/10.1089/jir.2020.0066.

Eichler, T., Bender, K., Murtha, M.J., Schwartz, L., Metheny, J., Solden, L., et al., 2019. Ribonuclease 7 shields the kidney and bladder from invasive uropathogenic *escherichia coli* infection. Journal of the American Society of Nephrology 30 (8), 1385–1397. Available at https://doi.org/10.1681/ASN.2018090929, https://jasn.asnjournals.org/content/30/8/1385.full.pdf.

Eichler, T.E., Becknell, B., Easterling, R.S., Ingraham, S.E., Cohen, D.M., Schwaderer, A.L., et al., 2016. Insulin and the phosphatidylinositol 3-kinase signaling pathway regulate Ribonuclease 7 expression in the human urinary tract. Kidney International 90 (3), 568–579. Available at https://doi.org/10.1016/j.kint.2016.04.025, https://www.journals.elsevier.com/kidney-international.

Eilers, R.E., Gandhi, M., Patel, J.D., Mulcahy, M.F., Agulnik, M., Hensing, T., et al., 2010. Dermatologic infections in cancer patients treated with epidermal growth factor receptor inhibitor therapy. JNCI Journal of the National Cancer Institute 102 (1), 47–53. Available at https://doi.org/10.1093/jnci/djp439.

Eller, C.H., Raines, R.T., 2020. Antimicrobial synergy of a ribonuclease and a peptide secreted by human cells. ACS Infectious Diseases 6 (11), 3083–3088. Available at https://doi.org/10.1021/acsinfecdis.0c00594, http://pubs.acs.org/journal/aidcbc.

Eriksson, J., Reimert, C.M., Kabatereine, N.B., Kazibwe, F., Ireri, E., Kadzo, H., et al., 2007. The 434(G>C) polymorphism within the coding sequence of eosinophil cationic protein (ECP) correlates with the natural course of Schistosoma mansoni infection. International Journal for Parasitology 37 (12), 1359–1366. Available at https://doi.org/10.1016/j.ijpara.2007.04.001.

Firat, Y.H., Simanski, M., Rademacher, F., Schröder, L., Brasch, J., Harder, J., 2014. Infection of keratinocytes with trichophytum rubrum induces epidermal growth factor-dependent RNase 7 and human beta-defensin-3 expression. PLoS One 9 (4). Available at https://doi.org/10.1371/journal.pone.0093941, http://www.plosone.org/article/fetchObject.action?uri=info%3Adoi%2F10.1371%2Fjournal.pone.0093941&representation=PDF.

Freund, I., Buhl, D.K., Boutin, S., Kotter, A., Pichot, F., Marchand, V., et al., 2019. 2′-O-methylation within prokaryotic and eukaryotic tRNA inhibits innate immune activation by endosomal Toll-like receptors but does not affect recognition of whole organisms. RNA (New York, N.Y.) 27 (5), 869–880. Available at https://doi.org/10.1261/rna.070243.118, https://rnajournal.cshlp.org/content/25/7/869.full.pdf+html.

Fritz, P., Beck-Jendroschek, V., Brasch, J., 2012. Inhibition of dermatophytes by the antimicrobial peptides human β-defensin-2, ribonuclease 7 and psoriasin. Medical Mycology 50 (6), 579–584. Available at https://doi.org/10.3109/13693786.2012.660203.

Garofalo, R., Kimpen, J.L.L., Welliver, R.C., Ogra, P.L., 1992. Eosinophil degranulation in the respiratory tract during naturally acquired respiratory syncytial virus infection. The Journal of Pediatrics 120 (1), 28–32. Available at https://doi.org/10.1016/S0022-3476(05)80592-X.

Gläser, R., Navid, F., Schuller, W., Jantschitsch, C., Harder, J., Schröder, J.M., et al., 2009. UV-B radiation induces the expression of antimicrobial peptides in human keratinocytes in vitro and in vivo. Journal of Allergy and Clinical Immunology 123 (5), 1117–1123. Available at https://doi.org/10.1016/j.jaci.2009.01.043.

Goo, S.M., Cho, S., 2013. The expansion and functional diversification of the mammalian ribonuclease a superfamily epitomizes the efficiency of multigene families at generating biological novelty. Genome Biology and Evolution 5 (11), 2124–2140. Available at https://doi.org/10.1093/gbe/evt161, http://gbe.oxfordjournals.org/.

Greulich, W., Wagner, M., Gaidt, M.M., Stafford, C., Cheng, Y., Linder, A., et al., 2019. TLR8 Is a sensor of RNase T2 degradation products. Cell 179 (6), 1264. Available at https://doi.org/10.1016/j.cell.2019.11.001.

Gutiérrez-Peña, E.J., Knab, J., Büttner, D.W., 1998. Immunoelectron microscopic evidence for release of eosinophil granule matrix protein onto microfilariae of *Onchocerca volvulus* in the skin after exposure to amocarzine. Parasitology Research 84 (8), 607–615. Available at https://doi.org/10.1007/s004360050459.

Gyan, B., Boakye, D., Ayi, I., van der Puije, W., Asuming-Brempong, E., Bimi, L., et al., 2015. Relationship between eosinophil cationic protein and infection intensity in a schistosomiasis endemic community in Ghana. Research and Reports in Tropical Medicine 1. Available at https://doi.org/10.2147/RRTM.S51713.

Haisma, E.M., Rietveld, M.H., Breij, A., Van Dissel, J.T., Ghalbzouri, A.E., Nibbering, P.H., 2013. Inflammatory and antimicrobial responses to methicillin-resistant *Staphylococcus aureus* in an in vitro wound infection model. PLoS One 8 (12). Available at https://doi.org/10.1371/journal.pone.0082800Netherlands, http://www.plosone.org/article/fetchObject.action?uri=info%3Adoi%2F10.1371%2Fjournal.pone.0082800&representation=PDF.

Hamann, K.J., Barker, R.L., Loegering, D.A., Gleich, G.J., 1987. Comparative toxicity of purified human eosinophil granule proteins for newborn larvae of *Trichinella spiralis*. The Journal of Parasitology 73 (3), 523–529. Available at https://doi.org/10.2307/3282130.

Hamann, K.J., Gleich, G.J., Checkel, J.L., Loegering, D.A., Mccall, J.W., Barker, R.L., 1990. In vitro killing of microfilariae of *Brugia pahangi* and*Brugia malayi* by eosinophil granule proteins. Journal of Immunology 144 (8), 3166–3173. Available at http://www.jimmunol.org/.

Harder, J., Schröder, J.M., 2002. RNase 7, a novel innate immune defense antimicrobial protein of healthy human skin. Journal of Biological Chemistry 277 (48), 46779–46784. Available at https://doi.org/10.1074/jbc.M207587200.

Harris, P., Johannessen, K.M., Smolenski, G., Callaghan, M., Broadhurst, M.K., Kim, K., et al., 2010. Characterisation of the anti-microbial activity of bovine milk ribonuclease4 and ribonuclease5 (angiogenin). International Dairy Journal 20 (6), 400–407. Available at https://doi.org/10.1016/j.idairyj.2009.12.018.

Harrison, A.M., Bonville, C.A., Rosenberg, H.F., Domachowske, J.B., 1999. Respiratory syncytical virus-induced chemokine expression in the lower airways: eosinophil recruitment and degranulation. American Journal of Respiratory and Critical Care Medicine 159 (6), 1918–1924. Available at https://doi.org/10.1164/ajrccm.159.6.9805083, http://ajrccm.atsjournals.org/.

Hooper, L.V., Stappenbeck, T.S., Hong, C.V., Gordon, J.I., 2003. Angiogenins: a new class of microbicidal proteins involved in innate immunity. Nature Immunology 4 (3), 269–273. Available at https://doi.org/10.1038/ni888.

Hotz, J.F., Kaczirek, K., Stremitzer, S., Waneck, F., Auer, H., Perkmann, T., et al., 2022. Evaluation of eosinophilic cationic protein as a marker of alveolar and cystic *Echinococcosis*. Pathogens 11 (2). Available at https://doi.org/10.3390/pathogens11020261, https://www.mdpi.com/2076-0817/11/2/261/pdf.

Huang, Y.-C., Lin, Y.-M., Chang, T.-W., Wu, S.-J., Lee, Y.-S., Chang, M.D.-T., et al., 2007. The flexible and clustered lysine residues of human ribonuclease 7 are critical for membrane permeability and antimicrobial activity. Journal of Biological Chemistry 282 (7), 4626–4633. Available at https://doi.org/10.1074/jbc.m607321200.

Ireland, D.D.C., Stohlman, S.A., Hinton, D.R., Kapil, P., Silverman, R.H., Atkinson, R.A., et al., 2009. RNase L mediated protection from virus induced demyelination. PLoS Pathogens 5 (10), e1000602. Available at https://doi.org/10.1371/journal.ppat.1000602.

Kaneva, E.M., Harizanov, R.N., Rainova, I.G., Kaftandjiev, I.T., Tsvetkova, N.D., 2020. Study of eosinophil cationic protein serum levels in patients with toxocariasis. Iberoamerican Journal of Medicine 2 (3), 178–182. Available at https://doi.org/10.53986/ibjm.2020.0032.

Kim, C.K., Seo, J.K., Ban, S.H., Fujisawa, T., Kim, D.W., Callaway, Z., 2013. Eosinophil-derived neurotoxin levels at 3 months post-respiratory syncytial virus bronchiolitis are a predictive biomarker of recurrent wheezing. Biomarkers: Biochemical Indicators of Exposure, Response, and Susceptibility to Chemicals 18 (3), 230–235. Available at https://doi.org/10.3109/1354750X.2013.773078.

Koller, D.Y., Wojnarowski, C., Herkner, K.R., Weinlander, G., Raderer, M., Eichler, I., et al., 1997. High levels of eosinophil cationic protein in wheezing infants predict the development of asthma. Journal of Allergy and Clinical Immunology 99 (6), 752–756. Available at https://doi.org/10.1016/S0091-6749(97)80007-3, http://www.elsevier.com/inca/publications/store/6/2/3/3/6/8/index.htt.

Kopfnagel, V., Wagenknecht, S., Harder, J., Hofmann, K., Kleine, M., Buch, A., et al., 2018. RNase 7 strongly promotes TLR9-mediated dna sensing by human plasmacytoid dendritic cells. Journal of Investigative Dermatology 138 (4), 872–881. Available at https://doi.org/10.1016/j.jid.2017.09.052.

Kopfnagel, V., Dreyer, S., Baumert, K., Stark, M., Harder, J., Hofmann, K., et al., 2020. RNase 7 promotes sensing of self-DNA by human keratinocytes and activates an antiviral immune response. Journal of Investigative Dermatology 140 (8), 1589. Available at https://doi.org/10.1016/j.jid.2019.09.029.

Kosgey, J.C., Jia, L., Nyamao, R.M., Zhao, Y., Xue, T., Yang, J., et al., 2020. RNase 1, 2, 5 & 8 role in innate immunity: strain specific antimicrobial activity. International Journal of Biological Macromolecules 160, 1042–1049. Available at https://doi.org/10.1016/j.ijbiomac.2020.06.001, www.elsevier.com/locate/ijbiomac.

Krutskikh, A., Poliandri, A., Cabrera-Sharp, V., Dacheux, J.L., Poutanen, M., Huhtaniemi, I., 2012. Epididymal protein Rnase10 is required for posttesticular sperm maturation and male fertility. FASEB Journal 26 (10), 4198–4209. Available at https://doi.org/10.1096/fj.12-205211, http://www.fasebj.org/content/26/10/4198.full.pdf+html.

Lee, Y.S., Chen, C.H., Chao, A., Chen, E.S., Wei, M.L., Chen, L.K., et al., 2005. Molecular signature of clinical severity in recovering patients with severe acute respiratory syndrome coronavirus (SARS-CoV). BMC Genomics 6. Available at https://doi.org/10.1186/1471-2164-6-132Taiwan, http://www.biomedcentral.com/1471-2164/6/132.

Lee-Huang, S., Huang, P.L., Sun, Y., Huang, P.L., Kung, H.F., Blithe, D.L., et al., 1999. Lysozyme and RNases as anti-HIV components in β-core preparations of human chorionic gonadotropin. Proceedings of the National Academy of Sciences of the United States of America 96 (6), 2678–2681. Available at https://doi.org/10.1073/pnas.96.6.2678, www.pnas.org.

Lehrer, R.I., Szklarek, D., Barton, A., Ganz, T., Hamann, K.J., Gleich, G.J., 1989. Antibacterial properties of eosinophil major basic protein and eosinophil cationic protein. Journal of Immunology 142 (12), 4428–4434.

Leutscher, P.D.C., Reimert, C.M., Vennervald, B.J., Ravaoalimalala, V.E., Ramarokoto, C.E., Serieye, J., et al., 2000. Morbidity assessment in urinary schistosomiasis infection through ultrasonography and measurement of eosinophil cationic protein (ECP) in urine. Tropical Medicine and International Health 5 (2), 88–93. Available at https://doi.org/10.1046/j.1365-3156.2000.00522.x.

Li, J., Boix, E., 2021. Host defence RNases as antiviral agents against enveloped single stranded RNA viruses. Virulence 12 (1), 444–469. Available at https://doi.org/10.1080/21505594.2021.1871823.

Li, Y., Zhao, Y., Liu, J., Huang, Y., Liu, Z., Xue, C., 2010. A promising alternative anti-HBV agent: the targeted ribonuclease. International Journal of Molecular Medicine 26 (1), 51–56. Available at https://doi.org/10.3892/ijmm_00000434, http://www.spandidos-publications.com/serveFile/ijmm_26_1_51_PDF.pdf?type=article&article_id=ijmm_26_1_51&item=PDF.

Li, Z., Ender, C., Meister, G., Moore, P.S., Chang, Y., John, B., 2012. Extensive terminal and asymmetric processing of small RNAs from rRNAs, snoRNAs, snRNAs, and tRNAs. Nucleic Acids Research 40 (14), 6787–6799. Available at https://doi.org/10.1093/nar/gks307.

Lichtenberger, B.M., Gerber, P.A., Holcmann, M., Buhren, B.A., Amberg, N., Smolle, V., et al., 2013. Epidermal EGFR controls cutaneous host defense and prevents inflammation. Science Translational Medicine 5 (199). Available at https://doi.org/10.1126/scitranslmed.3005886Austria, http://stm.sciencemag.org/content/5/199/199ra111.full.pdf.

Lin, Y.M., Wu, S.J., Chang, T.W., Wang, C.F., Suen, C.S., Hwang, M.J., et al., 2010. Outer membrane protein I of pseudomonas aeruginosa is a target of cationic antimicrobial peptide/protein. Journal of Biological Chemistry 285 (12), 8985–8994. Available at https://doi.org/10.1074/jbc.M109.078725Taiwan, http://www.jbc.org/content/285/12/8985.full.pdf+html.

Lin, R.J., Chien, H.L., Lin, S.Y., Chang, B.L., Yu, H.P., Tang, W.C., et al., 2013. MCPIP1 ribonuclease exhibits broad-spectrum antiviral effects through viral RNA binding and degradation. Nucleic Acids Research 41 (5), 3314–3326. Available at https://doi.org/10.1093/nar/gkt019, https://academic.oup.com/nar/issue.

Lindsley, A.W., Schwartz, J.T., Rothenberg, M.E., 2020. Eosinophil responses during COVID-19 infections and coronavirus vaccination. Journal of Allergy and Clinical Immunology 146 (1), 1–7. Available at https://doi.org/10.1016/j.jaci.2020.04.021, http://www.elsevier.com/inca/publications/store/6/2/3/3/6/8/index.htt.

Lippi, G., Henry, B.M., 2020. Eosinophil count in severe coronavirus disease 2019. QJM: An International Journal of Medicine 113 (7), 511–512. Available at https://doi.org/10.1093/qjmed/hcaa137.

Liu, J., Li, J.Y., Wang, H.Y., Zhang, C.L., Li, N., Lin, Y.Q., et al., 2008. Cloning, expression and location of RNase9 in human epididymis. BMC Research Notes 1. Available at https://doi.org/10.1186/1756-0500-1-111.

Liu, F., Xu, A., Zhang, Y., Xuan, W., Yan, T., Pan, K., et al., 2020. Patients of COVID-19 may benefit from sustained lopinavir-combined regimen and the increase of eosinophil may predict the outcome of COVID-19 progression. International Journal of Infectious Diseases 95, 183–191. Available at https://doi.org/10.1016/j.ijid.2020.03.013, https://www.journals.elsevier.com/international-journal-of-infectious-diseases.

Lomax, J.E., Eller, C.H., Raines, R.T., 2017. Comparative functional analysis of ribonuclease 1 homologs: molecular insights into evolving vertebrate physiology. Biochemical Journal 474 (13), 2219–2233. Available at https://doi.org/10.1042/BCJ20170173, http://www.biochemj.org/content/ppbiochemj/474/13/2219.full.pdf.

Lu, L., Li, J., Moussaoui, M., Boix, E., 2018. Immune modulation by human secreted RNases at the extracellular space. Frontiers in Immunology 9 (MAY). Available at https://doi.org/10.3389/fimmu.2018.01012, https://www.frontiersin.org/articles/10.3389/fimmu.2018.01012/full.

Lu, L., Arranz-Trullén, J., Prats-Ejarque, G., Pulido, D., Bhakta, S., Boix, E., 2019. Human antimicrobial RNases inhibit intracellular bacterial growth and induce autophagy in mycobacteria-infected macrophages. Frontiers in Immunology 10. Available at https://doi.org/10.3389/fimmu.2019.01500, https://www.frontiersin.org/journals/immunology#.

Lu, L., Wei, R.L., Prats-Ejarque, G., Goetz, M., Wang, G., Torrent, M., et al., 2021. Human RNase3 immune modulation by catalytic-dependent and independent modes in a macrophage-cell line infection model. Cellular and Molecular Life Sciences 78 (6), 2963–2985. Available at https://doi.org/10.1007/s00018-020-03695-5, http://link.springer.de/link/service/journals/00018/index.htm.

Lu, L., Li, J., Wei, R., Guidi, I., Cozzuto, L., Ponomarenko, J., et al., 2022. Selective cleavage of ncRNA and antiviral activity by RNase2/EDN in THP1-induced macrophages. Cellular and Molecular Life Sciences 79 (4). Available at https://doi.org/10.1007/s00018-022-04229-x, http://link.springer.de/link/service/journals/00018/index.htm.

Madhry, D., Malvankar, S., Phadnis, S., Srivastava, R.K., Bhattacharyya, S., Verma, B., 2023. Synergistic correlation between host angiogenin and dengue virus replication. RNA Biology 20 (1), 805–816. Available at https://doi.org/10.1080/15476286.2023.2264003, http://www.tandfonline.com/toc/krnb20/current.

Magnaval, J.-F., Faufingue, J.-H., Morassin, B., Fabre, R., 2006. Eosinophil cationic protein, specific IgE and IgG4 in human toxocariasis. Journal of Helminthology 80 (4), 417–423. Available at https://doi.org/10.1017/joh2006369.

Mahmoud, A.A., Warren, K.S., Peters, P.A., 1975. A role for the eosinophil in acquired resistance to *Schistosoma mansoni* infection as determined by antieosinophil serum. The Journal of Experimental Medicine 142 (4), 805–813. Available at https://doi.org/10.1084/jem.142.4.805.

Malathi, K., Saito, T., Crochet, N., Barton, D.J., Gale, M., Silverman, R.H., 2010. RNase L releases a small RNA from HCV RNA that refolds into a potent PAMP. RNA (New York, N.Y.) 16 (11), 2108–2119. Available at https://doi.org/10.1261/rna.2244210, http://rnajournal.cshlp.org/content/16/11/2108.full.pdf+html.

Malathi, K., Dong, B., Gale, M., Silverman, R.H., 2007. Small self-RNA generated by RNase L amplifies antiviral innate immunity. Nature 448 (7155), 816–819. Available at https://doi.org/10.1038/nature06042, http://www.nature.com/nature/index.html.

Manivannan, P., Siddiqui, M.A., Malathi, K., 2020. RNase l amplifies interferon signaling by inducing protein kinase r-mediated antiviral stress granules. Journal of Virology 94 (13). Available at https://doi.org/10.1128/JVI.00205-20, https://journals.asm.org/doi/10.1128/JVI.00205-20.

McLaren, D.J., McKean, J.R., Olsson, I., Venge, P., Kay, A.B., 1981. Morphological studies on the killing of schistosomula of*Schistosoma mansoni* by human eosinophil and neutrophil cationic proteins in vitro. Parasite Immunology 3 (4), 359–373. Available at https://doi.org/10.1111/j.1365-3024.1981.tb00414.x.

Mohammed, I., Yeung, A., Abedin, A., Hopkinson, A., Dua, H.S., 2011. Signalling pathways involved in ribonuclease-7 expression. Cellular and Molecular Life Sciences 68 (11), 1941–1952. Available at https://doi.org/10.1007/s00018-010-0540-2.

Molina, H.A., Kierszenbaum, F., Hamann, K.J., Gleich, G.J., 1988. Toxic effects produced or mediated by human eosinophil granule components on*Trypanosoma cruzi*. American Journal of Tropical Medicine and Hygiene 38 (2), 327–334. Available at https://doi.org/10.4269/ajtmh.1988.38.327.

Moreau, J.M., Dyer, K.D., Bonville, C.A., Nitto, T., Vasquez, N.L., Easton, A.J., et al., 2003. Diminished expression of an antiviral ribonuclease in response to pneumovirus infection in vivo. Antiviral Research 59 (3), 181–191. Available at https://doi.org/10.1016/S0166-3542(03)00111-6, www.elsevier.com/locate/antiviral.

Munitz, A., Levi-Schaffer, F., 2004. Eosinophils: 'new' roles for 'old' cells. Allergy 59 (3), 268–275. Available at https://doi.org/10.1111/j.1398-9995.2003.00442.x.

Murtha, M.J., Eichler, T., Bender, K., Metheny, J., Li, B., Schwaderer, A.L., et al., 2018. Insulin receptor signaling regulates renal collecting duct and intercalated cell antibacterial defenses. Journal of Clinical Investigation 128 (12), 5634–5646. Available at https://doi.org/10.1172/JCI98595, https://www.jci.org/articles/view/122383/pdf.

Nadal, A., Martell, M., Lytle, J.R., Lyons, A.J., Robertson, H.D., Cabot, B., et al., 2002. Specific cleavage of hepatitis C virus RNA genome by human RNase P. Journal of Biological Chemistry 277 (34), 30606–30613. Available at https://doi.org/10.1074/jbc.M203595200.

Nitto, T., Dyer, K.D., Mejia, R.A., Byström, J., Wynn, T.A., Rosenberg, H.F., 2004. Characterization of the divergent eosinophil ribonuclease, mEar 6, and its expression in response to*Schistosoma mansoni* infection in vivo. Genes & Immunity 5 (8), 668–674. Available at https://doi.org/10.1038/sj.gene.6364143.

Nunes, I.V., Breitenbach, L., Pawusch, S., Eigenbrod, T., Ananth, S., Schad, P., et al., 2024. Bacterial RNA sensing by TLR8 requires RNase 6 processing and is inhibited by RNA 2'O-methylation. EMBO Reports 25 (11), 4674–4692. Available at https://doi.org/10.1038/s44319-024-00281-9, https://www.embopress.org/page/journal/14693178/about.

Onorato, J., Esposito, S., Scovena, E., Morandi, B., Morelli, M., Pizzi, M., et al., 1999. Eosinophil involvement and serum IgE level in HIV-1–infected children. Journal of Allergy and Clinical Immunology 104 (1), 245–247. Available at https://doi.org/10.1016/s0091-6749(99)70146-6.

O'Reilly, M.A., Yee, M., Buczynski, B.W., Vitiello, P.F., Keng, P.C., Welle, S.L., et al., 2012. Neonatal oxygen increases sensitivity to influenza a virus infection in adult mice by suppressing epithelial expression of Ear1. American Journal of Pathology 181 (2), 441–451. Available at https://doi.org/10.1016/j.ajpath.2012.05.005.

Ostendorf, T., Zillinger, T., Andryka, K., Schlee-Guimaraes, T.M., Schmitz, S., Marx, S., et al., 2020. Immune sensing of synthetic, bacterial, and protozoan RNA by toll-like receptor 8 requires coordinated processing by RNase T2 and RNase 2. Immunity 52 (4), 591. Available at https://doi.org/10.1016/j.immuni.2020.03.009, www.immunity.com.

Paganelli, R., Fanales-Belasio, E., Scala, E., Carmini, D., Mezzaroma, I., Pinter, E., et al., 1991. Serum eosinophil cationic protein (ECP) in human immunodeficiency virus (HIV) infection. Journal of Allergy and Clinical Immunology 88 (3), 416–418. Available at https://doi.org/10.1016/0091-6749(91)90107-y.

Pearson, R.D., Uydess, I.L., Chapman, S.W., Steigbigel, R.T., 1987. Interaction of human eosinophils with *Leishmania donovani*. Annals of Tropical Medicine and Parasitology 81 (6), 735–739. Available at https://doi.org/10.1080/00034983.1987.11812179.

Pizzo, E., D'Alessio, G., 2007. The success of the RNase scaffold in the advance of biosciences and in evolution. Gene 406 (1-2), 8–12. Available at https://doi.org/10.1016/j.gene.2007.05.006.

Prats-Ejarque, G., Li, J., Ait-Ichou, F., Lorente, H., Boix, E., 2019. Testing a human antimicrobial rnase chimera against bacterial resistance. Frontiers in Microbiology 10. Available at https://doi.org/10.3389/fmicb.2019.01357.

Pulido, D., Torrent, M., Andreu, D., Nogues, M.V., Boix, E., 2013. Two human host defense ribonucleases against mycobacteria, the eosinophil cationic protein (RNase 3) and RNase 7. Antimicrobial Agents and Chemotherapy 57 (8), 3797–3805. Available at https://doi.org/10.1128/AAC.00428-13Spain, http://aac.asm.org/content/57/8/3797.full.pdf.

Pulido, D., Arranz-Trullén, J., Prats-Ejarque, G., Velázquez, D., Torrent, M., Moussaoui, M., et al., 2016a. Insights into the antimicrobial mechanism of action of human RNase6: structural determinants for bacterial cell agglutination and membrane permeation. International Journal of Molecular Sciences 17 (4), 552. Available at https://doi.org/10.3390/ijms17040552.

Pulido, D., Prats-Ejarque, G., Villalba, C., Albacar, M., González-López, J.J., Torrent, M., et al., 2016b. A novel RNase 3/ECP peptide for Pseudomonas aeruginosa biofilm eradication that combines antimicrobial, lipopolysaccharide binding, and cell-agglutinating activities. Antimicrobial Agents and Chemotherapy 60 (10), 6313–6325. Available at https://doi.org/10.1128/AAC.00830-16, http://aac.asm.org/content/60/10/6313.full.pdf+html.

Rademacher, F., Dreyer, S., Kopfnagel, V., Gläser, R., Werfel, T., Harder, J., 2019. The antimicrobial and immunomodulatory function of RNase 7 in skin. Frontiers in Immunology 10. Available at https://doi.org/10.3389/fimmu.2019.02553.

Rademacher, F., Scheel, A., Gläser, R., Schröder, L., Heinemann, N., Bartels, J., et al., 2024. Inhibition of RNase 7 by RNase inhibitor promotes inflammation and*Staphylococcus aureus* growth: implications for atopic dermatitis. Allergy 79 (6), 1573–1583. Available at https://doi.org/10.1111/all.16125.

Rajamanickam, A., Munisankar, S., Bhootra, Y., Dolla, C.K., Nutman, T.B., Babu, S., 2018. Elevated systemic levels of eosinophil, neutrophil, and mast cell granular proteins in *Strongyloides stercoralis* infection that diminish following treatment. Frontiers in Immunology 9. Available at https://doi.org/10.3389/fimmu.2018.00207, https://www.frontiersin.org/articles/10.3389/fimmu.2018.00207/full.

Ramos, Discipio, Ferreira, 2006. Eosinophil cationic protein damages protoscoleces in vitro and is present in the hydatid cyst. Parasite Immunology 28 (8), 347–355. Available at https://doi.org/10.1111/j.1365-3024.2006.00842.x.

Rendl, M., Mayer, C., Weninger, W., Tschachler, E., 2001. Topically applied lactic acid increases spontaneous secretion of vascular endothelial growth factor by human reconstructed epidermis. British Journal of Dermatology 145 (1), 3–9. Available at https://doi.org/10.1046/j.1365-2133.2001.04274.x.

Rodrigo-Muñoz, J.M., Sastre, B., Cañas, J.A., Gil-Martínez, M., Redondo, N., Del Pozo, V., 2021. Eosinophil response against classical and emerging respiratory viruses: Covid-19. Journal of Investigational Allergology and Clinical Immunology 31 (2), 94–107. Available at https://doi.org/10.18176/jiaci.0624, http://www.jiaci.org/revistas/vol31issue2_1.pdf.

Rosenberg, H.F., 1995. Recombinant human eosinophil cationic protein: ribonuclease activity is not essential for cytotoxicity. Journal of Biological Chemistry 270 (14), 7876–7881. Available at https://doi.org/10.1074/jbc.270.14.7876, http://www.jbc.org/content/by/year.

Rosenberg, H.F., 2008. RNase A ribonucleases and host defense: an evolving story. Journal of Leukocyte Biology 83 (5), 1079–1087. Available at https://doi.org/10.1189/jlb.1107725, http://www.jleukbio.org/cgi/reprint/83/5/1079.

Rosenberg, H.F., Dyer, K.D., Tiffany, H.L., Gonzalez, M., 1995. Rapid evolution of a unique family of primate ribonuclease genes. Nature Genetics 10 (2), 219–223. Available at https://doi.org/10.1038/ng0695-219.

Rudolph, B., Podschun, R., Sahly, H., Schubert, S., Schröder, J.M., Harder, J., 2006. Identification of RNase 8 as a novel human antimicrobial protein. Antimicrobial Agents and Chemotherapy 50 (9), 3194–3196. Available at https://doi.org/10.1128/AAC.00246-06.

Rugeles, M.T., Trubey, C.M., Bedoya, V.I., Pinto, L.A., Oppenheim, J.J., Rybak, S.M., et al., 2003. Ribonuclease is partly responsible for the HIV-1 inhibitory effect activated by HLA alloantigen recognition. AIDS (London, England) 17 (4), 481–486. Available at https://doi.org/10.1097/00002030-200303070-00002.

Ruiz-Rosado, J.D.D., Cortado, H., Kercsmar, M., Li, B., Ballash, G., Cotzomi-Ortega, I., et al., 2023. Human ribonuclease 6 has a protective role during experimental urinary tract infection. Journal of Innate Immunity 15 (1), 865–875. Available at https://doi.org/10.1159/000534736, http://content.karger.com/ProdukteDB/produkte.asp?Aktion=JournalHome&ProduktNr=234234&ContentOnly=false.

Salazar, V.A., Arranz-Trullén, J., Navarro, S., Blanco, J.A., Sánchez, D., Moussaoui, M., et al., 2016. Exploring the mechanisms of action of human secretory RNase 3 and RNase 7 against Candida albicans. MicrobiologyOpen 5 (5), 830–845. Available at https://doi.org/10.1002/mbo3.373, http://onlinelibrary.wiley.com/journal/10.1002/(ISSN)2045-8827.

Salazar, V.A., Arranz-Trullén, J., Prats-Ejarque, G., Torrent, M., Andreu, D., Pulido, D., et al., 2019. Insight into the antifungal mechanism of action of human RNase N-terminus derived peptides. International Journal of Molecular Sciences 20 (18), 4558. Available at https://doi.org/10.3390/ijms20184558.

Saxena, S.K., Gravell, M., Wu, Y.N., Mikulski, S.M., Shogen, K., Ardelt, W., et al., 1996. Inhibition of HIV-1 production and selective degradation of viral RNA by an amphibian ribonuclease. Journal of Biological Chemistry 271 (34), 20783–20788. Available at https://doi.org/10.1074/jbc.271.34.20783.

Schein, C.H., Haugg, M., 1995. Deletions at the C-terminus of interferon γ reduce RNA binding and activation of double-stranded-RNA cleavage by bovine seminal ribonuclease. Biochemical Journal 307 (1), 123–127. Available at https://doi.org/10.1042/bj3070123, www.biochemj.org.

Schein, C.H., Haugg, M., Benner, S.A., 1990. Interferon-γ activates the cleavage of double-stranded RNA by bovine seminal ribonuclease. FEBS Letters 270 (1-2), 229–232. Available at https://doi.org/10.1016/0014-5793(90)81275-S.

Shaffer, J.F., Gupta, A., Kharkwal, G., Linares, E.E., Holmes, A.D., Swartz, J.R., et al., 2024. Epididymis-specific RNase A family genes regulate fertility and small RNA processing. Journal of Biological Chemistry 300 (12). Available at https://doi.org/10.1016/j.jbc.2024.107933.

Sikriwal, D., Seth, D., Parveen, S., Malik, A., Broor, S., Batra, J.K., 2012. An insertion in loop L7 of human eosinophil-derived neurotoxin is crucial for its antiviral activity. Journal of Cellular Biochemistry 113 (10), 3104–3112. Available at https://doi.org/10.1002/jcb.24187.

Silverman, R.H., 2007. Viral encounters with 2′,5′-oligoadenylate synthetase and RNase L during the interferon antiviral response. Journal of Virology 81 (23), 12720–12729. Available at https://doi.org/10.1128/JVI.01471-07.

Simanski, M., Rademacher, F., Hanna, L.S., Schumacher, M., Gläser, R., Harder, J., et al., 2013. IL-17A and IFN-γ synergistically induce RNase 7 expression via STAT3 in primary keratinocytes. PLoS One 8 (3), e59531. Available at https://doi.org/10.1371/journal.pone.0059531.

Singh, A., Batra, J.K., 2011. Role of unique basic residues in cytotoxic, antibacterial and antiparasitic activities of human eosinophil cationic protein. Biological Chemistry 392 (4), 337–346. Available at https://doi.org/10.1515/BC.2011.037.

Specht, S., Saeftel, M., Arndt, M., Endl, E., Dubben, B., Lee, N.A., et al., 2006. Lack of eosinophil peroxidase or major basic protein impairs defense against murine filarial infection. Infection and Immunity 74 (9), 5236–5243. Available at https://doi.org/10.1128/IAI.00329-06.

Stecher, C.W., Fofana, H.K.M., Madsen, H., Wilson, S., Keita, A.D., Landouré, A., et al., 2017. Inflammation dynamics after praziquantel treatment of *Schistosoma haematobium* reflected by urinary eosinophil cationic protein. Transactions of the Royal Society of Tropical Medicine and Hygiene 111 (5), 316–324. Available at https://doi.org/10.1093/TRSTMH/TRX057, http://trstmh.oxfordjournals.org/content/by/year/2013.

Suhasini, A.N., Sirdeshmukh, R., 2007. Onconase action on tRNALys3, the primer for HIV-1 reverse transcription. Biochemical and Biophysical Research Communications 363 (2), 304–309. Available at https://doi.org/10.1016/j.bbrc.2007.08.157.

Sultana, M.F., Suzuki, M., Yamasaki, F., Kubota, W., Takahashi, K., Abo, H., et al., 2022. Identification of crucial amino acid residues for antimicrobial activity of angiogenin 4 and its modulation of gut microbiota in mice. Frontiers in Microbiology 13. Available at https://doi.org/10.3389/fmicb.2022.900948, https://www.frontiersin.org/journals/microbiology#.

Sun, D., Bai, R., Zhou, W., Yao, Z., Liu, Y., Tang, S., et al., 2021. Angiogenin maintains gut microbe homeostasis by balancing α-Proteobacteria and Lachnospiraceae. Gut 70 (4), 666–676. Available at https://doi.org/10.1136/gutjnl-2019-320135, http://gut.bmj.com/content/by/year.

Sun, J., Chen, M., Hu, Z., Xu, N., Wang, W., Ping, Z., et al., 2024. Ribonuclease 4 functions as an intestinal antimicrobial protein to maintain gut microbiota and metabolite homeostasis. Nature Communications 15 (1). Available at https://doi.org/10.1038/s41467-024-50223-1.

Swidergall, M., Ernst, J.F., 2014. Interplay between Candida albicans and the antimicrobial peptide armory. Eukaryotic Cell 13 (8), 950–957. Available at https://doi.org/10.1128/EC.00093-14, http://ec.asm.org/content/13/8/950.full.pdf.

Torrent, M., Navarro, S., Moussaoui, M., Nogués, M.V., Boix, E., 2008. Eosinophil cationic protein high-affinity binding to bacteria-wall lipopolysaccharides and peptidoglycans. Biochemistry 47 (11), 3544–3555. Available at https://doi.org/10.1021/bi702065b.

Torrent, M., Sánchez, D., Buzón, V., Nogués, M.V., Cladera, J., Boix, E., 2009a. Comparison of the membrane interaction mechanism of two antimicrobial RNases: RNase 3/ECP and RNase 7. Biochimica et Biophysica Acta - Biomembranes 1788 (5), 1116–1125. Available at https://doi.org/10.1016/j.bbamem.2009.01.013.

Torrent, M., de la Torre, B.G., Nogués, V.M., Andreu, D., Boix, E., 2009b. Bactericidal and membrane disruption activities of the eosinophil cationic protein are largely retained in an N-terminal fragment. Biochemical Journal 421 (3), 425–434. Available at https://doi.org/10.1042/BJ20082330Spain, http://www.biochemj.org/bj/421/0425/4210425.pdf.

Torrent, M., Badia, M., Moussaoui, M., Sanchez, D., Nogués, M.V., Boix, E., 2010. Comparison of human RNase 3 and RNase 7 bactericidal action at the gram-negative and gram-positive bacterial cell wall. FEBS Journal 277 (7), 1713–1725. Available at https://doi.org/10.1111/j.1742-4658.2010.07595.x.

Torrent, M., Pulido, D., Nogués, M.V., Boix, E., Seifert, H.S., 2012. Exploring new biological functions of amyloids: bacteria cell agglutination mediated by host protein aggregation. PLoS Pathogens 8 (11), e1003005. Available at https://doi.org/10.1371/journal.ppat.1003005.

Torres-Juarez, F., Touqui, L., Leon-Contreras, J., Rivas-Santiago, C., Enciso-Moreno, J.A., Hernández-Pando, R., et al., 2018. RNase 7 but not psoriasin nor sPLA2-IIA associates with *Mycobacterium tuberculosis* during airway epithelial cell infection. Pathogens and disease 76 (2). Available at https://doi.org/10.1093/femspd/fty005.

Tran, T.T.Q., Narayanan, C., Loes, A.N., Click, T.H., Pham, N.T.H., Létourneau, M., et al., 2024. Ancestral sequence reconstruction dissects structural and functional differences among eosinophil ribonucleases. Journal of Biological Chemistry 300 (5). Available at https://doi.org/10.1016/j.jbc.2024.107280, https://www.sciencedirect.com/science/journal/00219258.

Tsai, P.W., Yang, C.Y., Chang, H.T., Lan, C.Y., 2011. Human antimicrobial peptide LL-37 inhibits adhesion of Candida albicans by interacting with yeast cell-wall carbohydrates. PLoS One 6 (3). Available at https://doi.org/10.1371/journal.pone.0017755Taiwan, http://www.plosone.org/article/fetchObjectAttachment.action?uri=info%3Adoi%2F10.1371%2Fjournal.pone.0017755&representation=PDF.

Venge, P., Strömberg, A., Braconier, J.H., Roxin, L.-E., Olsson, I., 1978. Neutrophil and eosinophil granulocytes in bacterial infection: sequential studies of cellular and serum levels of granule proteins. British Journal of Haematology 38 (4), 475–483. Available at https://doi.org/10.1111/j.1365-2141.1978.tb01072.x.

Venge, Byström, Carlson, Håkansson, Karawacjzyk, Peterson, et al., 1999. Eosinophil cationic protein (ECP): molecular and biological properties and the use of ECP as a marker of eosinophil activation in disease. Clinical & Experimental Allergy 29 (9), 1172–1186. Available at https://doi.org/10.1046/j.1365-2222.1999.00542.x.

Vert, A., Castro, J., Ribó, M., Benito, A., Vilanova, M., 2017. Activating transcription factor 3 is crucial for antitumor activity and to strengthen the antiviral properties of Onconase. Oncotarget 8 (7), 11692–11707. Available at https://doi.org/10.18632/oncotarget.14302.

Walker, C.R., Hautefort, I., Dalton, J.E., Overweg, K., Egan, C.E., Bongaerts, R.J., et al., 2013. Intestinal intraepithelial lymphocyte-enterocyte crosstalk regulates production of bactericidal angiogenin 4 by paneth cells upon microbial challenge. PLoS One 8 (12). Available at https://doi.org/10.1371/journal.pone.0084553, http://www.plosone.org/article/fetchObject.action?uri=info%3Adoi%2F10.1371%2Fjournal.pone.0084553&representation=PDF.

Wang, Q., Lee, I., Ren, J., Ajay, S.S., Lee, Y.S., Bao, X., 2013. Identification and functional characterization of tRNA-derived RNA fragments (tRFs) in respiratory syncytial virus infection. Molecular Therapy 21 (2), 368–379. Available at https://doi.org/10.1038/mt.2012.237, https://www.journals.elsevier.com/molecular-therapy.

Waters, L.S., Taverne, J., Tai, P.C., Spry, C.J., Targett, G.A., Playfair, J.H., 1987. Killing of *Plasmodium falciparum* by eosinophil secretory products. Infection and Immunity 55 (4), 877–881. Available at https://doi.org/10.1128/iai.55.4.877-881.1987.

Weiss, C.M., Trobaugh, D.W., Sun, C., Lucas, T.M., Diamond, M.S., Ryman, K.D., et al., 2018. The interferon-induced exonuclease ISG20 exerts antiviral activity through upregulation of type I interferon response proteins. mSphere 3 (5). Available at https://doi.org/10.1128/mSphere.00209-18, https://msphere.asm.org/content/msph/3/5/e00209-18.full.pdf.

Wildenburg, G., Darge, K., Knab, J., Tischendorf, F.W., Bonow, I., Buttner, D.W., 1994. Lymph nodes of onchocerciasis patients after treatment with ivermectin: reaction of eosinophil granulocytes and their cationic granule proteins. Tropical Medicine and Parasitology 45 (2), 87–96.

Wu, Y., Wen, J., Xiao, W., Zhang, B., 2019. Pregenomic RNA: how to assist the management of chronic hepatitis B? Reviews in Medical Virology 29 (4). Available at https://doi.org/10.1002/rmv.2051.

Yamada, K.J., Barker, T., Dyer, K.D., Rice, T.A., Percopo, C.M., Garcia-Crespo, K.E., et al., 2015. Eosinophil-associated ribonuclease 11 is a macrophage chemoattractant. Journal of Biological Chemistry 290 (14), 8863–8875. Available at https://doi.org/10.1074/jbc.M114.626648, http://www.jbc.org/content/290/14/8863.full.pdf+html.

Yang, D., Chen, Q., Shao, B.S., Zhang, P., Kurosaka, K., Caspi, R.R., et al., 2008. Eosinophil-derived neurotoxin acts as an alarmin to activate the TLR2-MyD88 signal pathway in dendritic cells and enhances Th2 immune responses. Journal of Experimental Medicine 205 (1), 79–90. Available at https://doi.org/10.1084/jem.20062027, http://www.jem.org/cgi/reprint/205/1/79.

Youle, R.J., Wu, Y.N., Mikulski, S.M., Shogen, K., Hamilton, R.S., Newton, D., et al., 1994. RNase inhibition of human immunodeficiency virus infection of H9 cells. Proceedings of the National Academy of Sciences of the United States of America 91 (13), 6012–6016. Available at https://doi.org/10.1073/pnas.91.13.6012, www.pnas.org.

Zanger, P., Holzer, J., Schleucher, R., Steffen, H., Schittek, B., Gabrysch, S., 2009. Constitutive expression of the antimicrobial peptide rnase 7 is associated with *Staphylococcus aureus* infection of the Skin. The Journal of Infectious Diseases 200 (12), 1907–1915. Available at https://doi.org/10.1086/648408.

Zechendorf, E., Beckers, C., Frank, N., Kraemer, S., Neu, C., Breuer, T., et al., 2023. A potential association between ribonuclease 1 dynamics in the blood and the outcome in COVID-19 patients. International Journal of Molecular Sciences 24 (15), 12428. Available at https://doi.org/10.3390/ijms241512428.

Zernecke, A., Preissner, K.T., 2016. Extracellular ribonucleic acids (RNA) enter the stage in cardiovascular disease. Circulation Research 118 (3), 469–479. Available at https://doi.org/10.1161/CIRCRESAHA.115.307961, http://circres.ahajournals.org.

Zhang, J., Dyer, K.D., Rosenberg, H.F., 2002. RNase 8, a novel RNase A superfamily ribonuclease expressed uniquely in placenta. Nucleic Acids Research 30 (5), 1169–1175. Available at https://doi.org/10.1093/nar/30.5.1169, https://academic.oup.com/nar/issue.

Zhang, J., Rosenberg, H.F., Nei, M., 1998. Positive Darwinian selection after gene duplication in primate ribonuclease genes. Proceedings of the National Academy of Sciences of the United States of America 95 (7), 3708–3713. Available at https://doi.org/10.1073/pnas.95.7.3708.

Zhou, J., Liu, S., Chen, Y., Fu, Y., Silver, A.J., Hill, M.S., et al., 2017. Identification of two novel functional tRNA-derived fragments induced in response to respiratory syncytial virus infection. Journal of General Virology 98 (7), 1600–1610. Available at https://doi.org/10.1099/jgv.0.000852, http://jgv.microbiologyresearch.org/deliver/fulltext/jgv/98/7/1600_vir000852.pdf?itemId=/content/journal/jgv/10.1099/jgv.0.000852&mimeType=pdf&isFastTrackArticle=.

CHAPTER 9

Ribonucleases, cellular signaling and inflammation

Laura Diaz Satizabal
Department of Biochemistry and Molecular Biology, Universitat Autònoma de Barcelona, Cerdanyola del Vallès, Spain

9.1 RNases, cellular signaling, and inflammation

This section summarizes evidence regarding human RNases involved in cellular signaling and inflammation, which have been shown to occur in both a dependent and independent manner of the catalytic activity of the RNases. An overview of changes in RNases' expression associated with inflammatory processes and cancer can be found in Table 9.1.

9.2 RNase1

RNase1 was described to enhance antitumor activity in the context of breast cancer by Wang et al. (2023). 4T1 mammary tumor cells expressing RNase1 or vector control were injected into immunocompetent mice. Mice injected with RNase1-expressing cells were observed to have less tumor volume and weight. Changes in the tumor microenvironment (TME) immunological profile were investigated by analyzing infiltrating cells. It was determined that RNase1-expressing tumors had increased frequency of natural killer (NK) cells, $CD8^+$ and $CD4^+$ T-cells, and decreased presence of inhibitory T cell activity cells, myeloid-derived suppressor cells (MDSCs). Moreover, among the T helper (Th) cell population, it was observed that in the presence of RNase1, there was an increased frequency of antitumor Th1 cells, an increase of Th17 cells (that can be protumorigenic but also antitumorigenic by increasing infiltration of NK and T cells to the TME), and no change in the Th2 protumor cells or immunosuppressive Treg cells. Using a coculture cell-engineered system, the group demonstrated CD3-mediated activation of T cells, which was significantly increased with the addition of recombinant RNase1. Furthermore, RNase1-mediated T cell activation resulted in increased tumor cell death compared to the control group. Interestingly, it was shown that in the absence of T cells, expression of RNase1 resulted in increased cancer cell growth, and, in the presence of T cells, expression of RNase1 resulted in decreased cancer cell proliferation and increased cell death.

Further experiments characterizing T-cell-mediated killing boosted by RNase1 determined that RNase1-binding to receptor tyrosine kinase (RTK) Epinephrin A4

RNases in Health and Diseases
DOI: https://doi.org/10.1016/B978-0-443-27312-4.00010-9

Table 9.1 Reports of RNase gene expression exhibiting a Log2-fold change ≥ 2 during inflammatory conditions and cancer. Search conducted using the EMBL-EBI Expression Atlas (George et al., 2024).

RNase	Log2-fold change	Adjusted *p*-value	Comparison	Reference
1	4.6	1.8×10^{-31}	Esophageal carcinoma versus normal	Maag et al. (2017)
	4	5.3×10^{-9}	Sepsis versus normal	Linsley et al. (2014)
	3.8	1.8×10^{-3}	Synovial fluid mononuclear cell versus peripheral blood mononuclear cell in spondyloarthropathy	BioStudies A-AFFY-1 (2022)
	3.7	3.4×10^{-11}	Hepatitis B virus-associated acute liver failure versus normal	BioStudies A-AFFY-44 (2022)
	2.8	1.2×10^{-3}	Myxosarcoma versus normal	Scalise et al. (2020)
	2.5	1.6×10^{-2}	Mucosa-associated lymphoid tissue lymphoma versus normal	Vicente-Dueñas et al. (2012)
	2.2	5.5×10^{-4}	Multisystem inflammatory syndrome in children versus normal	Jackson et al. (2023)
	4.6	7.3×10^{-3}	Psoriatic arthritis versus healthy in the synovial fluid	BioStudies A-AFFY-37 (2022)
	−2.3	3.5×10^{-5}	Ovarian cancer versus normal in epithelial cells	BioStudies A-AFFY-44 (2022)
	−3	1.9×10^{-5}	Small cell lung carcinoma versus normal	Jiang et al. (2016)
	−4.3	1.1×10^{-3}	Intraductal papillary-mucinous carcinoma versus normal	Hiraoka et al. (2011)
2	4	4.3×10^{-5}	COVID-19 versus normal lung	Wu et al. (2020)
	3.3	1.5×10^{-6}	Crohn's disease versus noninflammatory bowel disease	Haberman et al. (2014)
	2.9	1.8×10^{-8}	Sepsis versus normal whole blood	Linsley et al. (2014)
	2.4	1.1×10^{-3}	Glioblastoma versus normal	BioStudies A-AFFY-37 (2022)
	2	3.1×10^{-3}	Multisystem inflammatory syndrome in children versus normal	Jackson et al. (2023)
	−14.1	1.1×10^{-64}	Erythroleukemia (ZRSR2 shRNA) versus normal	Madan et al. (2015)
	−7.4	2.5×10^{-8}	Osteosarcoma versus normal	Jones et al. (2012)
	−3.2	1.2×10^{-7}	Squamous cell carcinoma versus normal	Hoang et al. (2017)
	−2.1	2.9×10^{-5}	Colon cancer versus normal colon fresh frozen tissue	Abdueva et al. (2010)

3	2.2	4.8×10^{-4}	Glioblastoma versus normal	Best et al. (2015)
	2.1	1.0×10^{-5}	Septic shock versus normal	Cvijanovich et al. (2008)
	−13.6	1.5×10^{-65}	Erythroleukemia (ZRSR2 shRNA) versus normal	Madan et al. (2015)
	−6.6	2.1×10^{-12}	Osteosarcoma versus normal	Jones et al. (2012)
	−3.2	2.6×10^{-4}	Squamous cell carcinoma versus normal	Hoang et al. (2017)
	−2.4	7.7×10^{-3}	Neutrophil sepsis versus normal	Linsley et al. (2014)
4	3.6	1.7×10^{-5}	Esophageal adenocarcinoma versus normal	Maag et al. (2017)
	−4.8	3.4×10^{-3}	Erythroleukemia (ZRSR2 shRNA) versus normal	Madan et al. (2015)
	−2.9	7.4×10^{-6}	Liver tumor versus adjacent nontumor tissue	Chan et al. (2014)
	−2.8	5.0×10^{-6}	Colorectal carcinoma versus normal	Best et al. (2015)
	−2.4	4.2×10^{-3}	Hepatobiliary carcinoma versus normal	Best et al. (2015)
	−2.2	1.8×10^{-3}	Tumoral versus paratumoral tissue in hepatocellular carcinoma	Ding et al. (2014)
	−2	2.1×10^{-3}	Breast carcinoma versus normal	Best et al. (2015)
5	2.7	1.9×10^{-11}	Esophageal adenocarcinoma versus normal	Maag et al. (2017)
	−5.8	5.4×10^{-52}	Neoplasm versus normal tissue adjacent	Sia et al. (2015)
	−4.4	1.7×10^{-12}	Colorectal carcinoma versus normal	Best et al. (2015)
	−3.7	3.5×10^{-9}	Liver tumor versus adjacent nontumor tissue	Chan et al. (2014)
	−3.6	1.8×10^{-11}	Tumoral versus paratumoral tissue in hepatocellular carcinoma	Ding et al. (2014)
	−3.6	6.7×10^{-3}	Hepatocellular carcinoma neoplasm versus normal liver	Barry et al. (2013)
	−3	2.8×10^{-6}	Pancreatic adenocarcinoma versus normal	Best et al. (2015)
	−2.8	4.4×10^{-7}	Breast carcinoma versus normal	Best et al. (2015)
	−2.5	2.9×10^{-2}	Metastatic prostate cancer versus benign prostate tumor	Varambally et al. (2005)
	−2.4	1.7×10^{-2}	Erythroleukemia (ZRSR2 shRNA) versus normal	Madan et al. (2015)
	−2.3	3.4×10^{-4}	Nonsmall cell lung carcinoma versus normal	Best et al. (2015)
	−2.2	1.3×10^{-5}	Lung cancer versus normal lung fresh frozen tissue	Abdueva et al. (2010)

(*Continued*)

Table 9.1 (Continued)

RNase	Log2-fold change	Adjusted *p*-value	Comparison	Reference
6	4	8.7×10^{-3}	Hepatitis B virus-associated acute liver failure versus normal	BioStudies A-AFFY-44 (2022)
	3.9	1.1×10^{-11}	Acute liver failure versus normal	Diaz et al. (2015)
	3.5	2.3×10^{-7}	Clear cell renal carcinoma pT3 versus normal adjacent tissue	BioStudies A-AFFY-44 (2022)
	3.1	6.0×10^{-5}	Glioblastoma versus normal	BioStudies A-AFFY-37 (2022)
	2.9	1.5×10^{-3}	Rheumatoid arthritis versus normal in the synovial membrane	BioStudies A-AFFY-44 (2022)
	−2.4	1.0×10^{-9}	Colorectal carcinoma versus normal	Best et al. (2015)
	−2.4	5.9×10^{-8}	Nonsmall cell lung carcinoma versus normal	Best et al. (2015)
	−2.2	6.5×10^{-7}	Ovarian cancer versus normal epithelial cells	BioStudies A-AFFY-44 (2022)
7	3.2	1.9×10^{-10}	Lesional psoriasis versus normal	Yao et al. (2008)
	2	1.3×10^{-2}	Vulvar intraepithelial neoplasia versus normal	Santegoets et al. (2007)
	−6.1	1.7×10^{-18}	Esophageal adenocarcinoma versus normal	Maag et al. (2017)
	−3.2	3.6×10^{-3}	HER2-positive breast carcinoma versus normal	Eswaran et al. (2012)
	−2.9	1.5×10^{-7}	Atopic eczema versus normal	Gittler et al. (2012)
8	3.8	3.0×10^{-2}	Breast carcinoma versus normal	Eswaran et al. (2012)
	−2.8	3.1×10^{-2}	Actinic keratosis versus normal	Hoang et al. (2017)

receptor (EphA4) (see Fig. 9.1) and further description in the RTK section of this chapter) is important for cancer cell killing, as preincubation with a neutralizing EphA4 antibody resulted in decreased killing (Wang et al., 2023).

The role of RNase1 in T-cell dysfunction in the context of cancer and T-cell immunotherapy was described by Yang et al. (2025). Decreased T-cell cytotoxicity from exposure to immunosuppressive factors can result in treatment failure, and thus, the authors identified inhibitors of T-cell activity secreted by cancer cells. mRNA levels of *interleukin-2* (IL-2) were used as a marker for T-cell activation of Jurkat T cells exposed to the conditioned media (CM) from cancer cell lines. CM from breast and cervical cancer cell lines, KPL4 and HeLa cells, respectively, were found to downregulate the mRNA levels of IL-2 in activated Jurkat cells. To further elucidate which proteins were responsible for IL-2 downregulation, KPL4-derived CM was fractioned based on molecular weight. It was observed that only the fraction with proteins under 30 kDa was able to lower mRNA levels of IL-2 in both activated Jurkat T cells and human peripheral blood mononuclear-derived T cells. Four differentially expressed proteins in KPL4-derived CM versus the control were found and identified as

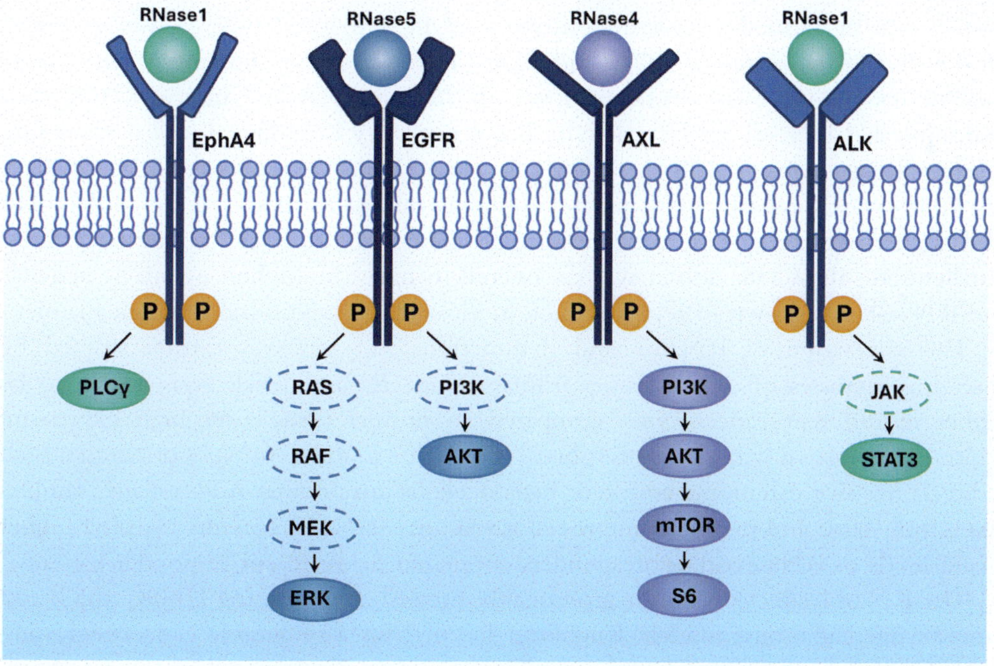

Figure 9.1 Schematic representation of some pathways induced by the RTK–RNase interaction. P represents phosphorylation, dotted lines represent known pathways, while the solid ones represent proteins analyzed in the publication: RNase1 (Lee et al., 2021; Liu et al., 2024), RNase5 (Wang et al., 2018), and RNase4 (Lee, 2024).

Tetranectin isoform 2 (TN), the hemoglobin subunit alpha, RNase1, and the retinol-binding protein 4 isoform b. Based on information from the Cancer Genome Atlas (TCGA) database, RNase1 was selected for the following analyses. Activated Jurkat T cells were treated with recombinant RNase1, and both *IL-2* mRNA levels and IL-2 protein concentration in the CM were measured. Upon treatment, a decrease in mRNA *IL-2* levels was observed along with a time-dependent decrease of IL-2 in the CM. Because a decrease in cytokines occurs in dysfunctional T cells, the effect of RNase1 was studied for other cytokines in activated T cells. A reduction in mRNA levels and protein concentration was observed for interferon gamma, interleukin-6 (IL-6), and granulocyte-macrophage colony-stimulating factor. Furthermore, immune checkpoint (IC) molecules, which are increased during T cell dysfunction, were found to be increased by RNase1 treatment in both activated Jurkat cells and peripheral blood mononuclear cell (PBMC)-derived T cells. Altogether suggesting that RNase1 can induce T cell dysfunction in vitro and may contribute to immunosuppression in the TME (Yang et al., 2025).

The effect of RNase1 was also analyzed for activated T cell subsets, $CD4^+$ and $CD8^+$, isolated from PBMC. An IC increase was observed in both subsets, and it appeared that $CD8^+$ T cells could uptake more RNase1 than $CD4^+$ T cells. Further studies were performed to determine if the catalytic activity of RNase1 was necessary for the T cell dysfunction of $CD4^+$ and $CD8^+$ T cells to occur. The use of an RNase1 catalytic-inactive revealed decreased levels of *IL-2* mRNA and increased ICs, thus indicating that RNase1 catalytic activity is not necessary to induce T cell dysfunction. Because dysfunctional T cells also exhibit decreased killing, the effect of RNase1 on T cell killing was assessed by silencing and knocking out *rnase1*. It was observed that both silencing and knocking out *rnase1* resulted in increased cancer cell killing by T cells. Furthermore, the $CD8^+$ subset was discovered to have the highest cytotoxic activity, and RNase1 was shown to repress killing in vitro in a catalytic-independent manner.

The association of RNase1 and dysfunctional T cells was further studied by analyzing databases of cancer tissues from patients. *RNase1* levels were found to be higher in different cancer types compared to normal tissue, and high expression positively correlated with the expression of several ICs in different types of cancers, such as breast invasive carcinoma and liver hepatocellular carcinoma. Additionally, samples taken from head and neck squamous cell carcinoma (HNSC) patients revealed higher serum levels of RNase1 in nonresponders compared to treatment-responder patients.

The RNase1 effect on tumor growth was assessed in vivo using HNSC and breast cancer syngeneic mouse models. Knocking down *RNase1* resulted in cancer regression and a higher proportion of $CD8^+$ T cells expressing effector cytokines and reduced expression of ICs. Furthermore, depletion of $CD8^+$ but not $CD4^+$ T cells resulted in overturning tumor regression, indicating that RNase1 contributes to tumorigenicity by increasing $CD8^+$ T cell dysfunction in vivo.

Further studies revealed that T cell dysfunction was not induced by binding to TKRs ephrin-A4 or anaplastic lymphoma kinase (ALK). It was found that RNase1 was internalized by T cells and accumulated in the cytoplasm, and inhibiting internalization stopped the decrease of *IL-2* mRNA levels. Immunoprecipitation assays and correlation analyses led to the identification of signal transducer and activator of transcription 1 (STAT1) as interacting with RNase1 and possibly mediating T cell dysregulation. Treatment with RNase1 led to a significant decrease in STAT1 phosphorylation and a shift in STAT1 localization from the nucleus to the cytoplasm. Mutagenesis of RNase1 to prevent STAT1 binding resulted in no decrease in phosphorylation of STAT1 and no inhibition of T cell killing. Furthermore, transferring $CD8^+$ T cells with the mutated RNase1 into mice resulted in no increase in tumor weight and volume.

Altogether, the data presented suggest that once internalized, RNase1 induces T cell dysfunction by preventing STAT1 activation and nuclear localization, increasing the expression of ICs, and impairing $CD8^+$ T cell cytotoxicity both in vitro and in vivo (Yang et al., 2025).

In 2018, research exploring the potential for the RNase1 variant QBI-139 as a cancer chemotherapeutic agent was published. QBI-139 was previously developed to have lower affinity for the human ribonuclease inhibitor (RNH1) in order to increase its availability to kill cancerous cells (Hoang et al., 2018). Research published shows the synergistic effect of using QBI-139 with small-molecule inhibitors of protein kinases. Treatment of the lung cancer cell line A549 with different kinase inhibitors and QBI-139 revealed that the combination with trametinib, a mitogen-activated protein kinase 1 (MEK1) and MEK2 inhibitor, was the most effective at killing the cancer cells. Using an inhibitor upstream of MEK on two different cancer cell lines suggested QBI-139 to be interacting with proteins in the ERK pathway. RI phosphorylation was shown to increase its affinity for RNase, and further experiments using ERK pathway kinase inhibitors revealed that these kinases are responsible for RI phosphorylation. Altogether, this publication demonstrated that the synergistic effect of QBI-139 and ERK pathway inhibitors is based on further lowering the affinity of the RI for the RNase variant and thus allowing for higher cytotoxicity (Hoang et al., 2018).

9.3 RNase2

RNase2 was already described as an alarmin with chemotactic properties at the beginning of the 21st century by Rosenberg and coworkers. However, the protein underlying the mechanism of action is still under study. Yang et al. (2003) reported the chemotactic properties of RNase2. In vitro studies showed that RNase2 induced chemotaxis of $CD34^+$ cell-derived immature dendritic cells (iDCs). Subsequent experiments revealed that induced migration was mediated by G protein-coupled receptors (GPCRs), as pretreatment with Pertussis toxin (PTX), a GPCR signaling inhibitor, completely inhibited cellular migration. Additionally, RNH1 inhibited migration of iDCs by RNase2 but did not inhibit migration by the stromal cell-derived factor 1α (SDF-1α[EB4]), a dendritic cell (DC) precursor chemoattractant. Furthermore, RNase2 was able to induce migration in monocyte-derived DCs and mature DCs, but failed to do so in monocytes, neutrophils, eosinophils, and T cells. Additionally, treatment of DCs with RNase2 induced phosphorylation of extracellular signal-regulated kinases 1/2 (ERK1/2), indicating activation of the MAPK/ERK signaling pathway (Yang et al., 2003).

Yang et al. (2008) reported RNase2 as an alarmin of the immune system. In vitro studies of human monocyte-derived DCs treated with RNase2 revealed that RNase2 induced full activation of DCs. Phenotypically, cell surface marker expression and chemokine receptor expression of treated DCs were consistent with those of mature DCs. Functionally, treated DCs became capable of antigen presentation, shown by stimulation of T cell proliferation. Additional studies identified RNase2 as a Toll-like receptor 2 (TLR2) ligand capable of inducing cytokine production in a MyD88-dependent manner. Finally, RNase2 was shown to enhance antigen-specific Th2 immune responses in a TLR2-dependent manner through Th2 cytokine production in immunized mice (Yang et al., 2008).

Recently, machine learning algorithms were used to screen genes from 2 microarray sets related to sepsis. Four common diagnostic genes were found: *IRAK3*, *CD177*, *RNASE2*, and *S100A12*. A two-step validation of these genes was done to assess their potential as biomarkers. First, the expressions of *IRAK3*, *CD177*, *RNASE2*, and *S100A12* were found to be significantly upregulated in both infective myocarditis and sepsis datasets. Second, single-cell-RNA sequencing datasets from sepsis were analyzed, and gene expression was found to be increased in the sepsis samples compared to healthy controls. Furthermore, different correlation analyses using microarray and RNA-seq sepsis and infective myocarditis data sets revealed the following: first, the four genes were associated. Second, the construction of an mRNA–miRNA network determined the genes to be associated with miRNAs involved in the immune response, inflammatory response, cell cycle regulation, and the development of tumors. And third, a protein–protein interaction network constructed with diagnostic genes and inflammatory cytokines found interleukin-10 (IL-10) to be significantly correlated. Altogether, *CD177* and *IRAK3* are proposed to be potential diagnostic genes for infective myocarditis, and *RNASE2* is a potential diagnostic gene in both infective myocarditis and sepsis (Yi et al., 2024).

9.4 RNase3

Since its isolation from patients with chronic myeloid leukemia in 1971 (Olsson & Venge, 1972, 1974), RNase3, named the eosinophil cationic protein (ECP), has been extensively characterized. Venge et al. (1999) proposed the use of ECP as a marker of eosinophil activation in disease. In 2011 this group published their work on the effect of ECP on classical Hodgkin lymphoma cell lines (Glimelius et al., 2011).

In 2007, Venge's group reported the role of ECP in the context of asthma (Zagai et al., 2007), a condition characterized by eosinophilic inflammation and the associated fibrotic remodeling of the tissue. In this condition, fibroblasts, being the main cell type found in connective tissue, play a key role in the inflammatory remodeling process by releasing mediator cytokines like transforming growth factor beta 1 (TGF-β1) and matrix proteins, such as collagen. In vitro studies showed that human lung fibroblasts exposed to RNase3 increased the release of TGF-β1 in a dose-dependent manner. Collagen gel contraction was used to model the contractile process in fibrotic tissue remodeling. Gels containing both fibroblasts and RNase3 showed an increase in the release of TGF-β1 and contracted more than those without RNase3. Lastly, it was observed that during coculture, fibroblasts increased the release of TGF-β1, and eosinophils increased the release of RNase3. Altogether, these results suggest a potential role for eosinophils in the fibrotic remodeling process during asthma, mediated by RNase3 stimulation of fibroblasts (Zagai et al., 2007).

In 2022, Chuang and colleagues reported their findings on RNase3 in systemic lupus erythematosus (SLE) (Chuang et al., 2022). In SLE patients, exosome levels correlate to disease severity, and thus, T cell-derived exosomes from SLE patients were isolated and characterized through mass spectrometry proteomics. RNase3 was found to be the highest-scoring protein within exosomes. Proteomics of T-cells revealed an enrichment of RNase3 in T cells from SLE patients. To further understand the role of RNase3, a transgenic mouse with T cells and T cell-derived exosomes overexpressing RNase3 was generated. Over time, transgenic mice were found to have increased levels of proinflammatory cytokines. These mice were also found to develop severe inflammation of different tissues where RNase3-containing exosomes were found. Exosomes derived from transgenic mice T cells were transferred to wild-type mice and were found to infiltrate different tissues and cause inflammation. Single-cell RNA sequencing from T cells isolated from transgenic mice showed increased levels of genes consistent with T cell activation and inflammation. Lastly, the levels of follicular helper T cells, plasma B cells, and autoantibodies increased in transgenic mice over time compared to the wild-type mice control (Chuang et al., 2022).

9.5 RNase4

RNase4 has been described in both cancer and infectious diseases. RNase4 was shown to be implicated in the occurrence of urinary tract infections (UTIs) in a publication investigating the importance of insulin and insulin receptor (IR)-mediated signaling in the context of urinary infections (Murtha et al., 2018). Previous work from the same group demonstrated that antimicrobial peptides (AMP) are induced in the kidney through insulin-mediated PI3K/AKT signaling (Eichler et al., 2016). In the kidney, AMPs are mostly produced by intercalated cells (ICs) located in the collecting tubules; therefore, the group hypothesized that insulin regulates AMP production through PI3K/AKT activity and impacts kidney defenses. A knockout mouse lacking IR in the collecting duct and ICs in the kidney was generated. Transgenic mice were observed to have higher uropathogenic *Escherichia coli* (UPEC) burden in the urine, bladder, and kidney compared to the control. IR deletion did not appear to affect IC population or urine acidification but to impact AMP production by ICs. Gene expression of two AMPs, *Rnase4* and *Lipocalin* (*Lcn2*), was found to be suppressed in infected and noninfected knockout mice. This observation was validated by measuring the concentration of the two peptides in the urine, suggesting that AMP production is IR-mediated.

AMP's role in UPEC survival was assessed. Urine from knockout and control mice was treated with RNase4 and Lcn2 antibodies. The addition of the RNase4 antibody resulted in significant suppression of UPEC killing in the control sample, while the addition of the Lcn2 antibody resulted in killing suppression that was not statistically significant. Furthermore, the antimicrobial activity of RNase4 against UPEC was assessed using a colony-count reduction assay. Antimicrobial activity was observed down to micromolar concentrations of RNase4 against 2 UPEC strains, including a multidrug-resistant one.

Insulin was shown to induce expression of *Rnase4* and *Lcn2* in mouse and human in vitro, and pretreatment of human renal cells with a PI3K/AKT antagonist resulted in reduced production of RNase4 and Lcn2. In vivo assays treating the kidney of mice with the PI3K/AKT antagonist revealed decreased expression of *Rnase4* and *Lcn2*. Furthermore, these mice also exhibited a higher UPEC burden after UPEC challenge compared to the control, confirming that AMP production in the kidney is PI3K/AKT-mediated (Murtha et al., 2018).

Lastly, urine concentrations of RNase4 and Lcn2 were found to be lower in children and adolescents with type 2 diabetes compared to the healthy control group. However, diabetic patients were under glycemic control, and the effect of therapy on urine AMP concentration was not assessed (Murtha et al., 2018).

The association between RNase4 and prostate cancer was initially reported in Vanli et al. (2022). The search for new biomarkers for disease type and aggressiveness led investigators to discover elevated levels of plasma RNase4 in prostate cancer patients. Furthermore, investigators discovered that RNase4 levels correlated with prostate-specific antigen (PSA), a limited biomarker and diagnosis tool for prostate cancer, suggesting the use of both together may enhance the accuracy of diagnosis. Interestingly, authors found RNase4 to be a better marker than RNase5, according to an ROC analysis performed with the same cohort of patients and correlation to poor prognosis, higher risk of metastasis, and biopsy grade. Additionally, and unlike PSA, RNase4 levels were able to differentiate between cancer patients and benign prostatic hyperplasia (BPH), with plasma protein levels appearing higher in cancer patients, but no differences were observed in protein levels between BPH and normal prostate individuals.

The authors also investigated the mechanism of action of RNase4 in prostate cancer and found RNase4 to induce phosphorylation of the AXL RTK and to stimulate prostate cancer cell proliferation by activating the PI3K-AKT-mTOR-S6K signaling pathway downstream of AXL (Fig. 9.1). Furthermore, *RNASE4* knockdown was shown to suppress tumor growth and cell proliferation, but these were restored when providing cells with exogenous RNase4. Interestingly, catalytically inactive RNase4 could not restore cell proliferation, indicating that enzymatic activity is required for this process. The effect of the *RNASE4* knockdown was also studied in a xenograft mouse, where it was observed that knockdown tumors had slower growth, less cell proliferation, and less AXL phosphorylation when compared to the control (Vanli et al., 2022).

Finally, the publication reported the therapeutic value of targeting RNase4 using a human monoclonal antibody (mAb). The RNase4 mAb was shown to inhibit cell proliferation, reduce the number of cells in G2/S/M phases, and increase G0/G1. The mAb was also tested against the angiogenic properties of RNase4. In vitro assays revealed a reduced tube formation in endothelial cells in a dose-dependent manner. In vivo studies were conducted to explore the activity of the mAb in a prophylactic setting and against established tumors. First, the prophylactic activity was tested in mice inoculated with cancercells. Treatment with the mAb resulted in an 80% mean reduction in tumor size, a 90% mean reduction in tumor weight, enhanced apoptosis, and reduced AXL phosphorylation. Second, the use of the mAb against established tumors in xenografted mice resulted in inhibition of tumor growth in a dose-dependent manner, increased apoptosis, and decreased AXL phosphorylation (Vanli et al., 2022).

The role of RNase4 in glioblastoma (GBM) was investigated and reported by Lee (2024). Bioinformatic analysis of the TCGA revealed an increasing upregulation of *RNASE4* through disease progression and an inverse correlation between high levels of RNase4 and overall and progression-free survival. These results matched immunohistochemistry of brain tissue in which increasing protein presence was relative to the severity of the disease, indicating a relation between RNase4 and disease stage. In vitro studies showed a decrease in cell proliferation, migration, and invasion capabilities in the *RNASE4*-knockdown DBTRG-05MG, a glial cell line isolated from a GBM patient. Upon exposure to exogenous RNase4, migration and invasion properties were significantly restored. Gene set enrichment analysis of TCGA data sets correlated the expression of *RNASE4* to marker genes of the epithelial-mesenchymal transitional (EMT) pathway, TNF-mediated nuclear factor kappa B (NF-κB) signaling, and hypoxia. Based on this finding, the group then investigated the role of RNase4 in these processes. It was observed that a hypoxic environment led to increased levels of RNase4, phosphorylation of the AXL RTK, and AKT. Further studies conducted using an RNase4-overexpressing cell line showed increased phosphorylation of AXL and AKT, and increased presence of EMT markers. RNase4 enhanced NF-κB activity using an NF-κB reporter system. Furthermore, cells treated with RNase4 led to an increased phosphorylation of IκBα, a key step for the degradation of the NF-κB inhibitor. Also, downstream effectors of NF-κB signaling involved in apoptosis inhibition were upregulated. Based on their findings, the group proposes that RNase4 promotes the progression of GBM by AXL RTK activation and downstream effectors (Lee, 2024).

9.6 RNase5

The molecular mechanism through which RNase5 promotes cellular proliferation was described in 2017 (Hoang & Raines, 2017). Upon RNase5 internalization by endocytosis, it was shown that phosphorylation by protein kinase C and cyclin-dependent kinase was required for RNase5 nuclear translocation. In the nucleus, RNase5 cleaves promoter-associated RNA (pRNA) in a specific manner and stops the repression of ribosomal DNA (rDNA) transcription by the nucleolar remodeling complex (NoRC). Repression of rDNA transcription occurs by pRNA-binding to NoRC's large subunit, the transcription termination factor I-interacting protein 5 (TIP5), which in turn associates with the rDNA promoter. Additionally, it was shown that RNase5 was capable of cleaving pRNA within a pRNA-TIP5 complex; however, cleavage occurred faster when pRNA was in an unbound state. Altogether, the authors demonstrate that RNase5 can promote cellular proliferation by derepressing the rDNA transcription factor through pRNA cleavage (Hoang & Raines, 2017).

In 2020 Plummer and colleagues published their work investigating the contribution of myofibroblasts in the development of colitis-associated cancer. Myofibroblasts are RNase5-secreting cells that have been implicated in human colon cancer. Myofibroblasts also represent an important source of stroma-derived cyclo-oxygenase-2 (COX-2), an enzyme that produces prostaglandins, which are involved in the inflammatory response in colitis and susceptibility to cancer. In vitro studies using RNase5-blocking antibody pretreatment followed by stimulation with a combination of proinflammatory mediators, TNFα and bradykinin, restricted upregulation of COX-2 and protein kinase D (PKD) phosphorylation, previously shown to be induced by the combination. These results were reproduced using myofibroblasts isolated from *RNASE5*-knockout mice and implicate RNase5 in the regulation of COX-2 through PKD in myofibroblasts (Plummer et al., 2020).

9.7 RNase6

Differentially expressed genes (DEGs) associated with immunity and inflammation were identified in a transcriptome data set of glomeruli samples from diabetic nephropathy

(DN) patients, a common complication of diabetes. Among the 120 DEGs identified, RNase6 was found to be the highest-scoring hub gene. Because renal leukocyte infiltration is a characteristic process during DN, the association of the expression of RNase6 with cellular infiltration was analyzed. RNase6 expression was positively correlated with the infiltration of both activated and conventional DCs, CD4+ effective memory T cells, memory B-cells, epithelial cells, and macrophages. Further studies performed using a DN rat model revealed a significantly higher mRNA expression of RNase6, the inflammatory cytokine tumor necrosis factor alpha (TNF-α), the chemokine MCP-1, and a marker of renal inflammation associated with disease progression. Based on their findings, the authors propose RNase6 as a DN diagnosis and prognosis biomarker and suggest that it could be contributing to the inflammatory response through the renal mononuclear phagocytic system (Li et al., 2022).

9.8 RNase7

The role of RNase7 in the regulation of the immune response during UTIs, specifically in a high-glucose environment and UPEC infection, was reported by Ho et al. (2022). High-glucose environments occur during diabetes and regulate the production of AMP and, therefore, impact the immune response. In vitro studies using bladder epithelial cells showed that treatment with RNase7 prior to UPEC coculture decreased the infection rate that was enhanced by the high-glucose concentration. Further studies revealed the expression of STAT1, STAT3, and TLR4 to be downregulated in bladder cells pretreated with RNase7 versus the nonpretreated control. Moreover, pretreatment with RNase7 was shown to regulate the JAK/STAT signaling pathway by increasing the inhibitor SOCS3, resulting in reduced levels of proinflammatory cytokines IL-6, IL-1β, and IL-8. Finally, blocking JAK resulted in increased expression of pSTAT1, TLR4, and IL-6 but did not result in reduced UPEC infection. Altogether, results suggest that RNase7 can reduce UPEC infection in high-glucose environments and acts as an antiinflammatory mediator by downregulating the JAK/STAT pathway (Ho et al., 2022).

9.9 RNases and receptor tyrosine kinases

RTKs are a family of cell surface receptors important for a wide array of cellular and metabolic processes. RTKs are activated upon ligand-binding, and the tyrosine kinase domain located in the cytoplasmic region of the receptor is phosphorylated to initiate a signaling cascade (Lemmon & Schlessinger, 2010; Maruyama, 2014). This section compiles reports of RNases as RTK ligands, following a combined set of criteria based on the chronological order of publication and the importance of findings. Some of these interactions have been discussed earlier in this chapter in some anti and protumoral RNase activities downstream of RTK binding.

In their efforts to identify a predictive biomarker for pancreatic ductal adenocarcinoma responsiveness to erlotinib treatment, Wang et al. (2018) published the first report of RNase5 as a ligand of epithelial growth factor receptor (EGFR) (Fig. 9.1). RNase5 was shown to induce EGFR phosphorylation and dimerization, and activation of its downstream pathways through phosphorylation of protein kinase B (AKT) and ERK1/2. Furthermore, EGFR dimerization was observed upon activation with RNase5, which is also observed during activation of EGFR with classical ligands. Mutagenesis of both RNase5 and EGFR revealed that glutamine 93 of RNase5 is necessary for EGFR-binding and activation, while domains I and III of EGFR are required for binding to RNase5. Studying the dissociation constant of RNase5 binding to EGFR, along with epithelial growth factor (EGF)-binding to EGFR, revealed RNase5 to behave as a high-affinity ligand of EGFR that competes with EGF for EGFR-binding. Finally, through mutagenesis of RNase5, it was shown that binding to EGFR occurs in an RNase5 catalytic-independent manner (Wang et al., 2018).

Liu et al. (2021) proposed RNase7-driven activation of ROS1 as a potential target in hepatocellular carcinoma. RNase7 was shown to induce phosphorylation and thus activation of ROS1 and downstream ROS1 signaling pathway proteins, ERK1/2, and AKT1/2/3. Additionally, the dissociation constant for the ROS1-RNase7 complex was estimated and determined to fall in the nanomolar range, within the classification of high-affinity ligands of RTKs, and binding was suggested to occur within the extracellular ROS1 domain N3-P1 and RNase7 residues K1, K107, and K111.

However, recent work by Jones et al. (2024) was unable to replicate Liu et al.'s (2021) findings, as phosphorylated ROS1 was not detected after RNase7 treatment. Furthermore, they were unable to detect ROS1 mRNA in the hepatocellular carcinoma cell lines previously reported by Liu et al. (2021) to abundantly express ROS1; this finding was confirmed by the absence of protein detection through immunoblotting. They then used RNase7 to treat SW-1353 chondrosarcoma cells, a cell line demonstrated to express high levels of ROS1, and found no phosphorylation of ROS1 (Jones et al., 2024).

In the report of the immunomodulatory activity of **RNase3** on a macrophage cell line, Lu et al. (2021) describe a possible interaction between RNase3 and EGFR. Transcriptome analyses were performed comparing THP-1-derived macrophages exposed to RNase3 at two different timepoints. Additionally, the ribonucleolytic activity of RNase3 in relation to the immunomodulatory capacity of the protein was explored by comparing THP-1-derived macrophages treated with RNase3 wild-type and RNase3 lacking its catalytic activity, obtained by a single amino acid substitution (RNase3-H15A). DEGs were identified in all treatment groups, with a high number of overlapping DEGs between cells treated with RNase3 and RNase3-H15A, indicating that both proteins were able to elicit an immune response. Correlation analyses led to the identification of genes involved in the immune response in signaling pathways independent of the RNase3 catalytic activity. Among the primary hub genes identified was the EGFR. Interestingly, *EGFR* was found to link the other hub genes and to upregulate mediators of the immune response. Therefore *EGFR* was identified as the main hub of the immune response elicited by RNase3 and RNase3-H15A. Besides, upregulation of proinflammatory genes observed at 4 hours was reduced at 12 hours. Additional work to understand the relation between RNase3 signaling and the EGFR receptor demonstrated the activation of the EGFR-MAPK signaling cascade in cells treated with RNase3 (Lu et al., 2021).

Transcriptome analyses comparing cells treated with RNase3 and RNase3-H15A revealed that the macrophage antiviral response appeared to occur in an RNase3 catalytic-dependent manner. Interestingly, changes in gene expression appeared stronger at 12 hours versus 4 hours, suggesting that signaling by RNase3 is independent and dependent on the catalytic activity and may complement each other, with the first being the main contributor in the early immune response and the second the main contributor in the late immune response (Lu et al., 2021).

Further studies were performed toward the understanding of RNase3 and EGFR signaling. Macrophage treatment with cetuximab, an antibody capable of blocking EGFR activation, confirmed the capacity of RNase3 to activate the EGFR signaling pathway. Additionally, molecular modeling revealed the main region of interaction in RNAse3 to share key amino acids with EGFR-binding regions of RNase5 and EGF, 2 known EGFR ligands. Additionally, RNase3's catalytically active residues were not involved in EGFR-binding, which supports the transcriptomic analyses results (Lu et al., 2021).

RNase1 was shown to bind and activate the RTK EphA4 in 2021 (Lee et al., 2021) (Fig. 9.1). Treatment of HeLa and BT-549 cells resulted in phosphorylation of EphA4, the hepatocyte growth factor receptor, together with the endothelial TEK RTK (belonging to the Tie-2 tyrosine kinase family), and EphA10, together with the RTK-like orphan receptor 2, respectively. However, because of the scope of the study, further experiments focused on the EphA4 pathway. RNase1 activation of EphA4 also resulted in the activation of the IKK/NF-κB and MEK/ERK pathways. The association between the ribonuclease and EphA4 was also shown through an immunoprecipitation assay with an EphA4 antibody of cells treated with RNase1, through a GST pull-down assay using a GST-tagged RNase1, and through a Duolink in situ proximity ligation assay. GST pull-down assays of several EphA4 and RNase1 mutants led to the observation that the binding of the two proteins seems to occur through the ligand-binding domain within the extracellular domain of EphA4 and the C-terminus of RNase1. It was also observed that activation with RNase1 led to dimerization and oligomerization of EphrA4, which is seen in the classical activation of Eph receptors with Eph family receptor-interacting proteins or ephrin ligands. Furthermore, the dissociation constant of the EphrA4-RNase1 complex was determined to be 16.5 times lower than that observed for EphA4 and its ligand ephrin A5, however still in the nanomolar range, which is observed in the complex formation of Eph receptors and their cognate ligands (Lee et al., 2021).

A second report of RNase1 signaling through an RTK was done by Liu et al. (2024), who studied the role of RNase1 and the ALK RTK (Fig. 9.1) in macrophage polarization in the context of hepatocellular carcinoma and tumor immunity. Transcript levels of RNase1 were positively correlated with infiltration of pro-tumorigenic or T2-like tumor-associated macrophages in tumor samples from patients diagnosed with hepatocellular carcinoma. In vitro studies on the polarization of THP-1 macrophages led to the observation that RNase1-treated macrophages exhibited a similar transcriptome change as those treated with interleukin-4 (IL-4). Furthermore, the upregulation of M2 macrophage markers CD200R, MRC1, VSIG4, and FCER2 was observed in both RNase1 and IL-4-treated macrophages. Meanwhile, macrophages treated with LPs/IFN-γ exhibited a profile consistent with that of M1 macrophages. RNA-seq and gene enrichment analysis were performed on stimulated THP-1 monocytes to find gene sets associated with RNase1. Enrichment of the ALK tyrosine kinase receptor gene set was observed only in RNase1-stimulated THP-1 monocytes. RNase1 was found to induce phosphorylation of ALK and downstream protein signal transducer and activator of transcription 3 (STAT3). Furthermore, the knockdown of ALK resulted in lower levels of phosphorylated ALK and STAT3 in RNase1 -treated macrophages, along with increased expression of T1 markers and reduced expression of T2 markers (Liu et al., 2024).

9.10 RNases and the human RNase inhibitor

The RNH1 is a leucin-rich repeat (LRR) protein conserved among mammals that evolved through exon duplication (Dickson et al., 2005). Although found mainly in the cytosol, it can also be located in the nucleus and the mitochondria (Dickson et al., 2005; Furia et al., 2011; Lee & Vallee, 1993). RNH1 is capable of binding with high affinity to different proteins of the pancreatic RNase superfamily, which are rendered catalytically inactive by becoming partly surrounded by the horseshoe shape of the RNH1 (Dickson et al., 2005; Kobe & Deisenhofer, 1993; Shapiro et al., 2000). The RNH1 is a key regulator of intracellular RNase activity, and here we compile information regarding the RNH1-RNase interaction grouped by RNase types.

9.10.1 RNase1

Zechendorf et al. (2020) reported on the role of RNase1 and RNH1 during septic cardiomyopathy and cardiac apoptosis. Levels of RNH1, RNase1, and extracellular RNA (eRNA) were found to be elevated in serum samples of septic patients compared to healthy volunteers. eRNA is a damage-associated molecular pattern released from tissue injury; it is associated with signaling during injury, inflammation, and repair, and it is degraded by extracellular RNases. In vitro studies using murine cardiomyocytes revealed that cells exposed to necrotic cardiomyocytes releasing eRNA showed increased levels of activated caspase-3, a mediator of apoptosis. Activation of caspase-3 was dampened upon cell exposure to RNase1. These results were confirmed with flow cytometry, showing that apoptosis was the main cell death type occurring in NC-treated cells and that RNase1 treatment reduced the number of apoptotic events occurring.

Using a murine model for polymicrobial sepsis, cecal ligation and puncture mice, the group demonstrated that RNase1 treatment led to a reduction of serum concentration of TNF and RNH1. Furthermore, RNase1 treatment palliated renal dysfunction and hepatocellular injury. RNase1 reduced the levels of the pro-apoptosis protein Bax and attenuated the associated levels of activation of caspase-9 and caspase-3. These results were confirmed by the observation that the number of apoptotic cells in the heart decreased in septic mice treated with RNase1. Based on their results, the group proposes that increased levels of eRNA are the result of increased RNH1, which in turn inhibits RNase1 activity. The increase of eRNA then leads to increased inflammation and apoptosis in the heart (Zechendorf et al., 2020).

A later study from the same research group reported the role of RNH1 during inflammation due to sepsis and iron homeostasis through its regulation of RNases1 and 5. RNH1 increased levels were found in plasma samples of septic patients, while it was not detectable in healthy volunteers. Furthermore, mortality was associated with higher levels of RNH1. RNH1 levels were also associated with decreased renal and hepatic functions, indicating organ failure. No changes in RNase1 concentration were detected, but RNase5 levels correlated with RNH1 concentration, suggesting a connection between RNH1 and RNase5 but not RNH1 and RNase1 during sepsis. In vitro studies demonstrated that LPS-activated PBMCs increased the production of RNase5 compared to untreated PBMCs, but RNase5 levels were reduced in PBMCs treated with LPS and RNH1. Treatment with exogenous RNH1 stimulated a reduction in the production of TNF-α and a reduction in the mRNA from a noncanonical inflammasome gene cluster. Based on their findings and previous reports, the group proposes that in septic patients, RNH1 elevated levels could be the result of RNH1–RNase5 dissociation to promote RNase5-mediated cell survival (Neu et al., 2024).

9.10.2 RNase2

RNase2 affinity to RNH1 was thoroughly characterized by early studies. The structural basis for the interaction between RNH1 and RNases was explored by the hotspot model. Single residue mutagenesis analysis of the RNH1 interface revealed that each hotspot residue had little impact on the affinity for RNase2, but that mutation of asparagine 435 led to a significant reduction of the protein-inhibitor affinity. However, mutagenesis of multiple residues of the RNH1 hotspot led to the observation of a superadditive effect, thus indicating the importance of the RNH1 C-terminus in the RNH1–RNase2 complex. Furthermore, mutagenesis of the tryptophan cluster of RNH1 led to decreased affinity for RNase2, as previously observed for RNase5 (Teufel et al., 2003).

9.10.3 RNase5

An equivalent approach was undertaken to analyze RNase5–RNH1 complex formation. Single amino acid mutagenesis in the RNH1–bovine pancreatic RNaseA and the RNH1–RNase5 complexes led to the observation that while the location of the hotspots was almost the same in both complexes, the interactions in the interfaces were very different from each other (Russo et al., 1997). Consequently, single and multisite mutagenesis of a region of complementarity between RNH1 and RNase5 observed outside the hotspot led to the identification of a tryptophan cluster that is functionally linked to the hotspot, providing additional binding energy. This region was different in RNaseA and had less complementarity to RNH1 (Shapiro et al., 2000). Taken together, these studies showed that the hotspot of RNH1 located in the C-terminus of the protein interacts with the enzymatically active site of RNaseA and RNase5. Specifically, asparagine 435 of RNH1 forms key hydrogen bonds with the lysine responsible for the intermediate stabilization during catalysis (Lys41 in RNaseA) (Russo et al., 1997; Shapiro et al., 2000).

9.10.4 RNase7

The role of the RNH1–RNase7 interaction was described in the maintenance of urinary tract sterility. qPCR analyses of samples collected from patients with or without kidney infections revealed decreased *RNH1* and increased *RNASE7* levels in the kidneys of patients with an ongoing infection. Additionally, both full-length and degraded RNH1 were found in the urine of patients infected with *E. coli*, but were not found in the urine of uninfected patients; RNH1 was found in the insoluble fraction of the urine, indicating it came from cells in the urinary sediment. In vitro studies revealed proteolysis of RNH1 occurred in the presence of neutrophil but not bacterial proteases; RNase7 proteolysis was not observed in the urine samples or the in vitro studies. Other analyses showed RNH1-binding to endogenous RNase7, with reduced binding occurring in infected kidney samples. Moreover, a recombinant RNase7 was shown to bind RNH1 present in infected urine samples but not in sterile ones. This was proposed to occur because oxidative stress may impact the interaction between the inhibitor and the RNase; previous in vitro studies demonstrated the inactivation of RNH1 and loss of binding to RNase in oxidative stress conditions (Blázquez et al., 1996; Fominaya & Hofsteenge, 1992). Lastly, RNH1 was shown to stop the antimicrobial activity of RNase7 by preventing it from binding to the bacterial cell wall (Spencer et al., 2014).

Another publication reporting the importance of the interaction between RNase7 and RNH1 was described in the context of atopic dermatitis and *Staphylococcus aureus* growth. In vitro studies showed that RNase7 inhibited the proinflammatory response mediated by poly I:C in keratinocytes. Moreover, the knockdown of *RNASE7* led to an increase in the production of proinflammatory mediators. Further studies indicated that knockdown of *Rnh1* led to a decrease in the production of proinflammatory mediators upon stimulation with poly I:C. Finally, RNH1 was found to be upregulated and released from biopsy samples of atopic dermatitis, and to inhibit the antimicrobial activity of RNase7 against *S. aureus*. Altogether, the authors conclude that RNase7 reduces the proinflammatory response in atopic dermatitis lesions, but increased presence of RNH1 results in inhibition of the antimicrobial properties of RNase7 and promotes inflammation (Rademacher et al., 2024).

Cell protection from invading ribonucleases, regulation of RNase activity, and monitoring cell oxidation are other roles proposed for the RNH1 (Dickson et al., 2005). However, its physiological function is not yet fully known.

There are very few reports about *RNH1* deficiencies in humans and their clinical implications. A case study characterizing two siblings with a genetic variant of *RNH1* was published in 2023. The older affected sibling passed away at 8 months of age, while the younger sister was alive but with severe complications. Both children presented congenital cataracts, global developmental delay, seizures, and macrocytic anemia, among other symptoms. These children were born to healthy first-cousin parents who had a total of seven kids and lost a gestational baby at 20 weeks. Whole genome sequencing revealed that the affected children carried the same homozygous splice-site variant in the *RNH1* gene. The parents were found to be healthy heterozygous carriers for the variant, and the healthy siblings were either heterozygous variant carriers or homozygous for the WT allele. Analyses from samples obtained from the living sibling showed *RNH1* exon 7, which encodes for part of the LRR domain, to be missing. Furthermore, neither full-length nor smaller RNH1 was detected through Western blot, thus suggesting that aberrant RNH1 is degraded. Functional analyses of fibroblasts isolated from the living sibling showed susceptibility to extracellular RNase. An increased apoptotic rate was seen in fibroblasts challenged with RNaseA versus the control groups. Foremost, this phenotype was reversed upon RNH1 complementation (Hedberg-Oldfors et al., 2023).

A second study of eight children from four families diagnosed with either acute necrotizing encephalopathy or acute demyelinating encephalomyelitis associated with

RNH1 variants was published later in 2023. Genome sequencing from all affected families revealed different mutations occurring within *RNH1*. The first family presented two heterozygous missense variants in trans with very low levels of protein detected in fibroblasts. In the second family, a homozygous nonsynonymous variant found in the affected child was found to be heterozygous in the parents and in the unaffected siblings. Like with the first family, this variant also results in barely detectable RNH1 in fibroblasts. The third family presented a truncated variant and a nonsynonymous variant that was also found in the affected child from family four and resulted in loss of function. Interestingly, all these mutations but one were located within the LRR region of *RNH1*. The non LRR mutation corresponded with the truncated variant found in family 3, and was located prior to the LRR domain (Shashi et al., 2023).

Different roles of RNH1 have been described using the mouse model. Chennupati et al. (2018) reported the role of RNH1 in erythropoiesis. The generation of *RNH1*-knockout mice resulted in embryonic death. Examination of the embryos revealed decreased numbers of erythroid cells with decreased levels of hemoglobin in the yolk sac and placenta. Further analyses showed these embryos to have normal numbers of hematopoietic stem/progenitor cells (HSPCs), but significantly low numbers of embryonic-primitive erythroid (EryP) cells, which support early embryonic development. Since HSPCs are progenitors to EryP, it was suggested that *RNH1* is required for the differentiation of these cells. Furthermore, *RNH1* heterozygous mice survived but showed decreased erythroblast maturation, further indicating the importance of RNH1 in erythropoiesis. Hematopoietic transcription factor target genes, but not transcription factors, were found to be downregulated through transcriptome and ChIP-sequencing analyses. Yolk sac polysomes were isolated, and associated mRNAs were profiled as a measure of translation, revealing *RNH1*-knockout mice had an overall decreased rate of translation that was further decreased for *Gata1*, an important transcription factor in hematopoiesis. In vitro studies using the human erythroleukemia K562 cell line corroborated results obtained from the knockout mice; *RNH1*-knockout K562 exhibited decreased levels of GATA1, while mRNA levels were similar to those of the control. Additionally, low levels of hemoglobin were detected in hemin-induced cells, indicating a deficiency in erythroid differentiation in the *RNH1*-knockout cell line. Furthermore, K562 overexpressing RNH1 showed increased transcription of *Gata1* and hemoglobin levels upon hemin induction. Finally, RNH1 was found to interact with the ribosome 40S subunit in K562 lysates, which was confirmed through immunoprecipitation of RNH1 and identification of ligands through mass spectrometry (Chennupati et al., 2018).

The role of RNH1 in inflammasome activation was reported by Bombaci and colleagues in 2022. Inflammasomes are sensors of the innate immune response that induce inflammation. Investigating RNH1 in the NOD-like receptor protein 3

(NLRP3) inflammasome activation, a sensor for microbial and endogenous danger signals, an *RNH1*-knockout THP-1 cell line was treated with NLRP3 agonists. Treated cells showed an increased production of IL-1β, caspase-1 cleavage, and cell death compared to WT. IL-1β production was reduced with RNH1 restoration in the knockout cell line and in overexpressing RNH1-THP-1 cells stimulated with NLRP3 agonists, implicating RNH1 in the reduction of inflammasome activation. Further experiments demonstrated RNH1 to be involved in the first step of the NLRP3 inflammasome, activating the NF-κB. Phosphorylation of the NF-κB inhibitor, IκBα, leading to activation of NF-κB, was increased in the *RNH1*-knockout cells compared to the WT. Activation of NF-κB was then confirmed by increased production of proinflammatory cytokines. Looking into other inflammasomes and using the *RNH1*-knockout THP-1 cell line, it was revealed that RNH1 also reduced AIM2, activated by double-stranded DNA, and NAIP/NLRC4, activated by bacterial proteins. Based on these observations of RNH1 lowering different inflammasome activation, the authors investigated the common inflammasome downstream effector caspase-1. qPCR analysis revealed no differences in *caspase-1* expression in the *RNH1*-knockout cell line versus the WT, and further experiments implicated RNH1 in increasing proteasomal degradation of caspase-1 (Bombaci et al., 2022).

9.11 Conclusions

Here, we have summarized evidence available on the different roles of RNases outside their ribonucleolytic activity. RNase1 has been shown to enhance both antitumor immunity and tumor growth in competent and immunocompromised models, respectively. RNases4 and 5 have been shown to stimulate cellular proliferation and promote different types of cancer. RNases3 and 6 have been linked to inflammatory diseases and have been shown to contribute to the inflammatory response, respectively. RNase2 was shown to act as a TLR2 ligand and an alarmin of the immune system. Last, RNase7 was shown to mediate an antiinflammatory response during urinary tract infection. Recent research revealed RNases to act as ligands for different RTKs and to initiate signaling cascades, as seen with RNases5 and 7 binding to EGFR, and RNase1 to EphA4 and ALK.

This chapter also highlighted the importance of RNH1 in regulating RNase activity and RNase-mediated inflammation, together with the devastating clinical implications of RNH1 deficiencies.

RNase research is an evolving field, and it is worth noting that throughout these reports, RNases are being proposed as disease biomarkers, such as RNase6 in diabetic nephropathy and RNase2 in sepsis, or as therapeutic targets, such as RNase4 in prostate cancer.

References

Abdueva, D., Wing, M., Schaub, B., Triche, T., Davicioni, E., 2010. Quantitative expression profiling in formalin-fixed paraffin-embedded samples by affymetrix microarrays. The Journal of Molecular Diagnostics 12 (4), 409–417. Available at https://doi.org/10.2353/jmoldx.2010.090155.

Barry, S., Chelala, C., Lines, K., Sunamura, M., Wang, A., Marelli-Berg, F.M., et al., 2013. S100P is a metastasis-associated gene that facilitates transendothelial migration of pancreatic cancer cells. Clinical & Experimental Metastasis 30 (3), 251–264. Available at https://doi.org/10.1007/s10585-012-9532-y.

Best, M.G., Sol, N., Kooi, I., Tannous, J., Westerman, B.A., Rustenburg, F., et al., 2015. RNA-seq of tumor-educated platelets enables blood-based pan-cancer, multiclass, and molecular pathway cancer diagnostics. Cancer Cell 28 (5), 666–676. Available at https://doi.org/10.1016/j.ccell.2015.09.018.

BioStudies A-AFFY-1 2022 2022. Affymetrix GeneChip Human Genome U95Av2 [HG_U95Av2]. Available from:https://www.ebi.ac.uk/biostudies/arrayexpress/arrays/A-AFFY-1.

BioStudies A-AFFY-37 2022 2022. Affymetrix GeneChip Human Genome U133A 2.0 [HG-U133A_2]. Available from: https://www.ebi.ac.uk/biostudies/arrayexpress/arrays/A-AFFY-37.

BioStudies A-AFFY-44 2022 2022. Affymetrix GeneChip Human Genome U133 Plus 2.0 [HG-U133_Plus_2]. Available from: https://www.ebi.ac.uk/biostudies/arrayexpress/arrays/A-AFFY-44.

Blázquez, M., Fominaya, J.M., Hofsteenge, J., 1996. Oxidation of sulfhydryl groups of ribonuclease inhibitor in epithelial cells is sufficient for its intracellular degradation. Journal of Biological Chemistry 271 (31), 18638–18642. Available at https://doi.org/10.1074/jbc.271.31.18638.

Bombaci, G., Sarangdhar, M.A., Andina, N., Tardivel, A., Yu, E.C.-W., Mackie, G.M., et al., 2022. LRR-protein RNH1 dampens the inflammasome activation and is associated with COVID-19 severity. Life Science Alliance 5 (6), e202101226. Available at https://doi.org/10.26508/lsa.202101226.

Chan, T.H.M., Lin, C.H., Qi, L., Fei, J., Li, Y., Yong, K.J., et al., 2014. A disrupted RNA editing balance mediated by ADARs (Adenosine DeAminases that act on RNA) in human hepatocellular carcinoma. Gut 63 (5), 832–843. Available at https://doi.org/10.1136/gutjnl-2012-304037.

Chennupati, V., Veiga, D.F.T., Maslowski, K.M., Andina, N., Tardivel, A., Yu, E.C.-W., et al., 2018. Ribonuclease inhibitor 1 regulates erythropoiesis by controlling GATA1 translation. Journal of Clinical Investigation 128 (4), 1597–1614. Available at https://doi.org/10.1172/JCI94956.

Chuang, H.-C., Chen, M.-H., Chen, Y.-M., Ciou, Y.-R., Hsueh, C.-H., Tsai, C.-Y., et al., 2022. Induction of Interferon-γ and tissue inflammation by overexpression of eosinophil cationic protein in T cells and exosomes. Arthritis & Rheumatology 74 (1), 92–104. Available at https://doi.org/10.1002/art.41920.

Cvijanovich, N., Shanley, T.P., Lin, R., Allen, G.L., Thomas, N.J., Checchia, P., et al., 2008. Validating the genomic signature of pediatric septic shock. Physiological Genomics 34 (1), 127–134. Available at https://doi.org/10.1152/physiolgenomics.00025.2008.

Diaz, G., Zamboni, F., Tice, A., Farci, P., 2015. Integrated ordination of miRNA and mRNA expression profiles. BMC Genomics 16 (1), 767. Available at https://doi.org/10.1186/s12864-015-1971-9.

Dickson, K.A., Haigis, M.C., Raines, R.T., 2005. Ribonuclease inhibitor: structure and function, pp. 349–374. https://doi.org/10.1016/S0079-6603(05)80009-1.

Ding, X., Yang, Y., Han, B., Du, C., Xu, N., Huang, H., et al., 2014. Transcriptomic characterization of hepatocellular carcinoma with CTNNB1 mutation. PLoS One 9 (5), e95307. Available at https://doi.org/10.1371/journal.pone.0095307.

Eichler, T.E., Becknell, B., Easterling, R.S., Ingraham, S.E., Cohen, D.M., Schwaderer, A.L., et al., 2016. Insulin and the phosphatidylinositol 3-kinase signaling pathway regulate Ribonuclease 7 expression in the human urinary tract. Kidney International 90 (3), 568–579. Available at https://doi.org/10.1016/j.kint.2016.04.025.

Eswaran, J., Cyanam, D., Mudvari, P., Reddy, S.D.N., Pakala, S.B., Nair, S.S., et al., 2012. Transcriptomic landscape of breast cancers through mRNA sequencing. Scientific Reports 2 (1), 264. Available at https://doi.org/10.1038/srep00264.

Fominaya, J.M., Hofsteenge, J., 1992. Inactivation of ribonuclease inhibitor by thiol-disulfide exchange. The Journal of Biological Chemistry 267 (34), 24655–24660.

Furia, A., Moscato, M., Calì, G., Pizzo, E., Confalone, E., Amoroso, M.R., et al., 2011. Giuseppe D·Alessio, the ribonuclease/angiogenin inhibitor is also present in mitochondria and nuclei. FEBS Letters 585 (4), 613–617. Available at https://doi.org/10.1016/j.febslet.2011.01.034.

George, N., Fexova, S., Fuentes, A.M., Madrigal, P., Bi, Y., Iqbal, H., et al., 2024. Expression Atlas update: insights from sequencing data at both bulk and single cell level. Nucleic Acids Research 52 (D1), D107. Available at https://doi.org/10.1093/nar/gkad1021.

Gittler, J.K., Shemer, A., Suárez-Fariñas, M., Fuentes-Duculan, J., Gulewicz, K.J., Wang, C.Q.F., et al., 2012. Progressive activation of TH2/TH22 cytokines and selective epidermal proteins characterizes acute and chronic atopic dermatitis. Journal of Allergy and Clinical Immunology 130 (6), 1344–1354. Available at https://doi.org/10.1016/j.jaci.2012.07.012.

Glimelius, I., Rubin, J., Fischer, M., Molin, D., Amini, R., Venge, P., et al., 2011. Effect of eosinophil cationic protein (ECP) on Hodgkin lymphoma cell lines. Experimental Hematology 39 (8), 850–858. Available at https://doi.org/10.1016/j.exphem.2011.05.006.

Haberman, Y., Tickle, T.L., Dexheimer, P.J., Kim, M.-O., Tang, D., Karns, R., et al., 2014. Pediatric Crohn disease patients exhibit specific ileal transcriptome and microbiome signature. Journal of Clinical Investigation 124 (8), 3617–3633. Available at https://doi.org/10.1172/JCI75436.

Hedberg-Oldfors, C., Mitra, S., Molinaro, A., Visuttijai, K., Fogelstrand, L., Oldfors, A., et al., 2023. Ribonuclease inhibitor 1 (RNH1) deficiency cause congenital cataracts and global developmental delay with infection-induced psychomotor regression and anemia. European Journal of Human Genetics 31 (8), 887–894. Available at https://doi.org/10.1038/s41431-023-01327-7.

Hiraoka, N., Yamazaki–Itoh, R., Ino, Y., Mizuguchi, Y., Yamada, T., Hirohashi, S., et al., 2011. CXCL17 and ICAM2 Are associated with a potential anti-tumor immune response in early intraepithelial stages of human pancreatic carcinogenesis. Gastroenterology 140 (1), 310. Available at https://doi.org/10.1053/j.gastro.2010.10.009.

Ho, C.-H., Liao, P.-W., Fan, C.-K., Liu, S.-P., Cheng, P.-C., 2022. RNase 7 inhibits uropathogenic *Escherichia coli*-induced inflammation in bladder cells under a high-glucose environment by regulating the JAK/STAT signaling pathway. International Journal of Molecular Sciences 23 (9), 5156. Available at https://doi.org/10.3390/ijms23095156.

Hoang, T.T., Raines, R.T., 2017. Molecular basis for the autonomous promotion of cell proliferation by angiogenin. Nucleic Acids Research 45 (2), 818–831. Available at https://doi.org/10.1093/nar/gkw1192.

Hoang, V.L.T., Tom, L.N., Quek, X.-C., Tan, J.-M., Payne, E.J., Lin, L.L., et al., 2017. RNA-seq reveals more consistent reference genes for gene expression studies in human non-melanoma skin cancers. PeerJ 5, e3631. Available at https://doi.org/10.7717/peerj.3631.

Hoang, T.T., Tanrikulu, I.C., Vatland, Q.A., Hoang, T.M., Raines, R.T., 2018. A human ribonuclease variant and ERK-pathway inhibitors exhibit highly synergistic toxicity for cancer cells. Molecular Cancer Therapeutics 17 (12), 2622–2632. Available at https://doi.org/10.1158/1535-7163.MCT-18-0724.

Jackson, H.R., Miglietta, L., Habgood-Coote, D., D'Souza, G., Shah, P., Nichols, S., et al., 2023. Diagnosis of multisystem inflammatory syndrome in children by a whole-blood transcriptional signature. Journal of the Pediatric Infectious Diseases Society 12 (6), 322–331. Available at https://doi.org/10.1093/jpids/piad035.

Jiang, L., Huang, J., Higgs, B.W., Hu, Z., Xiao, Z., Yao, X., et al., 2016. Genomic landscape survey identifies SRSF1 as a key oncodriver in small cell lung cancer. PLoS Genetics 12 (4), e1005895. Available at https://doi.org/10.1371/journal.pgen.1005895.

Jones, K.B., Salah, Z., Del Mare, S., Galasso, M., Gaudio, E., Nuovo, G.J., et al., 2012. miRNA signatures associate with pathogenesis and progression of osteosarcoma. Cancer Research 72 (7), 1865–1877. Available at https://doi.org/10.1158/0008-5472.CAN-11-2663.

Jones, K., Keddy, C., Jenkins, C., Nicholson, K., Shinde, U., Davare, M.A., 2024. Novel insight into mechanisms of ROS1 catalytic activation via loss of the extracellular domain. Scientific Reports 14 (1), 22191. Available at https://doi.org/10.1038/s41598-024-71687-7.

Kobe, B., Deisenhofer, J., 1993. Crystal structure of porcine ribonuclease inhibitor, a protein with leucine-rich repeats. Nature 366 (6457), 751–756. Available at https://doi.org/10.1038/366751a0.

Lee, H.-H., 2024. RNASE4 promotes malignant progression and chemoresistance in hypoxic glioblastoma via activation of AXL/AKT and NF-κB/cIAPs signaling pathways. American Journal of Cancer Research 14 (9), 4320–4336. Available at https://doi.org/10.62347/UDBJ5986.

Lee, F.S. and Vallee, B.L., 1993. Structure and action of mammalian ribonuclease (angiogenin) inhibitor, pp. 1–30. https://doi.org/10.1016/S0079-6603(08)60215-9.

Lee, H.-H., Wang, Y.-N., Yang, W.-H., Xia, W., Wei, Y., Chan, L.-C., et al., 2021. Human ribonuclease 1 serves as a secretory ligand of ephrin A4 receptor and induces breast tumor initiation. Nature Communications 12 (1), 2788. Available at https://doi.org/10.1038/s41467-021-23075-2.

Lemmon, M.A., Schlessinger, J., 2010. Cell signaling by receptor tyrosine kinases. Cell 141 (7), 1117–1134. Available at https://doi.org/10.1016/j.cell.2010.06.011.

Li, T., Wang, Y., Zhu, C., Yang, Y., Long, C., Chen, Q., 2022. Identification of ribonuclease 6 as an immunoinflammatory key gene associated with the glomerular injury in diabetic nephropathy. Scientific Reports 12 (1), 19709. Available at https://doi.org/10.1038/s41598-022-24289-0.

Linsley, P.S., Speake, C., Whalen, E., Chaussabel, D., 2014. Copy number loss of the interferon gene cluster in melanomas is linked to reduced T cell infiltrate and poor patient prognosis. PLoS One 9 (10), e109760. Available at https://doi.org/10.1371/journal.pone.0109760.

Liu, C., Zha, Z., Zhou, C., Chen, Y., Xia, W., Wang, Y.-N., et al., 2021. Ribonuclease 7-driven activation of ROS1 is a potential therapeutic target in hepatocellular carcinoma. Journal of Hepatology 74 (4), 907–918. Available at https://doi.org/10.1016/j.jhep.2020.09.030.

Liu, C., Zhou, C., Xia, W., Zhou, Y., Qiu, Y., Weng, J., et al., 2024. Targeting ALK averts ribonuclease 1-induced immunosuppression and enhances antitumor immunity in hepatocellular carcinoma. Nature Communications 15 (1), 1009. Available at https://doi.org/10.1038/s41467-024-45215-0.

Lu, L., Wei, R.L., Prats-Ejarque, G., Goetz, M., Wang, G., Torrent, M., et al., 2021. Human RNase3 immune modulation by catalytic-dependent and independent modes in a macrophage-cell line infection model. Cellular and Molecular Life Sciences 78 (6), 2963–2985. Available at https://doi.org/10.1007/s00018-020-03695-5.

Maag, J.L.V., Fisher, O.M., Levert-Mignon, A., Kaczorowski, D.C., Thomas, M.L., Hussey, D.J., et al., 2017. Novel aberrations uncovered in barrett's esophagus and esophageal adenocarcinoma using whole transcriptome sequencing. Molecular Cancer Research 15 (11), 1558–1569. Available at https://doi.org/10.1158/1541-7786.MCR-17-0332.

Madan, V., Kanojia, D., Li, J., Okamoto, R., Sato-Otsubo, A., Kohlmann, A., et al., 2015. Aberrant splicing of U12-type introns is the hallmark of ZRSR2 mutant myelodysplastic syndrome. Nature Communications 6 (1), 6042. Available at https://doi.org/10.1038/ncomms7042.

Maruyama, I., 2014. Mechanisms of activation of receptor tyrosine kinases: monomers or dimers. Cells 3 (2), 304–330. Available at https://doi.org/10.3390/cells3020304.

Murtha, M.J., Eichler, T., Bender, K., Metheny, J., Li, B., Schwaderer, A.L., et al., 2018. Insulin receptor signaling regulates renal collecting duct and intercalated cell antibacterial defenses. Journal of Clinical Investigation 128 (12), 5634–5646. Available at https://doi.org/10.1172/JCI98595.

Neu, C., Beckers, C., Frank, N., Thomas, K., Bartneck, M., Simon, T.-P., et al., 2024. Ribonuclease inhibitor 1 emerges as a potential biomarker and modulates inflammation and iron homeostasis in sepsis. Scientific Reports 14 (1), 14972. Available at https://doi.org/10.1038/s41598-024-65778-8.

Olsson, I., Venge, P., 1972. Cationic proteins of human granulocytes: I. isolation of the cationic proteins from the granules of leukaemic myeloid cells. Scandinavian Journal of Haematology 9 (1-6), 204–214. Available at https://doi.org/10.1111/j.1600-0609.1972.tb00932.x.

Olsson, I., Venge, P., 1974. Cationic proteins of human granulocytes. II. separation of the cationic proteins of the granules of leukemic myeloid cells. Blood 44 (2), 235–246. Available at https://doi.org/10.1182/blood.V44.2.235.235.

Plummer, R., Hu, G.-fu, Liu, T., Yoo, J., 2020. Angiogenin regulates PKD activation and COX-2 expression induced by TNF-α and bradykinin in the colonic myofibroblast. Biochemical and Biophysical Research Communications 525 (4), 870–876. Available at https://doi.org/10.1016/j.bbrc.2020.02.169.

Rademacher, F., Scheel, A., Gläser, R., Schröder, L., Heinemann, N., Bartels, J., et al., 2024. Inhibition of RNase 7 by RNase inhibitor promotes inflammation and Staphylococcus aureus growth: implications for atopic dermatitis. Allergy 79 (6), 1573–1583. Available at https://doi.org/10.1111/all.16125.

Russo, N., Shapiro, R., Vallee, B.L., 1997. 5′-Diphosphoadenosine 3′-phosphate is a potent inhibitor of bovine pancreatic ribonuclease A. Biochemical and Biophysical Research Communications 231 (3), 671–674. Available at https://doi.org/10.1006/bbrc.1997.6167.

Santegoets, L.A.M., van Seters, M., Helmerhorst, T.J.M., Heijmans-Antonissen, C., Hanifi-Moghaddam, P., Ewing, P.C., et al., 2007. HPV related VIN: highly proliferative and diminished responsiveness to extracellular signals. International Journal of Cancer 121 (4), 759–766. Available at https://doi.org/10.1002/ijc.22769.

Scalise, M., Torella, M., Marino, F., Ravo, M., Giurato, G., Vicinanza, C., et al., 2020. Atrial myxomas arise from multipotent cardiac stem cells. European Heart Journal 41 (45), 4332–4345. Available at https://doi.org/10.1093/eurheartj/ehaa156.

Shapiro, R., Ruiz-Gutierrez, M., Chen, C.-Z., 2000. Analysis of the interactions of human ribonuclease inhibitor with angiogenin and ribonuclease A by mutagenesis: importance of inhibitor residues inside versus outside the C-terminal "hot spot" 1 1Edited by J. Wells. Journal of Molecular Biology 302 (2), 497–519. Available at https://doi.org/10.1006/jmbi.2000.4075.

Shapiro, Robert, Ruiz-Gutierrez, Melisa, Chen, Chang-Zheng, 2000. Analysis of the interactions of human ribonuclease inhibitor with angiogenin and ribonuclease A by mutagenesis: importance of inhibitor residues inside versus outside the C-terminal "hot spot" 1 1Edited by J. Wells. Journal of Molecular Biology 302 (2), 497–519. Available at https://doi.org/10.1006/jmbi.2000.4075.

Shashi, V., Schoch, K., Ganetzky, R., Kranz, P.G., Sondheimer, N., Markert, M.L., et al., 2023. Biallelic variants in ribonuclease inhibitor (RNH1), an inflammasome modulator, are associated with a distinctive subtype of acute, necrotizing encephalopathy. Genetics in Medicine 25 (9), 100897. Available at https://doi.org/10.1016/j.gim.2023.100897.

Sia, D., Losic, B., Moeini, A., Cabellos, L., Hao, K., Revill, K., et al., 2015. Massive parallel sequencing uncovers actionable FGFR2–PPHLN1 fusion and ARAF mutations in intrahepatic cholangiocarcinoma. Nature Communications 6 (1), 6087. Available at https://doi.org/10.1038/ncomms7087.

Spencer, J.D., Schwaderer, A.L., Eichler, T., Wang, H., Kline, J., Justice, S.S., et al., 2014. An endogenous ribonuclease inhibitor regulates the antimicrobial activity of ribonuclease 7 in the human urinary tract. Kidney International 85 (5), 1179–1191. Available at https://doi.org/10.1038/ki.2013.395.

Teufel, D.P., Kao, R.Y.T., Acharya, K.R., Shapiro, R., 2003. Mutational analysis of the complex of human RNase inhibitor and human eosinophil-derived neurotoxin (RNase 2). Biochemistry 42 (6), 1451–1459. Available at https://doi.org/10.1021/bi026852o.

Vanli, N., Sheng, J., Li, S., Xu, Z., Hu, G.-F., 2022. Ribonuclease 4 is associated with aggressiveness and progression of prostate cancer. Communications Biology 5 (1), 625. Available at https://doi.org/10.1038/s42003-022-03597-1.

Varambally, S., Yu, J., Laxman, B., Rhodes, D.R., Mehra, R., Tomlins, S.A., et al., 2005. Integrative genomic and proteomic analysis of prostate cancer reveals signatures of metastatic progression. Cancer Cell 8 (5), 393–406. Available at https://doi.org/10.1016/j.ccr.2005.10.001.

Venge, P., Byström, J., Carlson, M., Håkansson, L., Karawacjzyk, M., Peterson, C., et al., 1999. Eosinophil cationic protein (ECP): molecular and biological properties and the use of ECP as a marker of eosinophil activation in disease. Clinical & Experimental Allergy 29 (9), 1172–1186. Available at https://doi.org/10.1046/j.1365-2222.1999.00542.x.

Vicente-Dueñas, C., Fontán, L., Gonzalez-Herrero, I., Romero-Camarero, I., Segura, V., Aznar, M.A., et al., 2012. Expression of MALT1 oncogene in hematopoietic stem/progenitor cells recapitulates the pathogenesis of human lymphoma in mice. Proceedings of the National Academy of Sciences 109 (26), 10534–10539. Available at https://doi.org/10.1073/pnas.1204127109.

Wang, Y.-N., Lee, H.-H., Chou, C.-K., Yang, W.-H., Wei, Y., Chen, C.-T., et al., 2018. Angiogenin/ribonuclease 5 Is an EGFR ligand and a serum biomarker for erlotinib sensitivity in pancreatic cancer. Cancer Cell 33 (4), 752. Available at https://doi.org/10.1016/j.ccell.2018.02.012.

Wang, Y.-N., Lee, H.-H., Jiang, Z., Chan, L.-C., Hortobagyi, G.N., Yu, D., et al., 2023. Ribonuclease 1 enhances antitumor immunity against breast cancer by boosting T cell activation. International Journal of Biological Sciences 19 (10), 2957–2973. Available at https://doi.org/10.7150/ijbs.84592.

Wu, M., Chen, Y., Xia, H., Wang, C., Tan, C.Y., Cai, X., et al., 2020. Transcriptional and proteomic insights into the host response in fatal COVID-19 cases. Proceedings of the National Academy of Sciences 117 (45), 28336–28343. Available at https://doi.org/10.1073/pnas.2018030117.

Yang, D., Rosenberg, H.F., Chen, Q., Dyer, K.D., Kurosaka, K., Oppenheim, J.J., 2003. Eosinophil-derived neurotoxin (EDN), an antimicrobial protein with chemotactic activities for dendritic cells. Blood 102 (9), 3396–3403. Available at https://doi.org/10.1182/blood-2003-01-0151.

Yang, D., Chen, Q., Su, S.B., Zhang, P., Kurosaka, K., Caspi, R.R., et al., 2008. Eosinophil-derived neurotoxin acts as an alarmin to activate the TLR2–MyD88 signal pathway in dendritic cells and enhances Th2 immune responses. The Journal of Experimental Medicine 205 (1), 79–90. Available at https://doi.org/10.1084/jem.20062027.

Yang, W.-H., Huang, B.-Y., Rao, H.-Y., Ye, P., Chen, B., Wang, H.-C., et al., 2025. Ribonuclease 1 induces T-cell dysfunction and impairs CD8+ T-cell cytotoxicity to benefit tumor growth through hijacking STAT1. Advanced Science 12 (13). Available at https://doi.org/10.1002/advs.202404961.

Yao, Y., Richman, L., Morehouse, C., de los Reyes, M., Higgs, B.W., Boutrin, A., et al., 2008. Type I interferon: potential therapeutic target for psoriasis? PLoS One 3 (7), e2737. Available at https://doi.org/10.1371/journal.pone.0002737.

Yi, C., Zhang, H., Yang, J., Chen, D., Jiang, S., 2024. Elucidating common pathogenic transcriptional networks in infective endocarditis and sepsis: integrated insights from biomarker discovery and single-cell RNA sequencing. Frontiers in Immunology 14. Available at https://doi.org/10.3389/fimmu.2023.1298041.

Zagai, U., Dadfar, E., Lundahl, J., Venge, P., Sköld, C.M., 2007. Eosinophil cationic protein stimulates TGF-β1 release by human lung fibroblasts in vitro. Inflammation 30 (5), 153–160. Available at https://doi.org/10.1007/s10753-007-9032-4.

Zechendorf, E., O'Riordan, C.E., Stiehler, L., Wischmeyer, N., Chiazza, F., Collotta, D., et al., 2020. Ribonuclease 1 attenuates septic cardiomyopathy and cardiac apoptosis in a murine model of polymicrobial sepsis. JCI Insight 5 (8). Available at https://doi.org/10.1172/jci.insight.131571.

CHAPTER 10

Ribonucleases in gut diseases

Jinghao Sheng
Zhejiang University School of Medicine, Hangzhou, Zhejiang, P.R. China

10.1 Introduction

Gut homeostasis represents a complex and dynamic equilibrium maintained through intricate interactions between the intestinal epithelium, immune system, and microbiota (Peterson and Artis, 2014; Zheng et al., 2020; Lee et al., 2022; Kayama et al., 2020). The intestinal epithelium serves as a critical barrier, comprising a single layer of specialized cells that balance selective permeability for nutrient absorption while maintaining defense against pathogens (Chelakkot et al., 2018). The maintenance of this epithelial barrier depends on intestinal stem cells located at the base of the crypts, which continuously generate all specialized epithelial cell types through tightly regulated proliferation and differentiation processes (Gehart and Clevers, 2019; Beumer and Clevers, 2021). Multiple mechanisms reinforce this barrier function, including tight junctions between epithelial cells, mucus layers produced by goblet cells, and antimicrobial peptides secreted by Paneth cells (Citi et al., 2024; Horowitz et al., 2023; Gustafsson and Johansson, 2022; Wallaeys et al., 2023). The gut immune system, encompassing both innate and adaptive components, maintains homeostasis by distinguishing between beneficial commensal microorganisms and potentially harmful pathogens, thus promoting tolerance while preserving defensive capabilities (Shao et al., 2023; Zhao et al., 2023). Additionally, the gut microbiota, comprising trillions of microorganisms, significantly contributes to host metabolism, immune development, and protection against pathogens through colonization resistance (Ma et al., 2024).

Disruption of this delicate balance can lead to various pathological conditions, such as inflammatory bowel diseases (IBD) and colorectal cancer (CRC) (Caruso et al., 2020; Janney et al., 2020). IBD, which includes ulcerative colitis (UC) and Crohn's disease, is a chronic and relapsing inflammatory disorder of the intestine (Colombel et al., 2019). Although the precise etiology and pathogenesis of IBD remain to be fully elucidated, genome-wide association studies have identified more than 200 IBD-associated susceptibility genes (Jostins et al., 2012; Liu et al., 2015). Many of these genes are known to be involved or implicated in barrier function, microbial defense, innate and adaptive immunity, autophagy, and cellular homeostasis, highlighting the complex interplay between these components in maintaining gut homeostasis. In addition to

RNases in Health and Diseases
DOI: https://doi.org/10.1016/B978-0-443-27312-4.00013-4

genetic factors, environmental influences play a crucial role in shaping the gut microbiome and maintaining intestinal homeostasis (Byndloss et al., 2018). The persistent inflammatory environment and sustained homeostatic disruption can ultimately contribute to CRC development through multiple mechanisms, including inflammation-induced DNA damage, altered stem cell dynamics, and dysbiosis-mediated carcinogenesis (Nadeem et al., 2020; Porter et al., 2021). These pathological changes create a permissive microenvironment that promotes tumor initiation, progression, and potential metastasis. Understanding the complex interplay between these various components of gut homeostasis has led to novel therapeutic approaches, including microbiota-based interventions, targeted immunotherapies, and strategies to enhance barrier function, offering new hope for treating various gastrointestinal and systemic diseases.

The ribonuclease A (RNaseA) superfamily is a group of vertebrate-specific enzymes with diverse functions (Sun et al., 2022). In humans, 13 members of the RNaseA superfamily have been identified, which are further divided into two subgroups: canonical RNases (1–8) and noncanonical RNases (9–13) (Lee et al., 2019). Interestingly, all of the RNase genes are clustered on chromosome 14q11.2, a region that has been identified as one of the susceptibility loci for IBD (IBD4) (Pierik et al., 2005; Duerr et al., 2000). This genomic location suggests a strong connection between these ribonucleases and the maintenance of gut homeostasis. These enzymes have been extensively studied and characterized, revealing their involvement in various biological processes, such as angiogenesis, neurotoxicity, immunomodulation, and host defense. Recent studies have shed light on the specific roles of some RNaseA superfamily members in maintaining gut homeostasis. Based on analysis of the expression patterns of RNaseA superfamily members in the gut, we have identified several enzymes that are highly expressed in this context (Sun et al., 2024). These include RNase1, 4, 5, and 6, which exhibit robust expression in the intestinal tissue. Additionally, eosinophil-derived neurotoxin (EDN/RNase2) and eosinophil cationic protein (ECP/RNase3) are highly expressed in eosinophils, immune cells that are known to play a role in gut homeostasis and disease (Rosenberg, 1998). Indeed, existing literature has reported the crucial roles of these RNases in maintaining gut homeostasis and their involvement in intestinal disorders.

Given the importance of these findings, this chapter will focus on elucidating the functional mechanisms of these specific RNaseA superfamily members in the context of gut homeostasis and disease. We will provide a comprehensive overview of the current understanding of their roles in maintaining a healthy gut environment and their potential involvement in the pathogenesis of intestinal disorders. Furthermore, we will discuss the implications of these findings and highlight the potential significance of targeting these enzymes for the development of novel therapeutic strategies.

10.2 RNase1

RNase1, a key digestive enzyme secreted by the pancreas, plays a crucial role in the digestion of food in the gut, particularly in the degradation of RNA (Garnett and Raines, 2022; Geng et al., 2021). Its expression levels vary greatly among vertebrates, with herbivorous mammals like ungulates, rodents, and marsupials exhibiting higher amounts than humans, dogs, cats, and lower vertebrates (Barnard, 1969). The elevated RNase1 levels in herbivorous mammals are due to their unique digestive systems and the high RNA load from their diets and microbial fermentation processes. In these species, RNase1 is crucial for maintaining digestive health and nutrient absorption by degrading bacterial RNA from gut flora. To process large amounts of RNA, these animals often have multiple RNase1 genes due to gene duplication (Barnard, 1969; McAllan, 1980). In contrast, species with lower RNase1 levels have a reduced need for the enzyme, accounting for only 0.5%–1% of pancreatic enzymes, reflecting the lower demand for RNA digestion in their digestive systems. In addition to its digestive role, RNase1 plays a vital role in regulating extracellular RNAs (exRNAs), a heterogeneous population of ribonucleic acids found in all biofluids (Sohail et al., 2022). High levels of exRNA can stimulate inflammatory pathways (Lu et al., 2018), while RNase1 acts as one of the main cleaners of exRNA, maintaining it at a safe level (Garnett and Raines, 2022; Lu et al., 2018; Nechooshtan et al., 2020). Moreover, RNase1 has been shown to modulate gut microbiota and metabolome following infections, such as *Aeromonas hydrophila* in blunt snout bream, by reducing inflammation and improving intestinal function (Geng et al., 2021; Chen et al., 2021). However, the specific role of RNase1 in maintaining the gut microbiome remains unclear.

In addition to its role in maintaining gut health, many studies have revealed that RNase1 is a risk gene for CRC, with significantly increased expression in tumor tissues (Gao et al., 2023; Wang, 2024). Moreover, Ji et al. analyzed survival data from the TCGA CRC cohort and showed that patients with high expression levels of RNase1 were strongly associated with a worse prognosis (Ji et al., 2024). As a result, the RNase1 level in serum may serve as a potential biomarker for CRC. Interestingly, RNase1 has been identified to have different roles in various types of cancer. In hepatocellular carcinoma (HCC), RNase1 has been found to mediate tumor resistance to immunotherapy and create an RNase1-mediated immunosuppressive tumor microenvironment, highlighting the potential of targeting RNase1 as a promising strategy for cancer immunotherapy in HCC (Liu et al., 2024). On the other hand, in breast cancer, RNase1 has been shown to have a tumor-suppressive role through adaptive immune response, providing a potential treatment strategy of combining RNase1 with cancer immunotherapies for immunocompetent patients (Wang et al., 2023). These findings suggest that the role of RNase1 in cancer may be context-dependent and vary across different types of malignancies. Despite the evidence implicating RNase1 as a risk gene in CRC, its precise role in the development and progression of this malignancy still requires further investigation.

10.3 RNase2 and RNase3

EDN (RNase2) and ECP (RNase3) are granule proteins released by activated eosinophils (Ackerman et al., 1983). Eosinophils are abundant in the lamina propria of the gut, where they play a crucial role in maintaining mucosal integrity and regulating immune responses (Salvo-Romero et al., 2022; Gurtner et al., 2021). This suggests that EDN and ECP are involved in maintaining gut homeostasis.

EDN primarily contributes to antiviral defense and immune modulation, while ECP possesses strong cytotoxic properties against pathogens. EDN has been shown to have in vitro antiviral activity, particularly against respiratory syncytial virus (Domachowske et al., 1998), marking it as a member of host antimicrobial proteins. Moreover, EDN acts as a potent chemoattractant for dendritic cells, enhancing adaptive immune responses (Yang et al., 2008, 2003, 2004). This interaction is crucial for managing the intestinal microbiota and responding to dietary antigens while preventing excessive inflammation. In contrast, ECP's potent cytotoxic activity against bacteria, parasites, and host cells contributes to its role in host defense and tissue damage in inflammatory conditions (Lee and Kim, 2022).

EDN and ECP have been implicated in various gastrointestinal disorders. In IBD, elevated levels of EDN and ECP have been observed in serum, feces, and colorectal mucus (Saitoh et al., 1999; Loktionov et al., 2017; Berstad et al., 1993). EDN has been proven to be a stable biomarker correlating with IBD disease activity (Saitoh et al., 1999), useful for evaluating response to treatment (Ling Lundström et al., 2023), and predicting flares in UC (Amcoff et al., 2019). Similarly, fecal ECP levels are upregulated in IBD patients, with acute phases exhibiting significantly higher levels than remission phases (Saitoh et al., 1999). These findings suggest that EDN and ECP could serve as noninvasive markers for diagnosing and monitoring disease activity in IBD. Furthermore, genetic polymorphisms in EDN (rs2073342) and ECP (rs2233860) have been associated with IBD susceptibility (Blom et al., 2012), highlighting their potential role in disease pathogenesis. In eosinophilic colitis, a rare condition characterized by extensive eosinophil accumulation in the colon, both EDN and ECP are highly expressed, and blood and urine EDN levels are significantly increased (Inamura et al., 2003). Additionally, EDN and ECP have been implicated in intestinal parasitic infections, such as hookworm and *Strongyloides stercoralis*, where they act as antimicrobial proteins to kill microbes and as alarmins to activate the immune system (Amoani et al., 2019; Clark et al., 2020; Rajamanickam et al., 2018).

Although research on EDN and ECP in the context of CRC has been limited, there are indications of potential roles for these proteins in tumor biology and patient prognosis. EDN has been associated with tumor-infiltrating eosinophils, which have been linked to improved survival outcomes in CRC patients (Reichman et al., 2019). A study highlighted that higher eosinophil infiltration in the tumor stroma correlates with better prognosis, suggesting that EDN may play a role in mediating this protective effect (Prizment et al., 2016). ECP has shown cytotoxic effects against various cancer cells (Krückel et al., 2019). While specific studies directly linking ECP levels to CRC outcomes are sparse, its role as a cytotoxic agent suggests it could influence tumor behavior. ECP's involvement in tissue remodeling and immune modulation indicates that it may also affect the tumor microenvironment, potentially impacting cancer progression. In summary, while both EDN and ECP are linked to eosinophilic activity in CRC (Varricchi et al., 2018), more research is needed to clarify their specific roles and potential as biomarkers for prognosis.

10.4 RNase4

RNase4 is a highly conserved protein among vertebrate species, with approximately 94% amino acid sequence identity observed across humans, bovines, mice, and pigs (Hofsteenge et al., 1998). This remarkable conservation suggests a fundamental biological role that is critical across different species. However, despite its potential importance, research on the functional mechanisms of RNase4 has been limited until recently. Recent studies have shed light on the antimicrobial properties of RNase4. It has been shown to exhibit potent antimicrobial activity against uropathogenic *Escherichia coli* (UPEC), a common cause of urinary tract infections (Bender et al., 2021; Schwartz et al., 2023). Our recent research has revealed that RNase4 is primarily expressed in intestinal epithelial cells, particularly in Paneth and goblet cells. Notably, RNase4 has demonstrated activity against certain gut bacteria, such as *Parasutterella*, which plays a role in maintaining gut microbiota homeostasis. This function is critical for intestinal health and may have implications in conditions like IBD (Sun et al., 2024).

In addition to its antimicrobial properties, RNase4 has been implicated in various aspects of cancer biology. It has been associated with angiogenesis, a process closely related to tumor growth (Li et al., 2013). Moreover, RNase4 has been linked to the aggressiveness and progression of prostate cancer and glioblastoma by influencing key tumorigenic processes such as cell proliferation, migration, invasion, and stemness properties (Lee, 2024; Vanli et al., 2022). The molecular mechanisms underlying these effects involve the activation of key signaling pathways such as receptor tyrosine kinase (AXL), protein kinase B (AKT), and nuclear factor kappa-light-chain-enhancer of activated B cells (NF-κB), as well as the modulation of epithelial-mesenchymal transition and the upregulation of inhibitors of apoptosis proteins. Our unpublished results indicate that RNase4 is highly expressed in CRC and may promote tumor growth by inhibiting CRC cell growth, migration, and metastasis. These findings suggest that RNase4 may act as a pro-tumorigenic factor in the context of CRC. The emerging evidence on the diverse functions of RNase4 highlights its potential importance in both gut health and CRC. However, there are still many aspects that require further research, particularly focusing on elucidating the molecular mechanisms underlying RNase4's diverse functions and exploring its potential as a biomarker or therapeutic target in these contexts.

10.5 RNase5

RNase5, also known as angiogenin (ANG), was first isolated from the conditioned media of a human colon adenocarcinoma cell line (HT-29) based on its angiogenic properties in 1985, suggesting a close relationship with the intestinal tract (Fett et al., 1985). Over the past four decades, our understanding of this protein in tumorigenesis and progression has expanded significantly. ANG functions through multiple mechanisms: (1) its canonical role in promoting angiogenesis through endothelial cell activation and basement membrane remodeling; (2) direct stimulation of cancer cell proliferation and survival; and (3) modulation of the tumor microenvironment through regulation of stromal cell functions (Garnett and Raines, 2022; Sheng and Xu, 2016). At the molecular level, ANG's functions are mediated through several distinct mechanisms. Extracellularly, ANG interacts with cell-surface proteins to activate multiple signaling cascades, including the ERK1/2 and AKT pathways (Liu et al., 2001; Peng et al., 2014; Ibaragi et al., 2009a; Miyake et al., 2015; Wang et al., 2018). Upon internalization, ANG accumulates in the nucleus where it stimulates ribosomal RNA (rRNA) transcription, thereby supporting the increased protein synthesis demands of rapidly proliferating cancer cells (Ibaragi et al., 2009a; Xu et al., 2002; Ibaragi et al., 2009b; Hu et al., 2009). Additionally, ANG generates tRNA-derived stress-induced small RNAs under stress conditions, which can modulate protein translation and promote stress response programs favoring tumor cell survival (Li et al., 2019; Tao et al., 2021; Saikia et al., 2014; Ivanov et al., 2011). These molecular mechanisms, initially identified across various cancer contexts, have been validated in CRC, where they contribute to tumor initiation, progression, and potentially therapeutic resistance.

Elevated ANG expression has been documented in various cancer types beyond CRC, including breast, prostate, and pancreatic cancers, correlating with poor prognosis (Yu et al., 2018, 2021). Furthermore, ANG has emerged as a promising therapeutic target, with several approaches under investigation, including neutralizing antibodies, small-molecule inhibitors, and aptamers targeting ANG's ribonucleolytic activity (Ibaragi et al., 2009b; Yu et al., 2017; Hirukawa et al., 2005; Olson et al., 1995; Nobile et al., 1998). Further research is needed to fully elucidate the safety, efficacy, and optimal therapeutic applications of ANG-targeted interventions before their translation into clinical practice.

Given its association with the intestinal tract and CRC, the question arises whether RNase5 also plays a role in gut homeostasis. In 2003 *Hooper* discovered that RNase5 can act as an antimicrobial peptide, regulating the gut microbiota by exhibiting antimicrobial activity against specific microbes, including *Enterococcus faecalis, Listeria monocytogenes*, *Streptococcus pneumoniae*, and *Candida albicans* (Hooper et al., 2003). Subsequent research has shown that ANG has varying degrees of bactericidal effects against viruses, bacteria, fungi, and even parasites, displaying broad-spectrum antimicrobial properties (Noschka et al., 2021; Forman et al., 2012; D'Elia et al., 2009; Cocchi et al., 2012). However, its expression and activity are subject to multilevel regulation by the host's physiological state and external environment. ANG employs diverse antimicrobial mechanisms, with its positively charged physicochemical properties disrupting microbial cell walls, while its ribonuclease activity is essential for its fungicidal effects. Our recent mouse model studies have revealed that ANG selectively disrupts the membranes of pathogenic bacteria, such as α-Proteobacteria, while promoting the growth of beneficial strains, such as Lachnospiraceae. This balance is crucial for preventing dysbiosis, which is linked to IBD (Sun et al., 2021).

Furthermore, ANG acts as a key factor in regulating host immune function. It has been discovered that macrophages in the intestinal lamina propria secrete ANG in response to gut microbiota stimulation. ANG directly acts on intestinal epithelial cells, regulating tRNA and rRNA metabolism through its membrane receptor plexin-B2 (PLXNB2) (Yu et al., 2017), thereby balancing cell apoptosis and proliferation levels and maintaining intestinal epithelial cell homeostasis (Bai et al., 2020; Yao et al., 2024).

ANG plays a multifaceted role in gut homeostasis and CRC. Its angiogenic properties, antimicrobial activity, and immunomodulatory effects highlight its importance in maintaining a balanced gut microbiome and regulating intestinal epithelial cell homeostasis. As research continues to unravel the complex functions of ANG, it may emerge as a potential therapeutic target for the prevention and treatment of gastrointestinal disorders and CRC.

10.6 RNase6

RNase6 is primarily known for its antimicrobial properties (Pulido et al., 2016). Studies indicate that RNase6 can reduce the burden of UPEC during experimental UTIs, thereby mitigating mucosal injury and inflammation (Ruiz-Rosado et al., 2023; Anguita et al., 2024; Cortado et al., 2024). Interestingly, RNase6 has also been detected in intestinal tissues, suggesting that it may contribute to the maintenance of gut homeostasis. Although there is currently limited direct evidence specifically linking RNase6 to gut health and disease, its well-established antimicrobial properties and role in immune modulation provide a strong rationale for its potential involvement in maintaining gut homeostasis and responding to gastrointestinal infections or inflammation. Further research is needed to clarify these connections and explore the therapeutic potential of RNase6 in gastrointestinal disorders.

10.7 Conclusion and future perspectives

Collectively, these results highlight the intricate complexity of RNase-mediated functions in intestinal homeostasis and disease (see Table 10.1). RNase1's primary role in RNA digestion and exRNA regulation has demonstrated significant correlations with CRC risk and clinical outcomes. The eosinophil-derived EDN and ECP exhibit dual functionality: orchestrating immune responses while serving as antimicrobial and alarmin-like molecules, with particular relevance as potential IBD biomarkers. RNase4 and RNase5 have emerged as crucial regulators of both microbial ecology and tumor biology. Although RNase6 remains relatively unexplored in the intestinal context, its established antimicrobial and immunomodulatory properties in the urinary tract suggest potentially parallel functions in gut homeostasis.

Despite substantial progress, critical knowledge gaps persist. The molecular determinants of substrate specificity and the complex interplay between ribonucleolytic activity and noncanonical functions remain incompletely understood. Furthermore, the integrated networks through which distinct RNases interact with each other and with various components of the intestinal ecosystem—including immune cells, epithelial stem cells, extracellular matrix, and microbiota—require

Table 10.1 Summary of Ribonuclease A superfamily members in gut homeostasis and disease.

RNase	Key gut functions	Gut disease associations
RNase1	RNA digestion; exRNA regulation; microbiota modulation	CRC risk gene; high expression correlates with poor CRC prognosis
RNase2	Antiviral defense; dendritic cell chemoattractant; alarmin function	IBD biomarker; polymorphism (rs2073342) linked to IBD susceptibility
RNase3	Cytotoxicity against pathogens; host defense	Elevated in IBD; polymorphism (rs2233860) linked to IBD susceptibility
RNase4	Antimicrobial activity; microbiota homeostasis	Highly expressed in CRC; potential pro-tumorigenic role
RNase5	Broad-spectrum antimicrobial activity; epithelial homeostasis	Promotes angiogenesis; elevated in CRC with poor prognosis
RNase6	Antimicrobial properties	Limited gut-specific evidence; potential role inferred from UTI studies

further investigation. Understanding these complex interactions at both cellular and molecular levels is essential for deciphering how their dysregulation contributes to pathologies such as IBD and CRC.

The translation of these findings into therapeutic applications faces several significant challenges. The development of RNase-based therapeutics encounters both technical and safety-related obstacles. Various therapeutic strategies, including small-molecule inhibitors, engineered RNases, gene-editing approaches, and microbiota-targeted interventions, require rigorous preclinical validation to ensure their specificity, stability, and safety profiles. Critical considerations include the potential immunogenicity of therapeutic RNase formulations and their long-term effects on the gut microbiome and the host immune system, necessitating careful evaluation through well-designed clinical trials. Addressing these challenges while leveraging the expanding understanding of RNase biology may ultimately yield novel therapeutic strategies for intestinal diseases.

References

Ackerman, S.J., Loegering, D.A., Venge, P., Olsson, I., Harley, J.B., Fauci, A.S., et al., 1983. Distinctive cationic proteins of the human eosinophil granule: major basic protein, eosinophil cationic protein, and eosinophil-derived neurotoxin. The Journal of Immunology 131 (6), 2977–2982. Available at https://doi.org/10.4049/jimmunol.131.6.2977.

Amcoff, K., Cao, Y., Zhulina, Y., Lampinen, M., Halfvarson, J., Carlson, M., 2019. Prognostic significance of faecal eosinophil granule proteins in inflammatory bowel disease. Scandinavian Journal of Gastroenterology 54 (10), 1237–1244. Available at https://doi.org/10.1080/00365521.2019.1670251.

Amoani, B., Adu, B., Frempong, M.T., Sarkodie-Addo, T., Nuvor, S.V., Wilson, M.D., et al., 2019. Levels of serum eosinophil cationic protein are associated with hookworm infection and intensity in endemic communities in Ghana. PLoS One 14 (9), e0222382. Available at https://doi.org/10.1371/journal.pone.0222382.

Anguita, R., Prats-Ejarque, G., Moussaoui, M., Becknell, B., Boix, E., 2024. A common polymorphism in RNASE6 impacts its antimicrobial activity toward uropathogenic *Escherichia coli*. International Journal of Molecular Sciences 25 (1), 604. Available at https://doi.org/10.3390/ijms25010604.

Bai, R., Sun, D., Chen, M., Shi, X., Luo, L., Yao, Z., et al., 2020. Myeloid cells protect intestinal epithelial barrier integrity through the angiogenin/plexin-B2 axis. The EMBO Journal 39 (13). Available at https://doi.org/10.15252/embj.2019103325.

Barnard, E.A., 1969. Biological function of pancreatic ribonuclease. Nature 221 (5178), 340–344. Available at https://doi.org/10.1038/221340a0.

Bender, K., Schwartz, L.L., Cohen, A., Vasquez, C.M., Murtha, M.J., Eichler, T., et al., 2021. Expression and function of human ribonuclease 4 in the kidney and urinary tract. American Journal of Physiology-Renal Physiology 320 (5), F972. Available at https://doi.org/10.1152/ajprenal.00592.2020.

Berstad, A., Borkje, B., Riedel, B., Elsayed, S., Berstad, A., 1993. Increased fecal eosinophil cationic protein in inflammatory bowel disease. Hepato-Gastroenterology 40 (3), 276–278.

Beumer, J., Clevers, H., 2021. Cell fate specification and differentiation in the adult mammalian intestine. Nature Reviews. Molecular Cell Biology 22 (1), 39–53. Available at https://doi.org/10.1038/s41580-020-0278-0.

Blom, K., Rubin, J., Halfvarson, J., Törkvist, L., Rönnblom, A., Sangfelt, P., et al., 2012. Eosinophil associated genes in the inflammatory bowel disease 4 region: correlation to inflammatory bowel disease revealed. World Journal of Gastroenterology 18 (44), 6409–6419. Available at https://doi.org/10.3748/wjg.v18.i44.6409, http://www.wjgnet.com/1007-9327/pdf/v18/i44/6409.pdf.

Byndloss, M.X., Pernitzsch, S.R., Bäumler, A.J., 2018. Healthy hosts rule within: ecological forces shaping the gut microbiota. Mucosal Immunology 11 (5), 1299–1305. Available at https://doi.org/10.1038/s41385-018-0010-y, http://www.nature.com/mi/lib/index.html.

Caruso, R., Lo, B.C., Núñez, G., 2020. Host–microbiota interactions in inflammatory bowel disease. Nature Reviews. Immunology 20 (7), 411–426. Available at https://doi.org/10.1038/s41577-019-0268-7, http://www.nature.com/nri/index.html.

Chelakkot, C., Ghim, J., Ryu, S.H., 2018. Mechanisms regulating intestinal barrier integrity and its pathological implications. Experimental & Molecular Medicine 50 (8), 1–9. Available at https://doi.org/10.1038/s12276-018-0126-x.

Chen, J., Huang, X., Geng, R., Zhu, D., Wang, W., Liu, H., 2021. Ribonuclease1 contributes to the antibacterial response and immune defense in blunt snout bream (*Megalobrama amblycephala*). International Journal of Biological Macromolecules 172, 309–320. Available at https://doi.org/10.1016/j.ijbiomac.2021.01.066.

Citi, S., Fromm, M., Furuse, M., González-Mariscal, L., Nusrat, A., Tsukita, S., et al., 2024. A short guide to the tight junction. Journal of Cell Science 137 (9). Available at https://doi.org/10.1242/jcs.261776.

Clark, E., Pritchard, H., Hemmige, V., Restrepo, A., Bautista, K., Damania, A., et al., 2020. Strongyloides stercoralis infection in solid organ transplant patients is associated with eosinophil activation and intestinal inflammation: a cross-sectional study. Clinical Infectious Diseases 71 (10), e580. Available at https://doi.org/10.1093/cid/ciaa233.

Cocchi, F., DeVico, A.L., Lu, W., Popovic, M., Latinovic, O., Sajadi, M.M., et al., 2012. Soluble factors from T cells inhibiting X4 strains of HIV are a mixture of β chemokines and RNases. Proceedings of the National Academy of Sciences 109 (14), 5411–5416. Available at https://doi.org/10.1073/pnas.1202240109.

Colombel, J.F., Shin, A., Gibson, P.R., 2019. AGA clinical practice update on functional gastrointestinal symptoms in patients with inflammatory bowel disease: expert review. Clinical Gastroenterology and Hepatology 17 (3), 380. Available at https://doi.org/10.1016/j.cgh.2018.08.001, http://www.elsevier.com/inca/publications/store/6/7/2/7/4/3/index.htt.

Cortado, H., Kercsmar, M., Li, B., Vasquez-Martinez, G., Gupta, S., Ching, C., et al., 2024. Murine ribonuclease 6 limits bacterial dissemination during experimental urinary tract infection. Journal of Innate Immunity 16 (1), 283–294. Available at https://doi.org/10.1159/000539177, http://content.karger.com/ProdukteDB/produkte.asp?Aktion=JournalHome&ProduktNr=234234&ContentOnly=false.

D'Elia, R., deSchoolmeester, M.L., Zeef, L.A.H., Wright, S.H., Pemberton, A.D., Else, K.J., 2009. Expulsion of *Trichuris muris* is associated with increased expression of angiogenin 4 in the gut and increased acidity of mucins within the goblet cell. BMC Genomics 10 (1), 492. Available at https://doi.org/10.1186/1471-2164-10-492.

Domachowske, J.B., Dyer, K.D., Bonville, C.A., Rosenberg, H.F., 1998. Recombinant human eosinophii-derived neurotoxin/RNase 2 functions as an effective antiviral agent against respiratory syncytial virus. Journal of Infectious Diseases 177 (6), 1458–1464. Available at https://doi.org/10.1086/515322, http://jid.oxfordjournals.org/content/current.

Duerr, R.H., Barmada, M.M., Zhang, L., Pfützer, R., Weeks, D.E., 2000. High-density genome scan in crohn disease shows confirmed linkage to chromosome 14q11-12. American Journal of Human Genetics 66 (6), 1857–1862. Available at https://doi.org/10.1086/302947, http://www.elsevier.com/wps/find/journaldescription.cws_home/713561/description#description.

Fett, J.W., Strydom, D.J., Lobb, R.R., Alderman, E.M., Bethune, J.L., Riordan, J.F., et al., 1985. Isolation and characterization of angiogenin, an angiogenic protein from human carcinoma cells. Biochemistry 24 (20), 5480–5486. Available at https://doi.org/10.1021/bi00341a030.

Forman, R.A., deSchoolmeester, M.L., Hurst, R.J.M., Wright, S.H., Pemberton, A.D., Else, K.J., et al., 2012. The goblet cell is the cellular source of the anti-microbial angiogenin 4 in the large intestine post *Trichuris muris* infection. PLoS One 7 (9), e42248. Available at https://doi.org/10.1371/journal.pone.0042248.

Gao, H., Ma, L., Zou, Q., Hu, B., Cai, K., Sun, Y., et al., 2023. Unraveling dynamic interactions between tumor-associated macrophages and consensus molecular subtypes in colorectal cancer: an integrative analysis of single-cell and bulk RNA transcriptome. Heliyon 9 (9), e19224. Available at https://doi.org/10.1016/j.heliyon.2023.e19224.

Garnett, E.R., Raines, R.T., 2022. Emerging biological functions of ribonuclease 1 and angiogenin. Critical Reviews in Biochemistry and Molecular Biology 57 (3), 244–260. Available at https://doi.org/10.1080/10409238.2021.2004577, http://www.tandfonline.com/loi/ibmg20.

Gehart, H., Clevers, H., 2019. Tales from the crypt: new insights into intestinal stem cells. Nature Reviews Gastroenterology & Hepatology 16 (1), 19–34. Available at https://doi.org/10.1038/s41575-018-0081-y.

Geng, R., Liu, H., Tan, K., Wang, Z., Wang, W., 2021. RNase1 can modulate gut microbiota and metabolome after *Aeromonas hydrophila* infection in blunt snout bream. Environmental Microbiology 23 (9), 5258–5272. Available at https://doi.org/10.1111/1462-2920.15564.

Gurtner, A., Gonzalez-Perez, I., Arnold, I.C., 2021. Intestinal eosinophils, homeostasis and response to bacterial intrusion. Seminars in Immunopathology 43 (3), 295–306. Available at https://doi.org/10.1007/s00281-021-00856-x, http://www.springer.com/east/home/generic/search/results?SGWID=5-40109-70-1042641-0.

Gustafsson, J.K., Johansson, M.E.V., 2022. The role of goblet cells and mucus in intestinal homeostasis. Nature Reviews Gastroenterology and Hepatology 19 (12), 785–803. Available at https://doi.org/10.1038/s41575-022-00675-x, https://www.nature.com/nrgastro/.

Hirukawa, S., Olson, K.A., Tsuji, T., Hu, G.F., 2005. Neamine inhibits xenografic human tumor growth and angiogenesis in athymic mice. Clinical Cancer Research 11 (24), 8745–8752. Available at https://doi.org/10.1158/1078-0432.CCR-05-1495.

Hofsteenge, J., Vicentini, A., Zelenko, O., 1998. Ribonuclease 4, an evolutionarily highly conserved member of the superfamily. Cellular and Molecular Life Sciences (CMLS) 54 (8), 804–810. Available at https://doi.org/10.1007/s000180050209.

Hooper, L.V., Stappenbeck, T.S., Hong, C.V., Gordon, J.I., 2003. Angiogenins: a new class of microbicidal proteins involved in innate immunity. Nature Immunology 4 (3), 269–273. Available at https://doi.org/10.1038/ni888.

Horowitz, A., Chanez-Paredes, S.D., Haest, X., Turner, J.R., 2023. Paracellular permeability and tight junction regulation in gut health and disease. Nature Reviews Gastroenterology and Hepatology 20 (7), 417–432. Available at https://doi.org/10.1038/s41575-023-00766-3, https://www.nature.com/nrgastro/.

Hu, M.G., Deshpande, A., Enos, M., Mao, D., Hinds, E.A., Hu, G.F., et al., 2009. A requirement for cyclin-dependent kinase 6 in thymocyte development and tumorigenesis. Cancer Research 69 (3), 810–818. Available at https://doi.org/10.1158/0008-5472.CAN-08-2473, http://cancerres.aacrjournals.org/cgi/reprint/69/3/810.

Ibaragi, S., Yoshioka, N., Kishikawa, H., Hu, J.K., Sadow, P.M., Li, M., et al., 2009a. Angiogenin-stimulated rRNA transcription is essential for initiation and survival of AKT-induced prostate intraepithelial neoplasia. Molecular Cancer Research 7 (3), 415–424. Available at https://doi.org/10.1158/1541-7786.mcr-08-0137.

Ibaragi, S., Yoshioka, N., Li, S., Hu, M.G., Hirukawa, S., Sadow, P.M., et al., 2009b. Neamine inhibits prostate cancer growth by suppressing angiogenin-mediated rRNA transcription. Clinical Cancer Research 15 (6), 1981–1988. Available at https://doi.org/10.1158/1078-0432.ccr-08-2593.

Inamura, H., Tomita, M., Okano, A., Kurosawa, M., 2003. Serial blood and urine levels of EDN and ECP in eosinophilic colitis. Allergy: European Journal of Allergy and Clinical Immunology 58 (9), 959–960. Available at https://doi.org/10.1034/j.1398-9995.2003.00135.x.

Ivanov, P., Emara, M.M., Villen, J., Gygi, S.P., Anderson, P., 2011. Angiogenin-induced tRNA fragments inhibit translation initiation. Molecular Cell 43 (4), 613–623. Available at https://doi.org/10.1016/j.molcel.2011.06.022.

Janney, A., Powrie, F., Mann, E.H., 2020. Host–microbiota maladaptation in colorectal cancer. Nature 585 (7826), 509–517. Available at https://doi.org/10.1038/s41586-020-2729-3, http://www.nature.com/nature/index.html.

Ji, T., Fu, H., Wang, L., Chen, J., Tian, S., Wang, G., et al., 2024. Single-cell RNA profiling reveals classification and characteristics of mononuclear phagocytes in colorectal cancer. PLoS Genetics 20 (2), e1011176. Available at https://doi.org/10.1371/journal.pgen.1011176.

Jostins, L., Ripke, S., Weersma, R.K., Duerr, R.H., McGovern, D.P., Hui, K.Y., et al., 2012. Host–microbe interactions have shaped the genetic architecture of inflammatory bowel disease. Nature 491 (7422), 119–124. Available at https://doi.org/10.1038/nature11582.

Kayama, H., Okumura, R., Takeda, K., 2020. Interaction between the microbiota, epithelia, and immune cells in the intestine. Annual Review of Immunology 38 (1), 23–48. Available at https://doi.org/10.1146/annurev-immunol-070119-115104.

Krückel, A., Moreira, A., Fröhlich, W., Schuler, G., Heinzerling, L., 2019. Eosinophil-cationic protein - a novel liquid prognostic biomarker in melanoma. BMC Cancer 19 (1). Available at https://doi.org/10.1186/s12885-019-5384-z, http://www.biomedcentral.com/bmccancer/.

Lee, J.H., Kim, T.B., 2022. Eosinophil granule proteins: what they can tell us about asthma. Thorax 77 (6), 532–533. Available at https://doi.org/10.1136/thoraxjnl-2021-218268, http://thorax.bmj.com/.

Lee, J.Y., Tsolis, R.M., Bäumler, A.J., 2022. The microbiome and gut homeostasis. Science (New York, N.Y.) 377 (6601). Available at https://doi.org/10.1126/science.abp9960, https://www.science.org/doi/10.1126/science.abp9960.

Lee, H.H., Wang, Y.N., Hung, M.C., 2019. Functional roles of the human ribonuclease a superfamily in RNA metabolism and membrane receptor biology. Molecular Aspects of Medicine 70, 106–116. Available at https://doi.org/10.1016/j.mam.2019.03.003, www.elsevier.com/locate/mam.

Lee, H.-H., 2024. RNASE4 promotes malignant progression and chemoresistance in hypoxic glioblastoma via activation of AXL/AKT and NF-κB/cIAPs signaling pathways. American Journal of Cancer Research 14 (9), 4320–4336. Available at https://doi.org/10.62347/udbj5986.

Ling Lundström, M., Peterson, C., Lampinen, M., Hedin, C.R.H., Keita, Å.V., Kruse, R., et al., 2023. Fecal biomarkers of neutrophil and eosinophil origin reflect the response to biological therapy and corticosteroids in patients with inflammatory bowel disease. Clinical and Translational Gastroenterology 14 (8), e00605. Available at https://doi.org/10.14309/ctg.0000000000000605.

Liu, S., Yu, D., Xu, Z.P., Riordan, J.F., Hu, G.F., 2001. Angiogenin activates Erk1/2 in human umbilical vein endothelial cells. Biochemical and Biophysical Research Communications 287 (1), 305–310. Available at https://doi.org/10.1006/bbrc.2001.5568, http://www.sciencedirect.com/science/journal/0006291X.

Liu, J.Z., Van Sommeren, S., Huang, H., Ng, S.C., Alberts, R., Takahashi, A., et al., 2015. Association analyses identify 38 susceptibility loci for inflammatory bowel disease and highlight shared genetic risk across populations. Nature Genetics 47 (9), 979–986. Available at https://doi.org/10.1038/ng.3359, http://www.nature.com/ng/index.html.

Liu, C., Zhou, C., Xia, W., Zhou, Y., Qiu, Y., Weng, J., et al., 2024. Targeting ALK averts ribonuclease 1-induced immunosuppression and enhances antitumor immunity in hepatocellular carcinoma. Nature Communications 15 (1). Available at https://doi.org/10.1038/s41467-024-45215-0.

Li, S., Sheng, J., Hu, J.K., Yu, W., Kishikawa, H., Hu, M.G., et al., 2013. Ribonuclease 4 protects neuron degeneration by promoting angiogenesis, neurogenesis, and neuronal survival under stress. Angiogenesis 16 (2), 387–404. Available at https://doi.org/10.1007/s10456-012-9322-9.

Li, S., Shi, X., Chen, M., Xu, N., Sun, D., Bai, R., et al., 2019. Angiogenin promotes colorectal cancer metastasis via tiRNA production. International Journal of Cancer 145 (5), 1395–1407. Available at https://doi.org/10.1002/ijc.32245.

Loktionov, A., Chhaya, V., Bandaletova, T., Poullis, A., 2017. Inflammatory bowel disease detection and monitoring by measuring biomarkers in non-invasively collected colorectal mucus. Journal of Gastroenterology and Hepatology 32 (5), 992–1002. Available at https://doi.org/10.1111/jgh.13627.

Lu, L., Li, J., Moussaoui, M., Boix, E., 2018. Immune modulation by human secreted RNases at the extracellular space. Frontiers in Immunology 9 (MAY). Available at https://doi.org/10.3389/fimmu.2018.01012, https://www.frontiersin.org/articles/10.3389/fimmu.2018.01012/full.

Ma, Z., Zuo, T., Frey, N., Rangrez, A.Y., 2024. A systematic framework for understanding the microbiome in human health and disease: from basic principles to clinical translation. Signal Transduction and Targeted Therapy 9 (1). Available at https://doi.org/10.1038/s41392-024-01946-6, https://www.nature.com/sigtrans.

McAllan, A.B., 1980. The degradation of nucleic acids in, and the removal of breakdown products from the small intestines of steers. British Journal of Nutrition 44 (1), 99–112. Available at https://doi.org/10.1079/bjn19800014.

Miyake, M., Goodison, S., Lawton, A., Gomes-Giacoia, E., Rosser, C.J., 2015. Angiogenin promotes tumoral growth and angiogenesis by regulating matrix metallopeptidase-2 expression via the ERK1/2 pathway. Oncogene 34 (7), 890–901. Available at https://doi.org/10.1038/onc.2014.2.

Nadeem, M.S., Kumar, V., Al-Abbasi, F.A., Kamal, M.A., Anwar, F., 2020. Risk of colorectal cancer in inflammatory bowel diseases. Seminars in Cancer Biology 64, 51–60. Available at https://doi.org/10.1016/j.semcancer.2019.05.001, http://www.elsevier.com/inca/publications/store/6/2/2/9/4/3/index.htt.

Nechooshtan, G., Yunusov, D., Chang, K., Gingeras, T.R., 2020. Processing by RNase 1 forms tRNA halves and distinct y RNA fragments in the extracellular environment. Nucleic Acids Research 48 (14), 8035–8049. Available at https://doi.org/10.1093/nar/gkaa526, https://academic.oup.com/nar/issue.

Nobile, V., Russo, N., Hu, G.F., Riordan, J.F., 1998. Inhibition of human angiogenin by DNA aptamers: nuclear colocalization of an angiogenin-inhibitor complex. Biochemistry 37 (19), 6857–6863. Available at https://doi.org/10.1021/bi973181z.

Noschka, R., Gerbl, F., Löffler, F., Kubis, J., Rodríguez, A.A., Mayer, D., Grieshober, M., et al., 2021. Unbiased identification of angiogenin as an endogenous antimicrobial protein with activity against virulent *Mycobacterium tuberculosis*. Frontiers in Microbiology 11. Available at https://doi.org/10.3389/fmicb.2020.618278.

Olson, K.A., Fett, J.W., French, T.C., Key, M.E., Vallee, B.L., 1995. Angiogenin antagonists prevent tumor growth in vivo. Proceedings of the National Academy of Sciences of the United States of America 92 (2), 442–446. Available at https://doi.org/10.1073/pnas.92.2.442, www.pnas.org.

Peng, Y., Li, L., Huang, M., Duan, C., Zhang, L., Chen, J., 2014. Angiogenin interacts with ribonuclease inhibitor regulating PI3K/AKT/mTOR signaling pathway in bladder cancer cells. Cellular Signalling 26 (12), 2782–2792. Available at https://doi.org/10.1016/j.cellsig.2014.08.021.

Peterson, L.W., Artis, D., 2014. Intestinal epithelial cells: regulators of barrier function and immune homeostasis. Nature Reviews. Immunology 14 (3), 141–153. Available at https://doi.org/10.1038/nri3608.

Pierik, M., Yang, H., Barmada, M.M., Cavanaugh, J.A., Annese, V., Brant, S.R., et al., 2005. The IBD international genetics consortium provides further evidence for linkage to IBD4 and shows gene-environment interaction. Inflammatory Bowel Diseases 11 (1), 1–7. Available at https://doi.org/10.1097/00054725-200501000-00001.

Porter, R.J., Arends, M.J., Churchhouse, A.M.D., Din, S., 2021. Inflammatory bowel disease-associated colorectal cancer:translational risks from mechanisms to medicines. Journal of Crohn's and Colitis 15 (12), 2131–2141. Available at https://doi.org/10.1093/ecco-jcc/jjab102, https://academic.oup.com/ecco-jcc/issue.

Prizment, A.E., Vierkant, R.A., Smyrk, T.C., Tillmans, L.S., Lee, J.J., Sriramarao, P., et al., 2016. Tumor eosinophil infiltration and improved survival of colorectal cancer patients: Iowa women's health study. Modern Pathology 29 (5), 516–527. Available at https://doi.org/10.1038/modpathol.2016.42, http://www.nature.com/modpathol/index.html.

Pulido, D., Arranz-Trullén, J., Prats-Ejarque, G., Velázquez, D., Torrent, M., Moussaoui, M., et al., 2016. Insights into the antimicrobial mechanism of action of human RNase6: structural determinants for bacterial cell agglutination and membrane permeation. International Journal of Molecular Sciences 17 (4), 552. Available at https://doi.org/10.3390/ijms17040552.

Rajamanickam, A., Munisankar, S., Bhootra, Y., Dolla, C.K., Nutman, T.B., Babu, S., 2018. Elevated systemic levels of eosinophil, neutrophil, and mast cell granular proteins in strongyloides stercoralis infection that diminish following treatment. Frontiers in Immunology 9. Available at https://doi.org/10.3389/fimmu.2018.00207.

Reichman, H., Itan, M., Rozenberg, P., Yarmolovski, T., Brazowski, E., Varol, C., et al., 2019. Activated eosinophils exert antitumorigenic activities in colorectal cancer. Cancer Immunology Research 7 (3), 388–400. Available at https://doi.org/10.1158/2326-6066.cir-18-0494.

Rosenberg, H.F., 1998. The eosinophil ribonucleases. Cellular and Molecular Life Sciences (CMLS) 54 (8), 795–803. Available at https://doi.org/10.1007/s000180050208.

Ruiz-Rosado, J.de D., Cortado, H., Kercsmar, M., Li, B., Ballash, G., Cotzomi-Ortega, I., et al., 2023. Human ribonuclease 6 has a protective role during experimental urinary tract infection. Journal of Innate Immunity 15 (1), 865–875. Available at https://doi.org/10.1159/000534736.

Saikia, M., Jobava, R., Parisien, M., Putnam, A., Krokowski, D., Gao, X.-H., et al., 2014. Angiogenin-cleaved tRNA halves interact with cytochrome c, protecting cells from apoptosis during osmotic stress. Molecular and Cellular Biology 34 (13), 2450–2463. Available at https://doi.org/10.1128/mcb.00136-14.

Saitoh, O., Kojima, K., Sugi, K., Matsuse, R., Uchida, K., Tabata, K., et al., 1999. Fecal eosinophil granule-derived proteins reflect disease activity in inflammatory bowel disease. American Journal of Gastroenterology 94 (12), 3513–3520. Available at https://doi.org/10.1111/j.1572-0241.1999.01640.x.

Salvo-Romero, E., Rodiño-Janeiro, B.K., Albert-Bayo, M., Lobo, B., Santos, J., Farré, R., et al., 2022. Eosinophils in the gastrointestinal tract: key contributors to neuro-immune crosstalk and potential implications in disorders of brain-gut interaction. Cells 11 (10), 1644. Available at https://doi.org/10.3390/cells11101644.

Schwartz, L., Bochter, M.S., Simoni, A., Bender, K., de Dios Ruiz Rosado, J., Cotzomi-Ortega, I., et al., 2023. Repurposing HDAC inhibitors to enhance ribonuclease 4 and 7 expression and reduce urinary tract infection. Proceedings of the National Academy of Sciences 120 (4). Available at https://doi.org/10.1073/pnas.2213363120.

Shao, T., Hsu, R., Rafizadeh, D.L., Wang, L., Bowlus, C.L., Kumar, N., et al., 2023. The gut ecosystem and immune tolerance. Journal of Autoimmunity 141, 103114. Available at https://doi.org/10.1016/j.jaut.2023.103114.

Sheng, J., Xu, Z., 2016. Three decades of research on angiogenin: a review and perspective. Acta Biochimica et Biophysica Sinica 48, 399–410. Available at https://doi.org/10.1093/abbs/gmv131.

Sohail, A.M., Khawar, M.B., Afzal, A., Hassan, A., Shahzaman, S., Ali, A., 2022. Multifaceted roles of extracellular RNAs in different diseases. Military Medical Research 9 (1). Available at https://doi.org/10.1186/s40779-022-00405-z, http://mmrjournal.biomedcentral.com/about.

Sun, D., Han, C., Sheng, J., 2022. The role of human ribonuclease A family in health and diseases: a systematic review. iScience 25 (11), 105284. Available at https://doi.org/10.1016/j.isci.2022.105284.

Sun, D., Bai, R., Zhou, W., Yao, Z., Liu, Y., Tang, S., et al., 2021. Angiogenin maintains gut microbe homeostasis by balancing α-Proteobacteria and Lachnospiraceae. Gut 70 (4), 666–676. Available at https://doi.org/10.1136/gutjnl-2019-320135.

Sun, J., Chen, M., Hu, Z., Xu, N., Wang, W., Ping, Z., et al., 2024. Ribonuclease 4 functions as an intestinal antimicrobial protein to maintain gut microbiota and metabolite homeostasis. Nature Communications 15 (1). Available at https://doi.org/10.1038/s41467-024-50223-1.

Tao, E.W., Wang, H.L., Cheng, W.Y., Liu, Q.Q., Chen, Y.X., Gao, Q.Y., 2021. A specific tRNA half, 5'tiRNA-His-GTG, responds to hypoxia via the HIF1α/ANG axis and promotes colorectal cancer progression by regulating LATS2. Journal of Experimental and Clinical Cancer Research 40 (1). Available at https://doi.org/10.1186/s13046-021-01836-7, http://www.jeccr.com/.

Vanli, N., Sheng, J., Li, S., Xu, Z., Hu, G.F., 2022. Ribonuclease 4 is associated with aggressiveness and progression of prostate cancer. Communications Biology 5 (1). Available at https://doi.org/10.1038/s42003-022-03597-1, https://www.springer.com/journal/11852.

Varricchi, G., Galdiero, M.R., Loffredo, S., Lucarini, V., Marone, G., Mattei, F., et al., 2018. Eosinophils: the unsung heroes in cancer? OncoImmunology 7 (2). Available at https://doi.org/10.1080/2162402x.2017.1393134.

Wallaeys, C., Garcia-Gonzalez, N., Libert, C., 2023. Paneth cells as the cornerstones of intestinal and organismal health: a primer. EMBO Molecular Medicine 15 (2). Available at https://doi.org/10.15252/emmm.202216427.

Wang, X., 2024. Multi-omics data-based analysis characterizes molecular alterations of the vesicle genes in human colorectal cancer. American Journal of Cancer Research 14 (3), 1402–1418. Available at https://doi.org/10.62347/qfkd1805.

Wang, Y.N., Lee, H.H., Chou, C.K., Yang, W.H., Wei, Y., Chen, C.T., et al., 2018. Angiogenin/ribonuclease 5 is an EGFR ligand and a serum biomarker for erlotinib sensitivity in pancreatic cancer.

Cancer Cell 33 (4), 752. Available at https://doi.org/10.1016/j.ccell.2018.02.012, https://www.journals.elsevier.com/cancer-cell.

Wang, Y.N., Lee, H.H., Jiang, Z., Chan, L.C., Hortobagyi, G.N., Yu, D., et al., 2023. Ribonuclease 1 enhances antitumor immunity against breast cancer by boosting T cell activation. International Journal of Biological Sciences 19 (10), 2957–2973. Available at https://doi.org/10.7150/ijbs.84592, https://www.ijbs.com/v19p2957.pdf.

Yang, D., Rosenberg, H.F., Chen, Q., Dyer, K.D., Kurosaka, K., Oppenheim, J.J., 2003. Eosinophil-derived neurotoxin (EDN), an antimicrobial protein with chemotactic activities for dendritic cells. Blood 102 (9), 3396–3403. Available at https://doi.org/10.1182/blood-2003-01-0151.

Yang, D., Chen, Q., Rosenberg, H.F., Rybak, S.M., Newton, D.L., Wang, Z.Y., et al., 2004. Human ribonuclease a superfamily members, eosinophil-derived neurotoxin and pancreatic ribonuclease, induce dendritic cell maturation and activation. The Journal of Immunology 173 (10), 6134–6142. Available at https://doi.org/10.4049/jimmunol.173.10.6134.

Yang, D., Chen, Q., Su, S.B., Zhang, P., Kurosaka, K., Caspi, R.R., et al., 2008. Eosinophil-derived neurotoxin acts as an alarmin to activate the TLR2–MyD88 signal pathway in dendritic cells and enhances Th2 immune responses. The Journal of Experimental Medicine 205 (1), 79–90. Available at https://doi.org/10.1084/jem.20062027.

Yao, Z., Bai, R., Liu, W., Liu, Y., Zhou, W., Xu, Z., et al., 2024. Activation of angiogenin expression in macrophages by lipopolysaccharide via the TLR4/NF-κB pathway in colitis. Acta Biochimica et Biophysica Sinica 56 (6), 857–865. Available at https://doi.org/10.3724/abbs.2024013, https://www.sciengine.com/abbs.

Yu, D., Cai, Y., Zhou, W., Sheng, J., Xu, Z., 2018. The potential of angiogenin as a serum biomarker for diseases: systematic review and meta-analysis. Disease Markers 2018, 1–15. Available at https://doi.org/10.1155/2018/1984718.

Yu, W., Goncalves, K.A., Li, S., Kishikawa, H., Sun, G., Yang, H., et al., 2017. Plexin-B2 mediates physiologic and pathologic functions of angiogenin. Cell 171 (4), 849. Available at https://doi.org/10.1016/j.cell.2017.10.005.

Yu, D., Sun, J., Weng, Y., Luo, L., Sheng, J., Xu, Z., 2021. Serum angiogenin as a potential biomarker for early detection of colorectal adenomas and colorectal cancer. Anti-Cancer Drugs 32 (7), 703–708. Available at https://doi.org/10.1097/cad.0000000000001047.

Xu, Z.P., Tsuji, T., Riordan, J.F., Hu, G.F., 2002. The nuclear function of angiogenin in endothelial cells is related to rRNA production. Biochemical and Biophysical Research Communications 294 (2), 287–292. Available at https://doi.org/10.1016/S0006-291X(02)00479-5.

Zhao, M., Chu, J., Feng, S., Guo, C., Xue, B., He, K., et al., 2023. Immunological mechanisms of inflammatory diseases caused by gut microbiota dysbiosis: a review. Biomedicine & Pharmacotherapy 164, 114985. Available at https://doi.org/10.1016/j.biopha.2023.114985.

Zheng, D., Liwinski, T., Elinav, E., 2020. Interaction between microbiota and immunity in health and disease. Cell Research 30 (6), 492–506. Available at https://doi.org/10.1038/s41422-020-0332-7.

CHAPTER 11

Human natural variants

Brian Becknell and Hanna Cortado
The Abigail Wexner Research Institute at Nationwide Children's Hospital, Columbus, OH, United States

11.1 Introduction

While the first human genome was sequenced over 20 years ago, the last two decades have illustrated the many ways in which genetic variation influences disease susceptibility in *Homo sapiens*. A growing number of monogenic disorders have been associated with causative mutations that disrupt protein function. Beyond these mutations, there is mounting evidence that single-nucleotide polymorphisms (SNPs)—distinguished from mutations based on their prevalence of 1% or higher in specific populations—contribute to human health and disease. Furthermore, structural genome variation has illustrated the extent of heterogeneity among humans in gene copy number, which carries significant implications for gene dosage and has been linked to pathological states.

In this chapter, we sought to uncover the role of genetic heterogeneity among human RNase-encoding genes. We organized our findings according to each gene, highlighting instances in which mutations in these genes appear to be responsible for human disease. We also paid attention to instances in which SNPs have been associated with functional differences in ribonucleases (RNases) and disease states. Along the way, we tried to highlight knowledge gaps and opportunities for future investigation. In an era of increasing access to genome-level information and personalized medicine, we anticipate that the functional and human-disease relevance of genetic variation in RNases will continue to expand.

RNASE1: Human pancreatic ribonuclease 1 (RNase1) is widely expressed, particularly by endothelial and epithelial cells, and exhibits high catalytic activity against single- and double-stranded RNA. When secreted, RNase1 serves roles in extracellular RNA degradation, which contributes to vascular homeostasis (Kramer et al., 2022; Fischer et al., 2011; Landré et al., 2002). Over a dozen SNPs have been identified in *RNASE1*, most of which result in amino acid substitutions (Ueki et al., 2008). However, the impact of these SNPs on RNase1 protein function and association with specific human disease states remains unknown.

RNASE2: Few studies have evaluated natural variants in *RNASE2*, which encodes RNase2, also known as eosinophil-derived neurotoxin (EDN) or eosinophil protein X. Zhang and Rosenberg investigated over 2.4 kilobases encompassing *RNASE2* and its close relative *RNASE3* to establish whether these genes were undergoing rapid

RNases in Health and Diseases
DOI: https://doi.org/10.1016/B978-0-443-27312-4.00009-2

diversification, as reported among paralogous eosinophil-associated ribonucleases in rodents. However, these investigators found similar nucleotide diversity at these loci to that reported for other human genes (0.06%–0.11%) in over 50 human subjects, thus rejecting the hypothesis that primates maintain numerous polymorphic *RNASE2* and *RNASE3* alleles (Zhang and Rosenberg, 2000). Jönsson and coworkers identified 405(G>C) rs2013109 as a common SNP in *RNASE2* that was associated with the cellular content of EDN protein in whole-blood extracts and in eosinophils from allergic and nonallergic subjects (Jönsson et al., 2014). These investigators noted that protein levels of RNase2 and RNase3 are reciprocally related and associated with the *RNASE2* 405(G>C) and *RNASE3* 562(G>C) polymorphisms. However, the relationship between the genotypes of these SNPs and allergic disorders remained unclear, as the genotypes of these SNPs did not differ between nonallergic and allergic subjects.

RNASE3: RNASE2 and *RNASE3* arose from a common genetic ancestry and are closely related in both structure and function (Zhang and Rosenberg, 2000). RNase3 is a highly basic protein originally purified from eosinophil granules and hence is more commonly known as eosinophil cationic protein (ECP) (Olsson et al., 1977). SNPs that impact ECP expression levels and function have been localized throughout the *RNASE3* gene, including the promoter, protein-coding, and 3' untranslated regions (3' UTR). Within the promoter, Noguchi and colleagues showed that, among children with asthma, there are lower serum levels of ECP in subjects with the -393CT and -393TT genotypes compared to -393CC. This SNP maps to a putative CEB/P transcription factor binding site, and reporter constructs made with -393T show lower promoter activity compared to -393C, substantiating the influence of this polymorphism on protein production (Noguchi et al., 2003).

One of the best studied polymorphisms in *RNASE3* is SNP rs2073342, located in exon 2, in which a nonsynonymous polymorphism alters the amino acid residue from arginine to threonine at position 124 in the ECP protein. The minor allele frequency for this and certain other SNPs varies geographically across the global population (Fig. 11.1). The variant alleles of the polymorphic gene at this site have been associated with different pathological conditions. In particular, the homozygous arginine variant was found to be more prevalent in allergic compared to nonallergic asthma patients in a small Swedish study (Jönsson et al., 2002). The same allele has also been associated with a lower prevalence of parasitic infections such as schistosomiasis and malaria (Adu et al., 2011; Eriksson et al., 2007). Despite having similar RNase activity, the arginine variant has higher cytotoxic potential compared to the threonine variant when native proteins were purified from the blood of healthy donors with the corresponding genotypes and tested against a lung carcinoma cell line. The lower cytotoxic activity of the threonine variant has been associated with worse clinical outcomes in tumor-associated tissue eosinophilia and other cancer pathologies in oral squamous cell carcinoma patients (Pereira et al., 2010). Mechanistically, the threonine variant has been associated with a

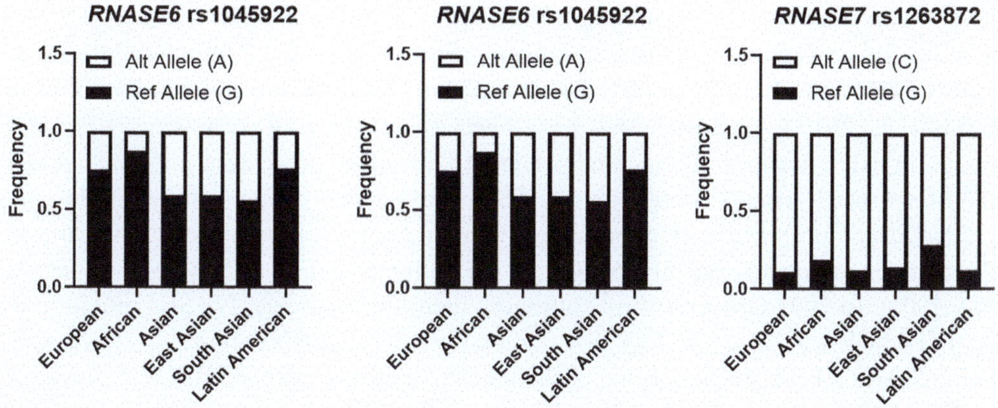

Figure 11.1 Geographic variation of common SNPs in *RNASE3*, *RNASE6*, and *RNASE7*. A query of dbSNP (https://www.ncbi.nlm.nih.gov/snp/ [accessed on January 17, 2026]) identified significant differences in each indicated SNP among major geographical subareas (*SNP*, single-nucleotide polymorphism).

higher degree of *N*-glycosylation, which limits cytotoxicity (Trulson et al., 2007). Glycosylation has also been inversely associated with the antimicrobial activity of ECP toward Gram-negative bacteria (Salazar et al., 2014), but it is not yet known whether this SNP is associated with susceptibility to bacterial infections in humans.

RNASE3 SNP alleles are adjacent to one another at the same locus and thus are inherited together as a set from each parent, called a haplotype. The G-G-T haplotype, encompassing the previously highlighted exon 2 SNP (rs2073342), along with two 3' UTR region SNPs (rs2233860 and rs8019343), has been implicated in susceptibility to severe malaria among Senegalese patients (Diop et al., 2018). Similarly, the C-G-G haplotype from the rs2233859, rs2073342, and rs2233860 polymorphisms was correlated significantly with severe malaria patients in India (Mukhi et al., 2020). Some investigators have proposed that polymorphisms in the *RNASE3* 3' UTR may affect interaction with transcription factors and therefore impact ECP cellular content (Jonsson et al., 2006).

RNASE4: RNASE4 variants have been investigated in amyotrophic lateral sclerosis (ALS), a fatal, progressive disorder that results in degenerative motoneurons in the brain and spinal cord, leading to a gradual onset of muscle weakness and eventual paralysis. ALS has a strong genetic basis with a complex, heterogeneous landscape involving over 40 genes (Nijs and Van Damme, 2024). Altered RNA processing has been implicated in ALS pathogenesis. As we will discuss in greater detail in the next section, *RNASE5* is one of the most frequently mutated genes in ALS (Greenway et al., 2004, 2006). The implication of missense mutations in *RNASE5* in ALS led investigators to focus on protein-coding polymorphisms in *RNASE4*, which shares the same locus and has a similar genetic structure as *RNASE5*. Li and coworkers identified an association between one such *RNASE4* polymorphism (rs3748338) and ALS (Li et al., 2013). Other investigators have found that certain rare, nonsynonymous *RNASE4* SNPs result in loss of RNase4 ribonuclease activity and capacity to undergo nuclear translocation (Padhi et al., 2019)—properties that have been associated with pathogenic *RNASE5* variants in ALS. Further studies are needed to establish the extent to which these *RNASE4* natural variants contribute to ALS and other neurodegenerative conditions.

RNASE5: Considerable evidence attests to a fundamental role for genetic variants in *RNASE5*, encoding angiogenin (ANG), in the pathogenesis of ALS (Greenway et al., 2004, 2006). In 2004 Greenway and colleagues identified a novel *RNASE5* mutation in two individuals with sporadic ALS that potentially inhibited ANG function (Greenway et al., 2004). These investigators subsequently identified seven *RNASE5* missense mutations in 15 individuals with familial and sporadic ALS, further attesting to a pivotal role for *RNASE5* in ALS pathogenesis (Greenway et al., 2006). At this point in time, over 30 heterozygous missense mutations in *RNASE5* have been associated with ALS onset. This topic has been recently reviewed in a comprehensive fashion by Gotte (2024). Most pathogenic variants in *RNASE5* lead to loss of ANG function, whether this is gauged on the basis of RNase activity, angiogenesis, or motoneuron survival.

RNASE6: RNase6 is expressed by monocytes, macrophages, and dendritic cells and exerts potent antimicrobial activity toward Gram-positive and Gram-negative bacteria (Lu et al., 2019; Rosenberg and Dyer, 1996; Tong et al., 2024; Cortado et al., 2024; Ruiz-Rosado et al., 2023; Becknell et al., 2015). This antimicrobial activity maps to the N-terminus of RNase6 and is associated with the ability of cationic lysine and arginine residues to associate with negatively charged bacterial cell wall components and agglutinate with Gram-negative bacteria, as well as amphipathic characteristics that enable disruption of microbial membranes (Pulido et al., 2016). In contrast, the enzymatic activity of RNase6 appears to be dispensable for its antimicrobial properties, as mutations in the catalytic triad of RNase6 render it inert as a ribonuclease without impacting its microbicidal activity (Lu et al., 2019).

Anguita and colleagues investigated the functional significance of the most common, nonsynonymous *RNASE6* SNP rs1045922 (Fig. 11.1), which results in an arginine (R) to glutamine (Q) transversion at position 66 in the mature protein sequence (R66Q) (Anguita et al., 2024). By structural analysis, these investigators observed that the R66Q substitution significantly reduced the positive electrostatic charge at the protein surface. When the antimicrobial activity of recombinant RNase6 R66 and -Q66 protein variants was assessed toward uropathogenic *Escherichia coli*, the most common cause of bacterial urinary tract infections (UTI), the R66 variant, encoded by the major SNP rs1045922 allele, exhibited superior microbicidal activity when compared to the Q66 variant. Increased bactericidal activity of the RNase6 R66 variant was associated with increased binding to bacterial cell wall components, such as lipopolysaccharide, as well as higher bacterial agglutination abilities, when compared to the Q66 variant. However, the variants did not differ in their enzymatic activity toward dinucleotide substrates (Anguita et al., 2024).

Further investigation is required to delineate the clinical implications of this polymorphism, which appears to be functionally significant from the standpoint of the antimicrobial properties of RNase6. In addition, as with other RNaseA superfamily members such as RNase3/ECP, it is plausible that RNase6 serves additional physiologic and pathologic roles apart from its antimicrobial activity and that nonsynonymous polymorphisms such as *RNASE6* rs1045922 may further impact these functional properties. Along these lines, it is intriguing that RNase6 has been recently implicated in processing bacterial RNA and generating oligonucleotides that serve as ligands for Toll-like receptor (TLR) 7 and TLR8 (Tong et al., 2024; Nunes et al., 2024). This newly discovered role for RNase6 raises the question of whether *RNASE6* variants such as rs1045922 influence TLR7/8 activation and sensing of pathogen RNA, which may have implications for pathogen detection by professional antigen-presenting cells and influence susceptibility to infectious and inflammatory diseases.

RNASE7: *RNASE7* encodes a potent antimicrobial protein expressed by epithelial cells in the skin, respiratory, and urinary tract (Harder and Schroder, 2002; Spencer et al., 2011). Like RNase6, the ribonuclease activity of RNase7 is independent of its antimicrobial activity (Huang et al., 2007). Pierce and colleagues evaluated the clinical and functional significance of a prevalent *RNASE7* SNP rs1263872 (Fig. 11.1), with a minor allele frequency of 0.07–0.19, in the context of bacterial UTI (Pierce et al., 2021). These investigators found that rs1263872 leads to an alanine (A) to proline (P) transversion at amino acid residue 103 in the mature RNase7 protein. RNase7 P103 exhibited increased antimicrobial activity toward uropathogenic bacteria when compared to RNase7 A103, along with augmented binding to bacterial cell wall components such as lipopolysaccharide. In addition, these investigators found that girls with recurrent UTIs were three times as likely to have biallelic inheritance of *RNASE7* alleles encoding the less potent RNase7 A103 variant than healthy controls. This impactful study established the significance of *RNASE7* rs1263872 as a common, nonsynonymous SNP impacting RNase7 protein function and UTI susceptibility in children.

DROSHA and *DICER1*: *DROSHA* and *DICER1* encode endoribonucleases that serve key roles in microRNA (miRNA) maturation. Specifically, Drosha converts primary miRNA to precursor miRNA, and Dicer cleaves premiRNA, yielding an miRNA duplex that is ~22 nucleotides in length (Lee et al., 2003). Biallelic inactivating mutations in *DICER1* were initially identified in sporadic pleuropulmonary blastoma, a rare pediatric malignancy (Seki et al., 2014), and subsequently associated with a spectrum of benign and malignant tumors (Schultz et al., 2024). These mutations occur in the RNaseIIIb domain of Dicer1 and impair its ability to generate mature miRNA. In most instances, the relationship between *DICER1* mutations and Dicer protein function is inferred but not experimentally tested. However, Wang and coworkers recently demonstrated that lineage-specific biallelic expression of *DICER1* mutations in the gynecologic tract of mice gives rise to cancer types that mirror *DICER1* syndrome-associated cancer (Wang et al., 2023). Likewise, Snuderi and coworkers identified biallelic mutations in *DROSHA* in children with pineoblastoma, a rare and highly aggressive brain cancer (Snuderl et al., 2018). These studies establish that germline and somatic mutations in miRNA processing machinery components predispose patients to cancer from a very young age.

RNASET2: *RNASET2* encodes an evolutionarily conserved endoribonuclease that is found in all human cells. Despite the broad degree of conservation that belies a foundational biological role, the physiologic properties of RNaseT2 are only recently coming to light. As is often the case, these insights have been facilitated by the discovery of genetic variants in *RNASET2*, specifically the association of biallelic inactivating mutations in *RNASET2* with cystic leukoencephalopathy, a rare disease that leads to neurological problems in infancy and resembles congenital cytomegalovirus brain infection (Henneke et al., 2009). Interestingly, RNaseT2 mutant zebrafish developed features of cystic leukoencephalopathy and implicated RNaseT2 in degrading ribosomal RNA (Haud et al., 2011). RNaseT2 has also been shown to localize to the lysosome and participate in the processing of pathogen and host RNAs into short oligoribonucleotides that serve as ligands for TLR7 and TLR8 (Ostendorf et al., 2020; Bérouti et al., 2024). This discovery may have important implications for future associations between *RNASET2* mutations and infectious and inflammatory disorders.

RNASEH2: *RNASEH2* encodes the RNaseH2, a heterotrimeric complex with critical roles in nucleic acid metabolism that prevents immune activation by processing RNA primers in Okazaki fragments during DNA replication, removing ribonucleotides misinserted by DNA polymerases, and eliminating RNA–DNA hybrids during cell death. The vital role of RNaseH2 in preventing human autoimmune disease was demonstrated by Crow and coworkers, who identified an association between biallelic inactivating *RNASEH2* mutations and Aicardi-Goutières syndrome, an autosomal recessive childhood disorder with neurological manifestations mirroring congenital viral brain infection and shared features with systemic lupus erythematosus (SLE) (Crow et al., 2006). Heterozygous carriers of *RNASEH2* mutations exhibit increased production of autoantibodies, a feature of systemic autoimmune disease. Furthermore, *RNASEH2* mutations that result in reduced RNaseH2 enzymatic activity were detected with increased frequency in patients with SLE, and the magnitude of RNaseH2 inhibition correlated directly with disease severity (Günther et al., 2015). These findings substantiate that genetic variation in *RNASEH2* impacts RNaseH2 activity and predisposes to human autoimmune disease.

11.2 Conclusions

Mounting evidence, collected over the past several decades, attests to considerable variation in genes encoding RNases, consisting of mutations and SNPs. At this point in time, there is no known evidence of copy number variation in RNase-encoding genes and an association with disease, but we surmise that it is only a matter of time before evidence emerges. In some instances, the functional implications of mutations and SNPs have been clarified and associated with disease status, but knowledge gaps persist in most cases. Further studies are needed to establish these structure–function relationships and their implications for human health and disease.

References

Adu, B., Dodoo, D., Adukpo, S., Gyan, B.A., Hedley, P.L., Goka, B., et al., 2011. Polymorphisms in the RNASE3 gene are associated with susceptibility to cerebral malaria in Ghanaian children. PLoS One 6 (12). Available at https://doi.org/10.1371/journal.pone.0029465Denmark, http://www.plosone.org/article/fetchObjectAttachment.action?uri=info%3Adoi%2F10.1371%2Fjournal.pone.0029465&representation=PDF.

Anguita, R., Prats-Ejarque, G., Moussaoui, M., Becknell, B., Boix, E., 2024. A common polymorphism in RNASE6 impacts its antimicrobial activity toward uropathogenic *Escherichia coli*. International Journal of Molecular Sciences 25 (1), 604. Available at https://doi.org/10.3390/ijms25010604.

Becknell, B., Eichler, T.E., Beceiro, S., Li, B., Easterling, R.S., Carpenter, A.R., et al., 2015. Ribonucleases 6 and 7 have antimicrobial function in the human and murine urinary tract. Kidney International 87 (1), 151–161. Available at https://doi.org/10.1038/ki.2014.268, https://www.journals.elsevier.com/kidney-international.

Bérouti, M., Lammens, K., Heiss, M., Hansbauer, L., Bauernfried, S., Stöckl, J., et al., 2024. Lysosomal endonuclease RNase T2 and PLD exonucleases cooperatively generate RNA ligands for TLR7 activation. Immunity 57 (7), 1482. Available at https://doi.org/10.1016/j.immuni.2024.04.010, https://www.sciencedirect.com/science/journal/10747613.

Cortado, H., Kercsmar, M., Li, B., Vasquez-Martinez, G., Gupta, S., Ching, C., et al., 2024. Murine ribonuclease 6 limits bacterial dissemination during experimental urinary tract infection. Journal of Innate Immunity 16 (1), 283–294. Available at https://doi.org/10.1159/000539177, http://content.karger.com/ProdukteDB/produkte.asp?Aktion=JournalHome&ProduktNr=234234&ContentOnly=false.

Crow, Y.J., Leitch, A., Hayward, B.E., Garner, A., Parmar, R., Griffith, E., et al., 2006. Mutations in genes encoding ribonuclease H_2 subunits cause Aicardi-Goutières syndrome and mimic congenital viral brain infection. Nature Genetics 38 (8), 910–916. Available at https://doi.org/10.1038/ng1842.

Diop, G., Derbois, C., Loucoubar, C., Mbengue, B., Ndao, B.N., Thiam, F., et al., 2018. Genetic variants of RNASE3 (ECP) and susceptibility to severe malaria in Senegalese population. Malaria Journal 17 (1). Available at https://doi.org/10.1186/s12936-018-2205-9, http://www.malariajournal.com/home/.

Eriksson, J., Reimert, C.M., Kabatereine, N.B., Kazibwe, F., Ireri, E., Kadzo, H., et al., 2007. The 434(G>C) polymorphism within the coding sequence of eosinophil cationic protein (ECP) correlates with the natural course of *Schistosoma mansoni* infection. International Journal for Parasitology 37 (12), 1359–1366. Available at https://doi.org/10.1016/j.ijpara.2007.04.001.

Fischer, S., Nishio, M., Dadkhahi, S., Gansler, J., Saffarzadeh, M., Shibamiyama, A., et al., 2011. Expression and localisation of vascular ribonucleases in endothelial cells. Thrombosis and Haemostasis 105 (2), 345–355. Available at https://doi.org/10.1160/TH10-06-0345Germany, http://www.schattauer.de/en/magazine/subject-areas/journals-a-z/thrombosis-and-haemostasis/issue/special/manuscript/14436/download.html.

Gotte, G., 2024. Effects of pathogenic mutants of the neuroprotective RNase 5-angiogenin in amyotrophic lateral sclerosis (ALS). Genes 15 (6), 738. Available at https://doi.org/10.3390/genes15060738.

Greenway, M.J., Alexander, M.D., Ennis, S., Traynor, B.J., Corr, B., Frost, E., et al., 2004. A novel candidate region for ALS on chromosome 14q11.2. Neurology 63 (10), 1936–1938. Available at https://doi.org/10.1212/01.WNL.0000144344.39103.F6, http://www.neurology.org.

Greenway, M.J., Andersen, P.M., Russ, G., Ennis, S., Cashman, S., Donaghy, C., et al., 2006. ANG mutations segregate with familial and 'sporadic' amyotrophic lateral sclerosis. Nature Genetics 38 (4), 411–413. Available at https://doi.org/10.1038/ng1742.

Günther, C., Kind, B., Reijns, M.A.M., Berndt, N., Martinez-Bueno, M., Wolf, C., et al., 2015. Defective removal of ribonucleotides from DNA promotes systemic autoimmunity. Journal of Clinical Investigation 125 (1), 413–424. Available at https://doi.org/10.1172/JCI78001, http://www.jci.org/articles/view/78001/version/2/pdf/render.

Harder, J., Schroder, J.M., 2002. RNase 7, a novel innate immune defense antimicrobial protein of healthy human skin. The Journal of Biological Chemistry 277 (48), 46779–46784. Available at https://doi.org/10.1074/jbc.M207587200.

Haud, N., Kar, F., Diekmann, S., Henneke, M., Willer, J.R., Hillwih, M.S., et al., 2011. Hurlstone, rnaset2 mutant zebrafish model familial cystic leukoencephalopathy and reveal a role for RNase T2 in degrading ribosomal RNA. Proceedings of the National Academy of Sciences of the United States of America 108 (3), 1099–1103. Available at https://doi.org/10.1073/pnas.1009811107, http://www.pnas.org/content/108/3/1099.full.pdf+html.

Henneke, M., Diekmann, S., Ohlenbusch, A., Kaiser, J., Engelbrecht, V., Kohlschütter, A., et al., 2009. RNASET2-deficient cystic leukoencephalopathy resembles congenital cytomegalovirus brain infection. Nature Genetics 41 (7), 773–775. Available at https://doi.org/10.1038/ng.398.

Huang, Y.C., Lin, Y.M., Chang, T.W., Wu, S.J., Lee, Y.S., Chang, M.D.T., et al., 2007. The flexible and clustered lysine residues of human ribonuclease 7 are critical for membrane permeability and antimicrobial activity. Journal of Biological Chemistry 282 (7), 4626–4633. Available at https://doi.org/10.1074/jbc.M607321200Taiwan, http://www.jbc.org/cgi/reprint/282/7/4626.

Jönsson, U.-B., Byström, J., Stålenheim, G., Venge, P., 2002. Polymorphism of the eosinophil cationic protein-gene is related to the expression of allergic symptoms. Clinical & Experimental Allergy 32 (7), 1092–1095. Available at https://doi.org/10.1046/j.1365-2222.2002.01410.x.

Jönsson, U.-B., Blom, K., Stålenheim, G., Douhan Håkansson, L., Venge, P., 2014. The production of the eosinophil proteins ECP and EPX / EDN are regulated in a reciprocal manner. APMIS: Acta Pathologica, Microbiologica, et Immunologica Scandinavica 122 (4), 283–291. Available at https://doi.org/10.1111/apm.12142.

Jonsson, U.B., Bystrom, J., Stalenheim, G., Venge, P., 2006. A (G->C) transversion in the 3' UTR of the human ECP (eosinophil cationic protein) gene correlates to the cellular content of ECP. Journal of Leukocyte Biology 79 (4), 846–851. Epub 20060124. Available at https://doi.org/10.1189/jlb.0904517, 16434694.

Kramer, T.J., Hubener, P., Pottker, B., Golz, C., Neulen, A., Pantel, T., et al., 2022. Ribonuclease-1 treatment after traumatic brain injury preserves blood-brain barrier integrity and delays secondary brain damage in mice. Scientific Reports 12 (1), 5731. Available at https://doi.org/10.1038/s41598-022-09326-2. Epub 20220406PubMed PMID: 35388024.

Landré, J.B.P., Hewett, P.W., Olivot, J.M., Friedl, P., Ko, Y., Sachinidis, A., et al., 2002. Human endothelial cells selectively express large amounts of pancreatic-type ribonuclease (RNase 1). Journal of Cellular Biochemistry 86 (3), 540–552. Available at https://doi.org/10.1002/jcb.10234.

Lee, Y., Ahn, C., Han, J., Choi, H., Kim, J., Yim, J., et al., 2003. The nuclear RNase III Drosha initiates microRNA processing. Nature 425 (6956), 415–419. Available at https://doi.org/10.1038/nature01957.

Li, S., Sheng, J., Hu, J.K., Yu, W., Kishikawa, H., Hu, M.G., et al., 2013. Ribonuclease 4 protects neuron degeneration by promoting angiogenesis, neurogenesis, and neuronal survival under stress. Angiogenesis 16 (2), 387–404. Available at https://doi.org/10.1007/s10456-012-9322-9.

Lu, L., Arranz-Trullen, J., Prats-Ejarque, G., Pulido, D., Bhakta, S., Boix, E., 2019. Human antimicrobial RNases inhibit intracellular bacterial growth and induce autophagy in mycobacteria-infected macrophages. Frontiers in Immunology 10, 1500. Available at https://doi.org/10.3389/fimmu.2019.01500. Epub 2019/07/18Epub 20220406PubMed PMID: 31312205.

Mukhi, B., Gupta, H., Wassmer, S.C., Anvikar, A.R., Ghosh, S.K., 2020. Haplotype of RNASE 3 polymorphisms is associated with severe malaria in an Indian population. Molecular Biology Reports 47 (11), 8841–8848. Available at https://doi.org/10.1007/s11033-020-05934-1, www.springer.com/journal/11033.

Nijs, M., Van Damme, P., 2024. The genetics of amyotrophic lateral sclerosis. Current Opinion in Neurology 37 (5), 560–569. Available at https://doi.org/10.1097/WCO.0000000000001294. Epub 20240822PubMed PMID: 38967083.

Noguchi, E., Iwama, A., Takeda, K., Takeda, T., Kamioka, M., Ichikawa, K., et al., 2003. The promoter polymorphism in the eosinophil cationic protein gene and its influence on the serum eosinophil cationic protein level. American Journal of Respiratory and Critical Care MedicineX 167 (2), 180–184. Available at https://doi.org/10.1164/rccm.200204-292oc.

Nunes, I.V., Breitenbach, L., Pawusch, S., Eigenbrod, T., Ananth, S., Schad, P., et al., 2024. Bacterial RNA sensing by TLR8 requires RNase 6 processing and is inhibited by RNA 2'O-methylation. EMBO Reports 25 (11), 4674–4692. Available at https://doi.org/10.1038/s44319-024-00281-9, https://www.embopress.org/page/journal/14693178/about.

Olsson, I., Venge, P., Spitznagel, J.K., Lehrer, R.I., 1977. Arginine rich cationic proteins of human eosinophil granules. Comparison of the constituents of eosinophilic and neutrophilic leukocytes. Laboratory Investigation 36 (5), 493–500.

Ostendorf, T., Zillinger, T., Andryka, K., Schlee-Guimaraes, T.M., Schmitz, S., Marx, S., et al., 2020. Immune sensing of synthetic, bacterial, and protozoan rna by toll-like receptor 8 requires coordinated processing by RNase T2 and RNase 2. Immunity 52 (4), 591. Available at https://doi.org/10.1016/j.immuni.2020.03.009, www.immunity.com.

Padhi, A.K., Narain, P., Dave, U., Satija, R., Patir, A., Gomes, J., 2019. Insights into the role of ribonuclease 4 polymorphisms in amyotrophic lateral sclerosis. Journal of Biomolecular Structure and Dynamics 37 (1), 116–130. Available at https://doi.org/10.1080/07391102.2017.1419147, http://www.tandfonline.com/loi/tbsd20.

Pereira, M.C., Oliveira, D.T., Olivieri, E.H.R., Rogatto, S.R., Carvalho, A.L., Landman, G., et al., 2010. The 434(G>C) polymorphism in the eosinophil cationic protein gene and its association with tissue eosinophilia in oral squamous cell carcinomas. Journal of Oral Pathology and Medicine 39 (1), 56–62. Available at https://doi.org/10.1111/j.1600-0714.2009.00795.x.

Pierce, K.R., Eichler, T., Mosquera Vasquez, C., Schwaderer, A.L., Simoni, A., Creacy, S., et al., 2021. Ribonuclease 7 polymorphism rs1263872 reduces antimicrobial activity and associates with pediatric urinary tract infections. Journal of Clinical Investigation 131 (22). Available at https://doi.org/10.1172/jci149807.

Pulido, D., Arranz-Trullén, J., Prats-Ejarque, G., Velázquez, D., Torrent, M., Moussaoui, M., et al., 2016. Insights into the antimicrobial mechanism of action of human RNase6: structural determinants for bacterial cell agglutination and membrane permeation. International Journal of Molecular Sciences 17 (4), 552. Available at https://doi.org/10.3390/ijms17040552.

Rosenberg, H.F., Dyer, K.D., 1996. , Molecular cloning and characterization of a novel human ribonuclease (RNase k6): increasing diversity in the enlarging ribonuclease gene family. Nucleic Acids Research 24 (18), 3507–3513. Available at https://doi.org/10.1093/nar/24.18.3507, https://academic.oup.com/nar/issue.

Ruiz-Rosado, J.D.D., Cortado, H., Kercsmar, M., Li, B., Ballash, G., Cotzomi-Ortega, I., et al., 2023. Human ribonuclease 6 has a protective role during experimental urinary tract infection. Journal of Innate Immunity 15 (1), 865–875. Available at https://doi.org/10.1159/000534736, http://content.karger.com/ProdukteDB/produkte.asp?Aktion=JournalHome&ProduktNr=234234&ContentOnly=false.

Salazar, V.A., Rubin, J., Moussaoui, M., Pulido, D., Nogués, M.V., Venge, P., et al., 2014. Protein post-translational modification in host defense: the antimicrobial mechanism of action of human eosinophil cationic protein native forms. FEBS Journal 281 (24), 5432–5446. Available at https://doi.org/10.1111/febs.13082, http://onlinelibrary.wiley.com/journal/10.1111/(ISSN)1742-4658.

Schultz, K.A.P., Nelson, A.T., Mallinger, P.H.R., Harris, A.K., Kamihara, J., Baldinger, S., et al., 2024. DICER1-related tumor predisposition: identification of at-risk individuals and recommended surveillance strategies. Clinical Cancer Research 30 (24), 5681–5692. Available at https://doi.org/10.1158/1078-0432.CCR-24-1532, https://aacrjournals.org/clincancerres/article/30/24/5666/750442/Spatial-Dynamics-of-T-and-B-Cell-Responses.

Seki, M., Yoshida, K., Shiraishi, Y., Shimamura, T., Sato, Y., Nishimura, R., et al., 2014. Biallelic DICER1 mutations in sporadic pleuropulmonary blastoma. Cancer Research 74 (10), 2742–2749. Available at https://doi.org/10.1158/0008-5472.CAN-13-2470, http://cancerres.aacrjournals.org/content/74/10/2742.full.pdf+html.

Snuderl, M., Kannan, K., Pfaff, E., Wang, S., Stafford, J.M., Serrano, J., et al., 2018. Recurrent homozygous deletion of DROSHA and microduplication of PDE4DIP in pineoblastoma. Nature Communications 9 (1). Available at https://doi.org/10.1038/s41467-018-05029-3, http://www.nature.com/ncomms/index.html.

Spencer, J.D., Schwaderer, A.L., Dirosario, J.D., McHugh, K.M., McGillivary, G., Justice, S.S., et al., 2011. Ribonuclease 7 is a potent antimicrobial peptide within the human urinary tract. Kidney International 80 (2), 174–180. Available at https://doi.org/10.1038/ki.2011.109, https://www.journals.elsevier.com/kidney-international.

Tong, A.J., Leylek, R., Herzner, A.M., Rigas, D., Wichner, S., Blanchette, C., et al., 2024. Nucleotide modifications enable rational design of TLR7-selective ligands by blocking RNase cleavage. Journal of Experimental Medicine 221 (2). Available at https://doi.org/10.1084/jem.20230341, https://rupress.org/jem/article/221/2/e20230341/276460/Nucleotide-modifications-enable-rational-design-of.

Trulson, A., Byström, J., Engström, Å., Larsson, R., Venge, P., 2007. The functional heterogeneity of eosinophil cationic protein is determined by a gene polymorphism and post-translational modifications. Clinical & Experimental Allergy 37 (2), 208–218. Available at https://doi.org/10.1111/j.1365-2222.2007.02644.x.

Ueki, M., Iida, R., Fujihara, J., Takeshita, H., Yuasa, I., Kominato, Y., et al., 2008. Development of genotyping methods for single nucleotide polymorphism in the human pancreatic ribonuclease gene (RNASE1) and their application to population studies. Biochemical Genetics 46 (3-4), 145–153. Available at https://doi.org/10.1007/s10528-007-9138-2.

Wang, Y., Chen, S.Y., Ta, M., Senz, J., Tao, L.V., Thornton, S., et al., 2023. Biallelic dicer1 mutations in the gynecologic tract of mice drive lineage-specific development of DICER1 syndrome–associated cancer. Cancer Research 83 (21), 3517–3528. Available at https://doi.org/10.1158/0008-5472.CAN-22-3620, http://www.aacrjournals.org/.

Zhang, J., Rosenberg, H.F., 2000. Sequence variation at two eosinophil-associated ribonuclease loci in humans. Genetics 156 (4), 1949–1958. Available at https://doi.org/10.1093/genetics/156.4.1949.

CHAPTER 12

Ribonucleases and drug design

Guillem Prats-Ejarque[1,2]
[1]Department of Biochemistry and Molecular Biology, Autonomous University of Barcelona, Cerdanyola del Vallès, Spain,
[2]ALBA synchrotron Light Source, Cerdanyola del Vallès, Spain

12.1 Introduction

The ribonuclease A superfamily is a vertebrate-specific family of proteins homologous to the bovine pancreatic ribonuclease (RNaseA). RNaseA was a reference protein in the pioneering biochemistry studies in the 1970s, being the first enzyme whose catalytic mechanism was determined before its 3D structure was solved. The proteins of this family share several key features. At the structural level, all of them have a high isoelectric point (between 9 and 11), a kidney-shaped structure maintained by three to four disulfide bridges, and a shared "catalytic triad," formed by two histidines and one lysine, which cleaves RNA by an acid–base catalysis (Cuchillo et al., 2011).

In addition to its catalytic function, evolutionary studies suggest that the RNaseA family started off as host-defense proteins (Pizzo and D'Alessio, 2007; Rosenberg, 2008a,b; Lomax et al., 2017). Antimicrobial and immune-regulatory properties were reported for several family members, and their presence in a variety of biological fluids is associated with a protective physiological role (Hooper et al., 2003; Lu et al., 2018; Goncalves et al., 2016; Xu et al., 2016; Koczera et al., 2016). Several authors have shown the importance of RNases in innate immunity due to their presence in mucosal secretions and their expression by innate immune cells, such as eosinophils and phagocytes. Moreover, in vitro studies have found that certain RNases possess antibacterial (Boix et al., 2012; Boix and Nogués, 2007), antiviral (Yang et al., 2003), antifungal (Salazar et al., 2019), antiparasitic (Rosenberg, 1998; Klion and Nutman, 2004), and even antitumoral activities (Ardelt et al., 2009; Gotte and Menegazzi, 2019).

These wide-ranging antipathogen activities do not share a common mechanism of action. In some cases, ribonuclease activity is essential for these activities (e.g., antiviral activity), while in other cases, it is completely unrelated. In certain scenarios, catalytic activity is not indispensable for the demise of the pathogen; however, it does serve a concomitant defensive role. This is exemplified by the antibacterial activity, where the death of the pathogen occurs via disruption of its membrane (Carreras et al., 2003; Torrent et al., 2008). However, ribonuclease activity could still be important for the clearance of free circulating pathogen RNAs released during the lysis of the bacterial cell. Furthermore, the existence of chemoattractant activity in certain RNases has been documented, implying a potential involvement in signaling pathways that regulate the response to pathogen infection.

RNases in Health and Diseases
DOI: https://doi.org/10.1016/B978-0-443-27312-4.00012-2

Regarding antitumoral activities, they have mainly been reported in onconase and bovine seminal RNase (BS-RNase) (Ardelt et al., 2009). Traditionally, this antitumoral activity has been related to the degradation of cellular RNAs (tRNA, rRNA, mRNA, and even siRNAs and microRNAs). This degradation of RNA would preclude protein synthesis and carry the cell to apoptosis (Iordanov et al., 2000) or autophagic cell death (Fiorini et al., 2014; Fiorini et al., 2015).

Considering this context, RNases are a good starting point for therapeutic agents for at least two of the most alarming issues in health sciences today: antimicrobial resistance and cancer. Several strategies are already being explored, including peptide-derived designs, the use of whole proteins as drugs, and the coadministration of approved drugs with RNases, among others. In this chapter, I am going to cover several of these potential therapeutic properties of RNases, aiming to provide a state-of-the-art view on each topic and explore new potential pharmacological uses.

12.2 RNases to fight antimicrobial resistance

12.2.1 Bacteria

Nowadays, antibiotic resistance is one of the main problems facing global public health. The 2016 UK government O'Neill Review projected 10 million AMR-related deaths annually by 2050 (O'Neill, 2016), with antimicrobial resistance infections being the main cause of death by that year. These findings have later been refined by the 2024 Lancet GRAM study to about 1.9 million direct AMR deaths (and over 8 million, including contributing cases) per year (Naghavi et al., 2024).

Thanks to their genetic plasticity, bacteria have an extraordinary ability to adapt to antimicrobial agents/antibiotic exposure. These resistance mechanisms can be classified into two major strategies: mutational resistance, in which genes related to the antibiotic mechanism of action are mutated, and horizontal gene transfer, in which bacteria acquire foreign DNA coding for the resistance determinant (Munita and Arias, 2016).

The mutations in the bacterial genome that result in antimicrobial resistance usually alter the antibiotic action through one of the following mechanisms: modifications of either the antimicrobial agent or the target, a decrease in drug uptake, the activation of efflux mechanisms, and global changes in important metabolic pathways via modulation of regulatory networks.

In this section, I will explain several RNase-based approaches to fight against bacterial antibiotic resistance.

12.2.2 Studies using canonical RNases

The RNaseA superfamily includes a series of proteins with antimicrobial and immune-modulatory activities and is considered to have emerged with an ancestral host-defense role (Rosenberg, 2008a,b; Lu et al., 2018; Boix and Nogués, 2007). In humans, there are 13 members of the RNaseA superfamily, of which RNases1–8 are known as the canonical ribonucleases, according to their structural, enzymatic, and biological properties (Sorrentino, 2010; Sorrentino and Libonati, 1997). Regarding RNases9–13, although they have a high percentage of identity, they do not present the catalytic triad, common in the rest of the family, and their function is still controversial (Cho et al., 2005; Goo and Cho, 2013).

Among the canonical RNases, only RNases3, 6, and 7 have been reported to display a significant bactericidal activity. RNase3 (also known as eosinophil cationic protein, ECP) deserves a special mention due to its main protagonism in the innate immune system. RNase3 is a secretory ribonuclease stored in the secondary granules of eosinophils and released at the focus of infection (Malik and Batra, 2012; Acharya and Ackerman, 2014), which displays the highest reported bactericidal activity (Rosenberg, 1998; Lehrer et al., 1989; Gupta et al., 2013; Torrent et al., 2012; Rosenberg et al., 2013) within the family with a broad-spectrum activity (Boix et al., 2012; Carreras et al., 2003; Malik and Batra, 2012; Torrent et al., 2012; Lien et al., 2013).

Interestingly, several studies (Carreras et al., 2003; Rosenberg, 1995; Torrent et al., 2007) indicate that the bactericidal activity of RNases is not directly related to their catalytic activity but is mainly due to their capacity to trigger the breakage of the bacterial cell wall. This bactericidal activity, retained at the N-terminal region of the protein (Torrent et al., 2010; Torrent et al., 2011b; Torrent et al., 2013), is similar to the mechanism that other antimicrobial proteins and peptides (AMPs) present. They share a marked cationicity that facilitates electrostatic interactions with the negatively charged bacterial surfaces (Torrent et al., 2011a). In addition, they show an abundance of hydrophobic residues and the presence of dynamic amphipathic modules that can adopt secondary structures upon interaction with the bacterial envelopes (Boix and Nogués, 2007; Boix et al., 2008). A depiction of this general mechanism of action is shown in Fig. 12.1.

Although the exact mechanism of the attack on the bacterial wall remains unclear, it has been proposed (Boix and Nogués, 2007; Torrent et al., 2008; Torrent et al., 2007) that the RNases bound by the cationic regions to lipopolysaccharides (LPS) or teichoic acids (TA) could form aggregates by interacting with other protein monomers by their hydrophobic regions. These large aggregates, bound to the bacterial wall, would inflict mechanical pressure on the membrane, causing the cell to burst. In addition, protein–protein aggregation favors cell agglutination. This bacterial agglutination would help the immune system cells to clear the infection focus. It has been observed that bacterial agglutination is mainly caused by a hydrophobic region of the N-terminus of the protein, in which Ile13 (numbered after RNase3) would play a key role (Torrent et al., 2012). Moreover, in Gram-negative strains, the competitive displacement of the cations that stabilize the LPS also contributes to this antibacterial activity (Boix and Nogués, 2007).

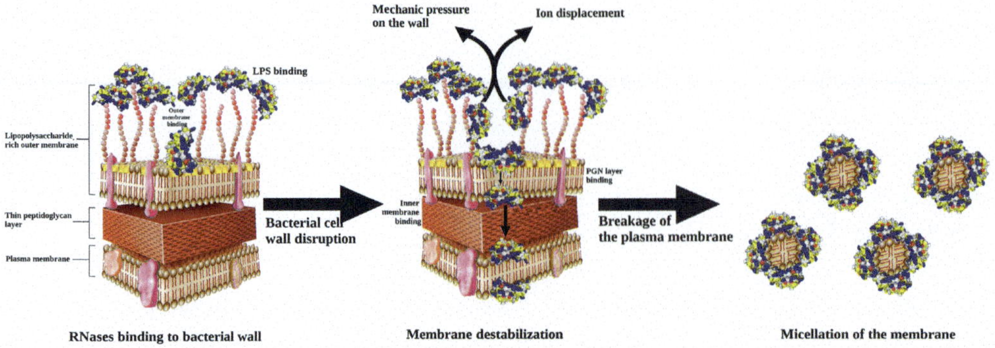

Figure 12.1 Breakage of the Gram-negative bacteria envelope by the action of RNases, from binding to the cell wall to disruption of the cell membranes. The RNases are represented usinga molecular surface view, where blue indicates cationic residues, red indicates anionic residues, and yellow indicates hydrophobic residues. In Gram-positive strains, the process is similar, with the difference that RNases do not need to break the outside membrane, and, instead of interacting with lipopolysaccharides (LPS), they bind to teichoic acids (TA) and then, to the peptidoglycan layer, until they reach the plasma membrane.

12.2.3 Using RNase-based antimicrobial peptides to fight antibiotic resistance

The mechanism of action of bactericidal RNases, such as RNase3, is similar to that of antimicrobial peptides, in that they bind the bacterial cell wall and the membrane through their cationic residues by electrostatic interactions. Then, the hydrophobic residues trigger the destablization of the membrane, permeabilizing it and inducing cell death. This mechanism is common for both Gram-negative and Gram-positive bacteria, although it may vary by peptide or pathogen type (Boix and Nogués, 2007; Hancock and Scott, 2000). Several studies indicate that the antibacterial activity of RNase3 is mostly retained in the first 45 residues of its N-terminus side (Torrent et al., 2009b), which comprises the first two helixes of the protein.

ECP's bacterial agglutination ability also corresponds to the first helix, particularly to the 8–16 residues, while the 33–36 residues are involved in destablization and breakage of the membranes. On the other hand, the LPS-binding region corresponds to the second helix, although some residues of the first helix, like Trp10 and Gln14, are also involved (Torrent et al., 2010). A combination of LPS-binding and agglutination promotes the protein's antibiofilm activity. It has been shown that ECP has high activity against bacterial biofilms of *Pseudomonas aeruginosa*, which is also retained at the protein N-terminus region (Pulido et al., 2016). After several minimization attempts based on the ECP(1–45) N-terminal peptide, the ECP(5–17P24–36) peptide was chosen (Torrent et al., 2011b) as the best pharmacophore of the parental protein, as depicted in Fig. 12.2.

One of the main challenges that antimicrobial peptides face is degradation by proteases, which is a known resistance mechanism against antimicrobial peptides (Joo et al., 2016).

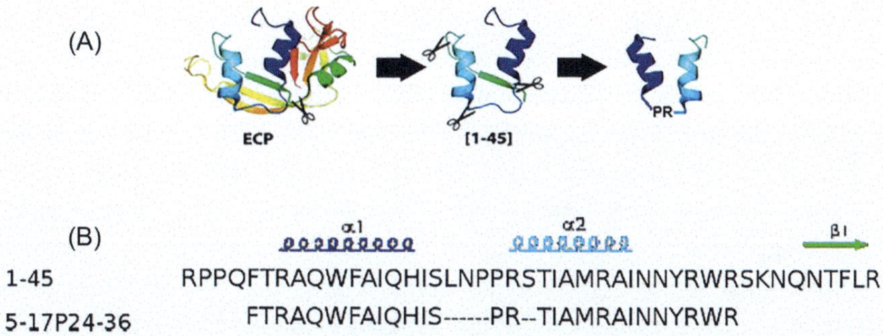

Figure 12.2 ECP(5–17P24–36) peptide design. (A) Structure representation of ECP, its N-terminal peptide, and the ECP(5–17P24–36) peptide. (B) ECP N-terminal and ECP(5–17P24–36) sequences. The secondary structure elements of the peptide are shown, based on the ECP structure.

To avoid it, a total D-enantiomer of the peptide was designed to protect the molecule against both bacterial and host proteases that retained an equivalent 2D structural fold and displayed even higher antimicrobial activity. However, the improvement in the peptide efficiency was to the detriment of a slight increase in its cytotoxicity, although it does not exceed the so-called therapeutic window, so this peptide was selected for further in vivo and serum stability studies (Salazar et al., 2019; Pulido et al., 2016; Pulido et al., 2018; Li et al., 2022; Sandín et al., 2021).

12.2.4 Using RNase catalytic activity to avoid bacterial resistance acquisition

As previously mentioned, RNases are dual-function proteins, meaning they also possess the capacity to degrade RNA. However, the enzymatic activity does not appear to be associated with their bactericidal activity (Rosenberg, 1995). Nevertheless, it is noteworthy that the conserved catalytic activity of this protein family, which is involved in the innate immune system, is not essential for its function.

With the aim of achieving maximal catalytic efficiency and bactericidal activity, a chimeric RNase was engineered. As previously explained, among the bactericidal ribonuclease (RNase) family, RNase3 is the most prominent member of the canonical RNases. Furthermore, RNase1 is the most catalytically active RNase in this group (Sorrentino and Libonati, 1994), despite the absence of reported bactericidal activity. Consequently, the integration of these two activities within a singular protein emerges as a promising avenue for exploration.

Some years ago, Torrent et al. (2009a) designed a novel bactericidal ribonuclease (PM5-YRWR) by replacing the 36–39 residues of RNase1 with the YRWR residues from the second helix of ECP, utilizing RNase1 as the scaffold. The present study demonstrated the viability of integrating both activities within a singular domain-engineered RNase.

Considering these previous results, a hybrid RNase was designed, called RNase3/1 (Prats-Ejarque et al., 2019). This RNase was designed to achieve a chimeric RNase that was able to combine, on one hand, the high catalytic activity of RNase1 and its affinity for the ribonuclease inhibitor (RI) and, on the other hand, the high antimicrobial activity of RNase3. The RNase3/1 conserves the RNase1 scaffold that provides the essential requirements for an elevated catalytic activity (Cuchillo et al., 2011; Nogués et al., 1998; Raines, 1998; Gagné et al., 2015; Doucet et al., 2009) and encompasses the key features of RNase3 antimicrobial activity (Boix et al., 2012; Carreras et al., 2003; Torrent et al., 2008; Torrent et al., 2007; Boix et al., 2008; Torrent et al., 2009b,c; Pulido et al., 2013). The protein nontoxicity for host cells in vivo is ensured thanks to the presence in the cytosol of all human tissue cells of the RNase inhibitor (RI), a horseshoe-shaped protein that binds to the human secretory RNase members (Lomax et al., 2014). The RI is ubiquitous and protects the host cells from the potential toxicity of secretory RNases (Dickson et al., 2005). Therefore this RNase chimera combines a high affinity for the bacterial cell wall together with an elevated catalytic efficiency and no toxicity to the host. A newer version of this chimeric protein (RNase3/1-v3) has been designed to improve the bactericidal activity by adding the cytotoxic loop 7 of RNase3 and shortening the long and flexible loop 1 of RNase1 to increase its activity, making it comparable to the RNase3 one, while it retains most of the catalytic activity of RNase1.

This hybrid RNase has proved to be effective in hindering antibiotic resistance against colistin by *Acinetobacter baumannii* by a reduction in the colistin minimal inhibitory concentration (MIC) by twofold after 330 generations (Prats-Ejarque et al., 2019). Not only that, a 1.5−5× reduction of the original colistin MIC value is observed when combined with the RNase. In addition to the therapeutic potential of the aforementioned data, these findings are noteworthy in light of the observation by Eller and Raines of the synergy between RNase1 and the antimicrobial peptide LL-37, which also demonstrated a correlation between bacterial membrane disruption and ribonucleolytic activity (Eller and Raines, 2020). In the presence of sublethal concentrations of LL-37, RNase1 is toxic to *Escherichia coli* cells at picomolar levels, similar to what has been observed with colistin and RNase3/1 with *A. baumannii*. The mechanism proposed by the authors, like what is observed in the case of colistin and RNase3/1, is that LL-37 would help to generate pores in the bacterial membrane, while RNase1, adhered to the membrane, would enter when the pore is created. The entrance of the RNase into the bacterial cell would degrade the cytosolic RNA and trigger bacterial cell death.

12.3 Parasites

Canonical RNases, particularly RNase3, have been reported to be active against several parasites like *Schistosoma mansoni* (Ackerman et al., 1983; Ackerman et al., 1985), helminths, and protozoa (Malik and Batra, 2012; Boix et al., 2008; Tischendorf et al., 1996; Singh and Batra, 2011). Of great interest is the reported capacity of RNase3 to kill *Plasmodium falciparum* (Waters et al., 1987). Several authors have pointed out an increase in expression of RNase3 during malaria infections (Kurtzhals et al., 1998) and that RNase3 polymorphisms could change the outcome of the disease (Adu et al., 2011; Mukhi et al., 2020; Diop et al., 2018).

12.4 RNases and cancer

The relationship between RNases and cancer is a double-edged sword. On one hand, and contrary to what happens with bactericidal activity, which is part of the physiological function of the protein family, antitumoral activity of RNases is possibly an "artifact" that we can use to treat cancer.

On the other hand, several authors are reporting that RNases could be potential targets to treat cancer by blocking its tumorigenic activity. Both approaches are discussed below.

12.4.1 Antitumoral RNases

As previously mentioned, some RNases are reported to present good antitumoral activity, directly related to their ribonucleolytic activity. Due to the high anionic properties of the cancer cells' membrane, RNases, with their high pI, preferentially enter the cancer cells, making them more specific against tumors.

This specificity is due to the cationic properties of RNases (Ardelt et al., 1991; Wu et al., 1993). Since the membrane potential of cancer cells is more negative than that of nonmalignant cells (Ran et al., 2002), cationic proteins such as RNases interact more strongly with cancer cells. As commented before, the mammalian cytosol presents a conserved protein, the RI, which has a high affinity for secretory ribonucleases (Dickson et al., 2005). Its constitutive expression among all cells makes it a sentry protein that protects them against internalized ribonucleases. The RI binds RNases in a 1:1 stoichiometry, engulfing and sequestering the protein, avoiding its cytotoxic action (Lomax et al., 2014; Rutkoski and Raines, 2008; Thomas et al., 2016). However, antitumoral RNases, unlike other members of the RNaseA family, have the ability to evade the RI (Rutkoski and Raines, 2008). This capacity to evade the RI, together with the aforementioned ability of RNases to bind preferentially to the tumoral anionic membranes, is the most important determinant of antitumoral activity.

The two main antitumor ribonucleases, the amphibian onconase and the dimeric BS-RNase, evade the RI, allowing them to exert high antitumoral activity. Besides, research is currently ongoing in engineered versions of the human pancreatic RNase1 that are able to evade the RI (Attery et al., 2018; Hoang et al., 2018; Noro et al., 2023).

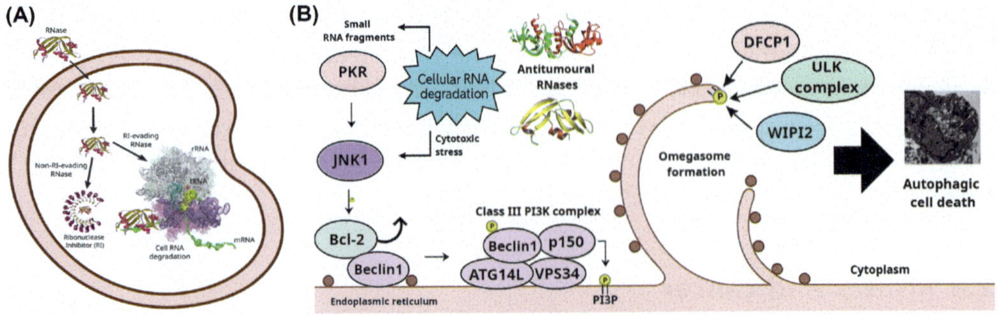

Figure 12.3 Schematic depiction of the antitumour RNase mechanism of action. (A) Entrance of the RNases into the cell and RNA degradation/RNase inhibition by RI. (B) Mechanism hypothesis of ribonuclease-induced autophagy, from induction by antitumoral cytotoxic RNases to the omegasome formation that finally triggers the autophagic death of the cancer cell. Abbreviations: Atg (Autophagy-related); Bcl-2 (B-cell lymphoma 2); Beclin1 (BCL-2 interacting myosin/moesin-like coiled-coil protein 1); DFCP1 (double FYVE domain-containing protein 1); JNK (c-Jun N-terminal kinase); PKR (dsRNA-dependent protein kinase); ULK (UNK-51-like kinase); Vps (vacuolar protein sorting); WIPI (WD repeat domain phosphoinositide-interacting). Source: *Own work, image of the onconase-induced autophagosomes from: Michaelis, Martin, Cinatl, Jaroslav, Anand, Puja, Rothweiler, Florian, Kotchetkov, Rouslan, Deimling, Andreas von, Doerr, Hans W., Shogen, Kuslima & Cinatl, Jindrich. (2007). Onconase induces caspase-independent cell death in chemoresistant neuroblastoma cells. Cancer Letters, 250(1), 107–116. https://doi.org/10.1016/j.canlet.2006.09.018.*

The advantage of these engineered proteins is that they would be less immunogenic than amphibian RNases, such as onconase or BS-RNase.

Studies with these antitumoral RNases indicate that they trigger caspase-independent cell death (autophagic cell death) in tumor cells (Fiorini et al., 2014, 2015; Michaelis et al., 2007). Supporting this hypothesis, experiments with RNaseL (Chakrabarti et al., 2012; Siddiqui and Malathi, 2012)—a dimeric ribonuclease from a different protein family—also demonstrated ribonuclease-mediated autophagy. In its latent form, RNaseL exists as a monomer that dimerizes and is activated during intracellular viral infections as part of the innate immune response (see further details in "viral infections" section in chapter 8). Although RNaseL shares no homology with onconase or BS-RNase, it serves as a valuable model for dissecting how ribonucleolytic stress induces cell death. The general antitumoral RNase mechanism of action is depicted in Fig. 12.3.

12.4.2 Onconase

Probably the most well-known antitumoral RNase, onconase (ONC), is an amphibian RNase isolated from the Northern Leopard Frog (*Rana pipiens*) (Ardelt et al., 1991). In physiological conditions, its function is probably related to the innate immunity of the frog and its oocytes. But in mammalian organisms, it has a powerful cytotoxic activity with a strong preference for cancer cells.

Onconase ability to evade the RI is caused by the smaller size of certain exposed loops that in mammalian RNases interact with the inhibitor, preventing the formation of the RI-protein complex (Lomax et al., 2014; Rutkoski and Raines, 2008). In RNaseA, 24 residues participate in interactions with the RI. Of these 24 residues, only three are conserved in onconase. In comparison with RNaseA, onconase has a shorter sequence (104 residues vs 123 residues) (Ardelt et al., 1991), which gives onconase a more compact structure. Besides, ONC also has a lower catalytic activity than RNaseA (Boix et al., 1996). Although sharing only 30% of sequence identity with RNaseA, it adopts the same kidney-shaped 3D structure. Another member of the amphibian ribonuclease family is amphinase (Singh et al., 2007), found in *Rana pipiens* oocytes, which also presents antitumoral action.

In 2007, autophagic cell death induced by onconase was observed in neuroblastoma cells (Michaelis et al., 2007) by transmission electron microscopy. Later, Fiorini and colleagues also reported an autophagic cell death induction by onconase in pancreatic cancer cells (Fiorini et al., 2015). They quantified cell death in pancreatic cancer cells (Panc1 and PaCa44) following onconase and compared it to the death of normal cells (fibroblasts and HaCaT). They found that onconase inhibits 60%–65% of the cancer cell growth, while in the case of the normal cells, they found an inhibition of 40%. This inhibition effect could be enhanced with coadministration of gemcitabine, nowadays the most used chemotherapeutic agent to treat pancreatic adenocarcinoma. In addition, it has been reported that this cell death could be reverted with the addition of autophagic inhibitors such as chloroquine or 3-methyladenine, suggesting that the mechanism of cell death is autophagy-dependent.

Several clinical trials were done with onconase, with some positive results, mainly an increase in median survival in the groups treated with onconase versus the groups treated with other chemotherapeutic agents (tested for nonsmall cell lung cancer (Mikulski et al., 1995) and unresectable malignant mesothelioma (Mikulski et al., 2002)). A full clinical trial with onconase was performed in 2009 on unresectable malignant mesothelioma patients, with an increase of median survival in the onconase-treated patient group (11.6 months vs 9,6 months). However, the clinical trials stopped because the high renal accumulation of onconase limited the effective doses due to renal toxicity. Currently, research on onconase is still ongoing, focusing on its oligomeric variants (Gotte et al., 2021) and its synergistic action together with other drugs (Raineri et al., 2019). Interestingly, onconase has been recently shown to downregulate specific oncogenic proteins (De Tomi et al., 2022).

12.4.3 Bovine seminal RNase

BS-RNase is another antitumoral ribonuclease that is part of the superfamily of RNaseA. BS-RNase is the only natural dimeric member of the superfamily, and, in solution, is present as a mixture of two different forms: a swapped (M × M) and a nonswapped form (M = M) (Rutkoski and Raines, 2008). I would not extend on this topic here because Chapter 6 will cover it thoroughly, so I will merely point out some aspects of its antitumoral activity related to its quaternary structure.

It could seem a bit surprising that, having 83% of sequence identity with RNaseA, which does not present antitumoral activity, BS-RNase does have antitumoral action. This is due to the dimeric conformation of BS-RNase, because this quaternary structure allows it to evade the RI. It has been reported that the dimeric structure of BS-RNase precludes the interaction with RI. The form M × M protects the BS-RNase from the reducing conditions of the cytosol, which could reduce the disulfide bonds that maintain the dimeric structure. Unlike the dimeric structure, the monomer of BS-RNase is sensitive to RI, so if the disulfide bonds are reduced, antitumoral activity will be lost.

In 2014 Fiorini et al. (2014) demonstrated that BS-RNase was also able to induce autophagic cell death in pancreatic cancer cells and that this induction was the cause of its antitumoral activity.

This study reported 60% inhibition of cancer cell growth (Panc1 strain), while in nontumorigenic fibroblasts, the inhibition of growth is about 40%. Another interesting aspect is the different response of cancer cells to conditions of serum starvation. The decrease in serum percentage in the culture reduces the growth inhibition of cancer cells treated with BS-RNase in comparison with untreated cells. Furthermore, the levels of autophagy decreased with the increase in starvation levels. These facts suggest that the antitumoral specificity of BS-RNase could also be related to the high proliferation level of the tumoral cells.

12.4.4 Targeting RNases to fight cancer

Besides the use of RNases as drugs to target cancer cells, in recent years an opposing strategy has been proposed by some authors, considering the possible role in vivo of RNases in cancer signaling. The role in cancer for RNase5, or angiogenin, has been extensively investigated in recent years, as detailed in Chapter 9. Angiogenin has been reported to directly bind to plexin-B2 (PLXNB2) (Yu et al., 2017) and the epidermal growth factor (Wang et al., 2018) receptors in solid and hematopoietic cancers.

Also, RNase4 has been shown to promote malignant progression in hypoxic glioblastoma through activation of AXL/AKT and NF-κB/cIAPs signaling pathways (Lee, 2024). The authors suggest that hypoxia-induced RNASE4-mediated upregulation of cellular inhibitor of apoptosis proteins (cIAPs) could represent a promising target for enhancing glioblastoma chemotherapy, as well as a prognostic biomarker.

Other reports from Lee et al. (2021) show that RNase1 could induce breast tumor initiation by enriching the stem-like cell population via activation of the ephrin A4 receptor (EphA4). This activation, by direct binding, is distinct from the classical EphA4-ephrin signaling, which occurs through contact-dependent cell–cell communication.

Recently, it also has been proposed that RNase7 could have oncogenic potential by inducing ROS1 activation in hepatocellular carcinoma (Liu et al., 2021). However, other authors were unable to reproduce these results, using either the cell line used in the original work or with other ROS1 cell lines (Jones et al., 2024). The oncogenic potential of RNase7 and its feasibility as an antitumoral target, then, remain to be unveiled.

To sum up, these discoveries suggest that potential treatment strategies could be designed to inhibit the RNase-induced oncogenic pathways, pending further investigation.

12.5 RNases to fight viruses

Antiviral activity is reported to be present in several RNases from several families (Ilinskaya and Mahmud, 2014; Li and Boix, 2021). In the RNaseA superfamily, RNase2 (named the eosinophil-derived neurotoxin; EDN) is the main representative, which is effective against several RNA viruses. RNase2 acts as a potent modulator of the host immune system and additionally displays a high antiviral activity against rhinovirus, adenovirus, respiratory syncytial virus, and human immunodeficiency virus, among others, in an RNase catalytic activity-dependent manner (Li and Boix, 2021; Domachowske et al., 1998; Rosenberg, 2008a,b; Rosenberg, 2015; Lee-Huang et al., 1999; Rugeles et al., 2003; Lu et al., 2022). Analogously, mouse eosinophil-associated RNases have also been found to reduce the infectivity of several rodent respiratory viruses (Moreau et al., 2003; Gaudreault and Gosselin, 2007; O'Reilly et al., 2012).

Apart from the RNaseA superfamily, other proteins with ribonuclease activity have been reported to be potent antiviral agents. One of the most studied cases is RNaseL, a ubiquitous intracellular endonuclease activated by 2′5′-oligoadenylates (2'5'-OA), specifically synthesized after IFN activation in response to viral infections (Chakrabarti et al., 2012; Silverman, 2007; Hornung et al., 2014; Banerjee et al., 2019). 2'5'-OA binds to the RNaseL monomer (the inactivated form) and induces its dimerization (the activated form) to cleave viral ssRNA, further enhancing the formation of stress granules, which include several antiviral proteins (Manivannan et al., 2020).

The monocyte chemoattractant protein-induced protein 1 (MCPIP1) is a zinc finger protein that also possesses a nuclease domain with ribonuclease activity, which has been shown to be antiviral (Lin et al., 2013). This protein is involved in the regulation of the inflammation of the cell, and it exerts potent antiviral activity in response to inflammation (Chen et al., 2018; Li et al., 2020). Interestingly, as it happens in RNaseL, the activity of MCPIP1 is dependent on its dimerization (Suzuki et al., 2011).

Another ribonuclease, which has been discovered at the end of the 20th century is the interferon-stimulated gene 20 kDa protein (ISG20) (Degols et al., 2007; Wu et al., 2019). ISG20 is able to target a broad spectrum of viruses. Its active site is composed of three aspartates and one glutamate, and it needs Mn^{2+} for its catalytic activity. Interestingly, as it happens with RNaseL, ISG20 activation is also induced by the IFN system. Several mechanisms of action for its antiviral activity have been proposed, but its precise mechanism has not been elucidated yet (Deymier et al., 2022).

Overall, ribonuclease activity is a promising strategy to tackle virus infection, and still more work is needed to design new therapeutic approaches based on this innate immune system strategy against viruses. Recently, a fusion protein between the human receptor ACE2 and RNaseA has been designed, with the goal of specifically attacking SARS-CoV-2 (Xu et al., 2021). Although this fusion protein has only been tested in vitro, the authors claim it as a proof-of-concept to use this antiviral activity therapeutically.

12.6 Design of artificial RNases

Taking all this into account, mimicking the catalytic activity of RNases in small molecules appears to be an interesting approach for drug synthesis. With this goal in mind, several researchers have designed artificial RNases, small molecules that mimic the catalytic mechanism of RNaseA (Helm et al., 2001; Fouace and Gaudin, 2004; Tamkovich et al., 2016) (or even other RNases (Mironova et al., 2007)), including imidazole groups that perform the acid–base catalysis. Currently, the main drawback of these molecules is their low catalytic activity. Some of these artificial RNases have proved to be effective in vitro against the influenza virus (Tamkovich et al., 2016). This strategy has been proposed to be applied for viral inactivation for vaccine development (Goncharova et al., 2009; Fedorova et al., 2012). Another artificial RNase strategy is focused on antiviral ribozymes that could specifically target viral RNA sequences. Some of them are currently in the clinical trials stage (Scarborough and Gatignol, 2015; Yu et al., 2020).

Altogether, it is clear that ribonuclease activity is a potential tool to address a broad range of diseases, and several tailored strategies have been proposed for each condition. With the current knowledge on RNases' antipathogenic activities, the possibility to design specific molecules with ribonuclease activity able to discriminate pathogenic RNAs might be closer than ever, opening a broad new field in the design of new therapeutic tools.

12.7 Future perspectives

As demonstrated in this chapter, RNases present high therapeutic potential in a broad range of applications. While antibacterial, antiviral, and antitumoral properties have been thoroughly studied, recent findings suggest key and very specialized roles of RNases in the immune system and cell regulation, as covered in other chapters of this book.

References

Acharya, K.R., Ackerman, S.J., 2014. Eosinophil granule proteins: form and function. Journal of Biological Chemistry 289 (25), 17406–17415. Available at https://doi.org/10.1074/jbc.R113.546218, http://www.jbc.org/content/289/25/17406.full.pdf+html.

Ackerman, S.J., Loegering, D.A., Venge, P., Olsson, I., Harley, J.B., Fauci, A.S., et al., 1983. Distinctive cationic proteins of the human eosinophil granule: major basic protein, eosinophil cationic protein, and eosinophil-derived neurotoxin. Journal of Immunology 131 (6), 2977–2982.

Ackerman, S.J., Gleich, G.J., Loegering, D.A., Richardson, B.A., Butterworth, A.E., 1985. Comparative toxicity of purified human eosinophil granule cationic proteins for schistosomula of *Schistosoma mansoni*. American Journal of Tropical Medicine and Hygiene 34 (4), 735–745. Available at https://doi.org/10.4269/ajtmh.1985.34.735.

Adu, B., Dodoo, D., Adukpo, S., Gyan, B.A., Hedley, P.L., Goka, B., et al., 2011. Polymorphisms in the RNASE3 gene are associated with susceptibility to cerebral malaria in Ghanaian children. PLoS ONE 6 (12), e29465. Available at https://doi.org/10.1371/journal.pone.0029465.

Ardelt, W., Mikulski, S.M., Shogen, K., 1991. Amino acid sequence of an anti-tumor protein from Rana pipiens oocytes and early embryos. Homology to pancreatic ribonucleases. Journal of Biological Chemistry 266 (1), 245–251. Available at https://doi.org/10.1016/s0021-9258(18)52427-3.

Ardelt, W., Ardelt, B., Darzynkiewicz, Z., 2009. Ribonucleases as potential modalities in anticancer therapy. European Journal of Pharmacology 625 (1-3), 181–189. Available at https://doi.org/10.1016/j.ejphar.2009.06.067.

Attery, A., Dey, P., Tripathi, P., Batra, J.K., 2018. A ribonuclease inhibitor resistant dimer of human pancreatic ribonuclease displays specific antitumor activity. International Journal of Biological Macromolecules 107, 1965–1970. Available at https://doi.org/10.1016/j.ijbiomac.2017.10.067, www.elsevier.com/locate/ijbiomac.

Banerjee, S., Gusho, E., Gaughan, C., Dong, B., Gu, X., Holvey-Bates, E., et al., 2019. OAS-RNase L innate immune pathway mediates the cytotoxicity of a DNA-demethylating drug. Proceedings of the National Academy of Sciences of the United States of America 116 (11), 5071–5076. Available at https://doi.org/10.1073/pnas.1815071116, https://www.pnas.org/content/pnas/116/11/5071.full.pdf.

Boix, E., Nogués, M.V., 2007. Mammalian antimicrobial proteins and peptides: overview on the RNase A superfamily members involved in innate host defence. Molecular BioSystems 3 (5), 317–335. Available at https://doi.org/10.1039/b617527a.

Boix, E., Wu, Y.N., Vasandani, V.M., Saxena, S.K., Ardelt, W., Ladner, J., et al., 1996. Role of the N terminus in RNase a homologues: differences in catalytic activity, ribonuclease inhibitor interaction and cytotoxicity. Journal of Molecular Biology 257 (5), 992–1007. Available at https://doi.org/10.1006/jmbi.1996.0218, https://www.journals.elsevier.com/journal-of-molecular-biology.

Boix, E., Torrent, M., Sánchez, D., Nogués, M.V., 2008. The antipathogen activities of eosinophil cationic protein. Current Pharmaceutical Biotechnology 9 (3), 141–152. Available at https://doi.org/10.2174/138920108784567353Spain, http://www.ingentaconnect.com/content/ben/cpb/2008/00000009/00000003/art00003.

Boix, E., Salazar, V.A., Torrent, M., Pulido, D., Nogués, M.V., Moussaoui, M., 2012. Structural determinants of the eosinophil cationic protein antimicrobial activity. Biological Chemistry 393 (8), 801–815. Available at https://doi.org/10.1515/hsz-2012-0160.

Carreras, E., Boix, E., Rosenberg, H.F., Cuchillo, C.M., Nogués, M.V., 2003. Both aromatic and cationic residues contribute to the membrane-lytic and bactericidal activity of eosinophil cationic protein. Biochemistry 42 (22), 6636–6644. Available at https://doi.org/10.1021/bi0273011.

Chakrabarti, A., Ghosh, P.K., Banerjee, S., Gaughan, C., Silverman, R.H., 2012. Rnase L triggers autophagy in response to viral infections. Journal of Virology 86 (20), 11311–11321. Available at https://doi.org/10.1128/JVI.00270-12, http://jvi.asm.org/content/86/20/11311.full.pdf.

Chen, X., Zhao, Q., Xie, Q., Xing, Y., Chen, Z., 2018. MCPIP1 negatively regulate cellular antiviral innate immune responses through DUB and disruption of TRAF3-TBK1-IKKε complex. Biochemical and Biophysical Research Communications 503 (2), 830–836. Available at https://doi.org/10.1016/j.bbrc.2018.06.083, http://www.sciencedirect.com/science/journal/0006291X.

Cho, S., Beintema, J.J., Zhang, J., 2005. The ribonuclease a superfamily of mammals and birds: identifying new members and tracing evolutionary histories. Genomics 85 (2), 208–220. Available at https://doi.org/10.1016/j.ygeno.2004.10.008.

Cuchillo, C.M., Nogués, M.V., Raines, R.T., 2011. Bovine pancreatic ribonuclease: fifty years of the first enzymatic reaction mechanism. Biochemistry 50 (37), 7835–7841. Available at https://doi.org/10.1021/bi201075b.

De Tomi, E., Campagnari, R., Orlandi, E., Cardile, A., Zanrè, V., Menegazzi, M., et al., 2022. Upregulation of miR-34a-5p, miR-20a-3p and miR-29a-3p by onconase in A375 melanoma cells correlates with the downregulation of specific onco-proteins. International Journal of Molecular Sciences 23 (3). Available at https://doi.org/10.3390/ijms23031647, https://www.mdpi.com/1422-0067/23/3/1647/pdf.

Degols, G., Eldin, P., Mechti, N., 2007. ISG20, an actor of the innate immune response. Biochimie 89 (6-7), 831–835. Available at https://doi.org/10.1016/j.biochi.2007.03.006.

Deymier, S., Louvat, C., Fiorini, F., Cimarelli, A., 2022. ISG20: an enigmatic antiviral RNase targeting multiple viruses. FEBS Open Biology 12 (6), 1096–1111. Available at https://doi.org/10.1002/2211-5463.13382, http://www.elsevier.com/wps/find/journaldescription.cws_home/726807/description#description.

Dickson, K.A., Haigis, M.C., Raines, R.T., 2005. Ribonuclease Inhibitor: structure and function. Progress in Nucleic Acid Research and Molecular Biology 80, 349–374. Available at https://doi.org/10.1016/S0079-6603(05)80009-1.

Diop, G., Derbois, C., Loucoubar, C., Mbengue, B., Ndao, B.N., Thiam, F., et al., 2018. Genetic variants of RNASE3 (ECP) and susceptibility to severe malaria in Senegalese population. Malaria Journal 17 (1). Available at https://doi.org/10.1186/s12936-018-2205-9.

Domachowske, J.B., Dyer, K.D., Bonville, C.A., Rosenberg, H.F., 1998. Recombinant human eosinophil-derived neurotoxin/RNase 2 functions as an effective antiviral agent against respiratory syncytial virus. The Journal of Infectious Diseases 177 (6), 1458–1464. Available at https://doi.org/10.1086/515322.

Doucet, N., Watt, E.D., Loria, J.P., 2009. The flexibility of a distant loop modulates active site motion and product release in ribonuclease A. Biochemistry 48 (30), 7160–7168. Available at https://doi.org/10.1021/bi900830g.

Eller, C.H., Raines, R.T., 2020. Antimicrobial synergy of a ribonuclease and a peptide secreted by human cells. ACS Infectious Diseases 6 (11), 3083–3088. Available at https://doi.org/10.1021/acsinfecdis.0c00594, http://pubs.acs.org/journal/aidcbc.

Fedorova, A.A., Goncharova, E.P., Kovpak, M.P., Vlassov, V.V., Zenkova, M.A., 2012. Influenza virus inactivated by artificial ribonucleases as a prospective killed virus vaccine. Vaccine 30 (19), 2973–2980. Available at https://doi.org/10.1016/j.vaccine.2012.02.051.

Fiorini, C., Gotte, G., Donnarumma, F., Picone, D., Donadelli, M., 2014. Bovine seminal ribonuclease triggers Beclin1-mediated autophagic cell death in pancreatic cancer cells. Biochimica et Biophysica Acta - Molecular Cell Research 1843 (5), 976–984. Available at https://doi.org/10.1016/j.bbamcr.2014.01.025.

Fiorini, C., Cordani, M., Gotte, G., Picone, D., Donadelli, M., 2015. Onconase induces autophagy sensitizing pancreatic cancer cells to gemcitabine and activates Akt/mTOR pathway in a ROS-dependent manner. Biochimica et Biophysica Acta - Molecular Cell Research 1853 (3), 549–560. Available at https://doi.org/10.1016/j.bbamcr.2014.12.016, www.elsevier.com/locate/bbamcr.

Fouace, S., Gaudin, C., Picard, S., Corvaisier, S., Renault, J., Carboni, B., Felden, B., 2004. Polyamine derivatives as selective RNaseA mimics. Nucleic Acids Research 32 (1), 151–157. Available at https://doi.org/10.1093/nar/gkh157.

Gagné, D., French, R.L., Narayanan, C., Simonović, M., Agarwal, P.K., Doucet, N., 2015. Perturbation of the conformational dynamics of an active-site loop alters enzyme activity. Structure 23 (12), 2256–2266. Available at https://doi.org/10.1016/j.str.2015.10.011, http://www.journals.elsevier.com/structure/.

Gaudreault, E., Gosselin, J., 2007. Leukotriene B 4 -mediated release of antimicrobial peptides against cytomegalovirus Is BLT1 dependent. Viral Immunology 20 (3), 407–420. Available at https://doi.org/10.1089/vim.2006.0099.

Goncalves, K.A., Silberstein, L., Li, S., Severe, N., Hu, M.G., Yang, H., et al., 2016. Angiogenin promotes hematopoietic regeneration by dichotomously regulating quiescence of stem and progenitor cells. Cell 166 (4), 894–906. Available at https://doi.org/10.1016/j.cell.2016.06.042.

Goncharova, E.P., Kovpak, M.P., Ryabchikova, E.I., Konevets, D.A., Sil'nikov, V.N., Zenkova, M.A., et al., 2009. Viral genome cleavage with artificial ribonucleases: a new method to inactivate RNA-containing viruses. Doklady Biochemistry and Biophysics 427 (1), 221–224. Available at https://doi.org/10.1134/s1607672909040140.

Goo, S.M., Cho, S., 2013. The expansion and functional diversification of the mammalian ribonuclease a superfamily epitomizes the efficiency of multigene families at generating biological novelty. Genome Biology and Evolution 5 (11), 2124–2140. Available at https://doi.org/10.1093/gbe/evt161, http://gbe.oxfordjournals.org/.

Gotte, G., Menegazzi, M., 2019. Biological activities of secretory RNases: focus on their oligomerization to design antitumor drugs. Frontiers in Immunology 10. Available at https://doi.org/10.3389/fimmu.2019.02626, https://www.frontiersin.org/journals/immunology#.

Gotte, G., Campagnari, R., Loreto, D., Bettin, I., Calzetti, F., Menegazzi, M., et al., 2021. The crystal structure of the domain-swapped dimer of onconase highlights some catalytic and antitumor activity features of the enzyme. International Journal of Biological Macromolecules 191, 560–571. Available at https://doi.org/10.1016/j.ijbiomac.2021.09.095, www.elsevier.com/locate/ijbiomac.

Gupta, S.K., Haigh, B.J., Griffin, F.J., Wheeler, T.T., 2013. The mammalian secreted RNases: mechanisms of action in host defence. Innate Immunity 19 (1), 86–97. Available at https://doi.org/10.1177/1753425912446955.

Hancock, R.E.W., Scott, M.G., 2000. The role of antimicrobial peptides in animal defenses. Proceedings of the National Academy of Sciences of the United States of America 97 (16), 8856–8861. Available at https://doi.org/10.1073/pnas.97.16.8856.

Helm, M., Kopka, M.L., Sharma, S.K., Lown, J.W., Giegé, R., 2001. Rnase activity of a DNA minor groove binder with a minimalist catalytic motif from RNase A. Biochemical and Biophysical Research Communications 281 (5), 1283–1290. Available at https://doi.org/10.1006/bbrc.2001.4503, http://www.sciencedirect.com/science/journal/0006291X.

Hoang, T.T., Tanrikulu, I.C., Vatland, Q.A., Hoang, T.M., Raines, R.T., 2018. A human ribonuclease variant and ERK-pathway inhibitors exhibit highly synergistic toxicity for cancer cells. Molecular Cancer Therapeutics 17 (12), 2622–2632. Available at https://doi.org/10.1158/1535-7163.mct-18-0724.

Hooper, L.V., Stappenbeck, T.S., Hong, C.V., Gordon, J.I., 2003. Angiogenins: a new class of microbicidal proteins involved in innate immunity. Nature Immunology 4 (3), 269–273. Available at https://doi.org/10.1038/ni888.

Hornung, V., Hartmann, R., Ablasser, A., Hopfner, K.-P., 2014. OAS proteins and cGAS: unifying concepts in sensing and responding to cytosolic nucleic acids. Nature Reviews Immunology 14 (8), 521–528. Available at https://doi.org/10.1038/nri3719.

Ilinskaya, O.N., Mahmud, R.S., 2014. Ribonucleases as antiviral agents. Molecular Biology 48 (5), 615–623. Available at https://doi.org/10.1134/s0026893314040050.

Iordanov, M.S., Ryabinina, O.P., Wong, J., Dinh, T.H., Newton, D.L., Rybak, S.M., et al., 2000. Molecular determinants of apoptosis induced by the cytotoxic ribonuclease onconase: evidence for cytotoxic mechanisms different from inhibition of protein synthesis. Cancer Research 60 (7), 1983–1994.

Jones, K., Keddy, C., Jenkins, C., Nicholson, K., Shinde, U., Davare, M.A., 2024. Novel insight into mechanisms of ROS1 catalytic activation via loss of the extracellular domain. Scientific Reports 14 (1). Available at https://doi.org/10.1038/s41598-024-71687-7, https://www.nature.com/srep/.

Joo, H.S., Fu, C.I., Otto, M., 2016. Bacterial strategies of resistance to antimicrobial peptides. Philosophical Transactions of the Royal Society B: Biological Sciences 371 (1695). Available at https://doi.org/10.1098/rstb.2015.0292, http://rstb.royalsocietypublishing.org/content/royptb/371/1695/20150292.full.pdf.

Klion, A.D., Nutman, T.B., 2004. The role of eosinophils in host defense against helminth parasites. Journal of Allergy and Clinical Immunology 113 (1), 30–37. Available at https://doi.org/10.1016/j.jaci.2003.10.050, http://www.elsevier.com/inca/publications/store/6/2/3/3/6/8/index.htt.

Koczera, P., Martin, L., Marx, G., Schuerholz, T., 2016. The ribonuclease a superfamily in humans: canonical RNases as the buttress of innate immunity. International Journal of Molecular Sciences 17 (8), 1278. Available at https://doi.org/10.3390/ijms17081278.

Kurtzhals, J.A.L., Reimert, C.M., Tette, E., Dunyo, S.K., Koram, K.A., Akanmori, B.D., et al., 1998. Increased eosinophil activity in acute *Plasmodium falciparum* infection—association with cerebral malaria. Clinical and Experimental Immunology 112 (2), 303–307. Available at https://doi.org/10.1046/j.1365-2249.1998.00586.x.

Lee, H.-H., 2024. RNASE4 promotes malignant progression and chemoresistance in hypoxic glioblastoma via activation of AXL/AKT and NF-κB/cIAPs signaling pathways. American Journal of Cancer Research 14 (9), 4320–4336. Available at https://doi.org/10.62347/udbj5986.

Lee, H.H., Wang, Y.N., Yang, W.H., Xia, W., Wei, Y., Chan, L.C., et al., 2021. Human ribonuclease 1 serves as a secretory ligand of ephrin A4 receptor and induces breast tumor initiation. Nature Communications 12 (1). Available at https://doi.org/10.1038/s41467-021-23075-2, http://www.nature.com/ncomms/index.html.

Lee-Huang, S., Huang, P.L., Sun, Y., Huang, P.L., Kung, H.-fu, Blithe, D.L., et al., 1999. Lysozyme and RNases as anti-HIV components in β-core preparations of human chorionic gonadotropin. Proceedings of the National Academy of Sciences 96 (6), 2678–2681. Available at https://doi.org/10.1073/pnas.96.6.2678.

Lehrer, R.I., Szklarek, D., Barton, A., Ganz, T., Hamann, K.J., Gleich, G.J., 1989. Antibacterial properties of eosinophil major basic protein and eosinophil cationic protein. Journal of Immunology 142 (12), 4428–4434.

Li, J., Boix, E., 2021. Host defence RNases as antiviral agents against enveloped single stranded RNA viruses. Virulence 12 (1), 444–469. Available at https://doi.org/10.1080/21505594.2021.1871823, http://www.tandfonline.com/loi/kvir20.

Li, Y., Que, L., Fukano, K., Koura, M., Kitamura, K., Zheng, X., et al., 2020. MCPIP1 reduces HBV-RNA by targeting its epsilon structure. Scientific Reports 10 (1). Available at https://doi.org/10.1038/s41598-020-77166-z, www.nature.com/srep/index.html.

Li, J., Prats-Ejarque, G., Torrent, M., Andreu, D., Brandenburg, K., Fernández-Millán, P., et al., 2022. In vivo evaluation of ECP peptide analogues for the treatment of *Acinetobacter baumannii* infection. Biomedicines 10 (2). Available at https://doi.org/10.3390/biomedicines10020386, https://www.mdpi.com/2227-9059/10/2/386/pdf.

Lien, P.-C., Kuo, P.-H., Chen, C.-J., Chang, H.-H., Fang, S.-lung, Wu, W.-S., et al., 2013. In silico prediction and in vitro characterization of multifunctional human RNase3. BioMed Research International 2013, 1–12. Available at https://doi.org/10.1155/2013/170398.

Lin, R.J., Chien, H.L., Lin, S.Y., Chang, B.L., Yu, H.P., Tang, W.C., et al., 2013. MCPIP1 ribonuclease exhibits broad-spectrum antiviral effects through viral RNA binding and degradation. Nucleic Acids Research 41 (5), 3314–3326. Available at https://doi.org/10.1093/nar/gkt019, https://academic.oup.com/nar/issue.

Liu, C., Zha, Z., Zhou, C., Chen, Y., Xia, W., Wang, Y.N., et al., 2021. Ribonuclease 7-driven activation of ROS1 is a potential therapeutic target in hepatocellular carcinoma. Journal of Hepatology 74 (4), 907–918. Available at https://doi.org/10.1016/j.jhep.2020.09.030, http://www.sciencedirect.com/science/journal/01688278.

Lomax, J.E., Bianchetti, C.M., Chang, A., Phillips, G.N., Fox, B.G., Raines, R.T., 2014. Functional evolution of ribonuclease inhibitor: insights from birds and reptiles. Journal of Molecular Biology 426 (17), 3041–3056. Available at https://doi.org/10.1016/j.jmb.2014.06.007, https://www.journals.elsevier.com/journal-of-molecular-biology.

Lomax, J.E., Eller, C.H., Raines, R.T., 2017. Comparative functional analysis of ribonuclease 1 homologs: molecular insights into evolving vertebrate physiology. Biochemical Journal 474 (13), 2219–2233. Available at https://doi.org/10.1042/BCJ20170173, http://www.biochemj.org/content/ppbiochemj/474/13/2219.full.pdf.

Lu, L., Li, J., Moussaoui, M., Boix, E., 2018. Immune modulation by human secreted RNases at the extracellular space. Frontiers in Immunology 9 (MAY). Available at https://doi.org/10.3389/fimmu.2018.01012, https://www.frontiersin.org/articles/10.3389/fimmu.2018.01012/full.

Lu, L., Li, J., Wei, R., Guidi, I., Cozzuto, L., Ponomarenko, J., et al., 2022. Selective cleavage of ncRNA and antiviral activity by RNase2/EDN in THP1-induced macrophages. Cellular and Molecular Life Sciences 79 (4). Available at https://doi.org/10.1007/s00018-022-04229-x, http://link.springer.de/link/service/journals/00018/index.htm.

Malik, A., Batra, J.K., 2012. Antimicrobial activity of human eosinophil granule proteins: involvement in host defence against pathogens. Critical Reviews in Microbiology 38 (2), 168–181. Available at https://doi.org/10.3109/1040841X.2011.645519.

Manivannan, P., Siddiqui, M.A., Malathi, K., 2020. RNase 1 amplifies interferon signaling by inducing protein kinase r-mediated antiviral stress granules. Journal of Virology 94 (13). Available at https://doi.org/10.1128/JVI.00205-20, https://journals.asm.org/doi/10.1128/JVI.00205-20.

Michaelis, M., Cinatl, J., Anand, P., Rothweiler, F., Kotchetkov, R., Deimling, Av, et al., 2007. Onconase induces caspase-independent cell death in chemoresistant neuroblastoma cells. Cancer Letters 250 (1), 107–116. Available at https://doi.org/10.1016/j.canlet.2006.09.018, www.elsevier.com/locate/canlet.

Mikulski, S., Chun, H., Mittelman, A., Panella, T., Puccio, C., Shogen, K., et al., 1995. Relationship between response rate and median survival in patients with advanced nonsmall cell lung-cancer - comparison of onconase(r) with other anticancer agents. International Journal of Oncology 6 (4), 889–897.

Mikulski, S.M., Costanzi, J.J., Vogelzang, N.J., McCachren, S., Taub, R.N., Chun, H., et al., 2002. K. Shogen, phase II trial of a single weekly intravenous dose of ranpirnase in patients with unresectable malignant mesothelioma. Journal of Clinical Oncology 20 (1), 274–281. Available at https://doi.org/10.1200/JCO.20.1.274.

Mironova, N.L., Pyshnyi, D.V., Shtadler, D.V., Fedorova, A.A., Vlassov, V.V., Zenkova, M.A., 2007. RNase T1 mimicking artificial ribonuclease. Nucleic Acids Research 35 (7), 2356–2367. Available at https://doi.org/10.1093/nar/gkm143.

Moreau, J.M., Dyer, K.D., Bonville, C.A., Nitto, T., Vasquez, N.L., Easton, A.J., et al., 2003. Diminished expression of an antiviral ribonuclease in response to pneumovirus infection in vivo. Antiviral Research 59 (3), 181–191. Available at https://doi.org/10.1016/S0166-3542(03)00111-6, www.elsevier.com/locate/antiviral.

Mukhi, B., Gupta, H., Wassmer, S.C., Anvikar, A.R., Ghosh, S.K., 2020. Haplotype of RNASE 3 polymorphisms is associated with severe malaria in an Indian population. Molecular Biology Reports 47 (11), 8841–8848. Available at https://doi.org/10.1007/s11033-020-05934-1, www.springer.com/journal/11033.

Munita, J.M., Arias, C.A., 2016. Mechanisms of antibiotic resistance. Microbiology Spectrum 4 (2). Available at https://doi.org/10.1128/microbiolspec.VMBF-0016-2015, http://www.asmscience.org/docserver/fulltext/microbiolspec/4/2/VMBF-0016-2015.pdf?expires=1486108180&id=id&accname=45052&checksum=43A031364F99D87142C9C34E0F0E2DC9.

Naghavi, M., Vollset, S.E., Ikuta, K.S., Swetschinski, L.R., Gray, A.P., Wool, E.E., et al., 2024. Global burden of bacterial antimicrobial resistance 1990–2021: a systematic analysis with forecasts to 2050. The Lancet 404 (10459), 1199–1226. Available at https://doi.org/10.1016/s0140-6736(24)01867-1.

Nogués, M.V., Moussaoui, M., Boix, E., Vilanova, M., Ribó, M., Cuchillo, C.M., 1998. The contribution of noncatalytic phosphate-binding subsites to the mechanism of bovine pancreatic ribonuclease A.

Cellular and Molecular Life Sciences 54 (8), 766–774. Available at https://doi.org/10.1007/s000180050205.

Noro, I., Bettin, I., Fasoli, S., Smania, M., Lunardi, L., Giannini, M., et al., 2023. Human RNase 1 can extensively oligomerize through 3D domain swapping thanks to the crucial contribution of its C-terminus. International Journal of Biological Macromolecules 249, 126110. Available at https://doi.org/10.1016/j.ijbiomac.2023.126110.

O'Neill, J., 2016. Tackling Drug-Resistant Infections Globally: Final Report and Recommendations.

O'Reilly, M.A., Yee, M., Buczynski, B.W., Vitiello, P.F., Keng, P.C., Welle, S.L., et al., 2012. Neonatal oxygen increases sensitivity to influenza a virus infection in adult mice by suppressing epithelial expression of Ear1. American Journal of Pathology 181 (2), 441–451. Available at https://doi.org/10.1016/j.ajpath.2012.05.005.

Pizzo, E., D'Alessio, G., 2007. The success of the RNase scaffold in the advance of biosciences and in evolution. Gene 406 (1-2), 8–12. Available at https://doi.org/10.1016/j.gene.2007.05.006.

Prats-Ejarque, G., Li, J., Ait-Ichou, F., Lorente, H., Boix, E., 2019. Testing a human antimicrobial RNase chimera against bacterial resistance. Frontiers in Microbiology 10. Available at https://doi.org/10.3389/fmicb.2019.01357, https://www.frontiersin.org/journals/microbiology#.

Pulido, D., Torrent, M., Andreu, D., Nogues, M.V., Boix, E., 2013. Two human host defense ribonucleases against mycobacteria, the eosinophil cationic protein (RNase 3) and RNase 7. Antimicrobial Agents and Chemotherapy 57 (8), 3797–3805. Available at https://doi.org/10.1128/AAC.00428-13Spain, http://aac.asm.org/content/57/8/3797.full.pdf.

Pulido, D., Prats-Ejarque, G., Villalba, C., Albacar, M., González-López, J.J., Torrent, M., et al., 2016. A novel RNase 3/ECP peptide for *Pseudomonas aeruginosa* biofilm eradication that combines antimicrobial, lipopolysaccharide binding, and cell-agglutinating activities. Antimicrobial Agents and Chemotherapy 60 (10), 6313–6325. Available at https://doi.org/10.1128/AAC.00830-16, http://aac.asm.org/content/60/10/6313.full.pdf+html.

Pulido, D., Prats-Ejarque, G., Villalba, C., Albacar, M., Moussaoui, M., Andreu, D., et al., 2018. Positional scanning library applied to the human eosinophil cationic protein/RNase3 N-terminus reveals novel and potent anti-biofilm peptides. European Journal of Medicinal Chemistry 152, 590–599. Available at https://doi.org/10.1016/j.ejmech.2018.05.012.

Raineri, A., Prodomini, S., Fasoli, S., Gotte, G., Menegazzi, M., 2019. Influence of onconase in the therapeutic potential of PARP inhibitors in A375 malignant melanoma cells. Biochemical Pharmacology 167, 173–181. Available at https://doi.org/10.1016/j.bcp.2019.06.006, www.elsevier.com/locate/biochempharm.

Raines, R.T., 1998. Ribonuclease A. Chemical Reviews 98 (3), 1045–1066. Available at https://doi.org/10.1021/cr960427h.

Ran, S., Downes, A., Thorpe, P.E., 2002. Increased exposure of anionic phospholipids on the surface of tumor blood vessels. Cancer Research 62 (21), 6132–6140.

Rosenberg, H.F., 1995. Recombinant human eosinophil cationic protein: ribonuclease activity is not essential for cytotoxicity. Journal of Biological Chemistry 270 (14), 7876–7881. Available at https://doi.org/10.1074/jbc.270.14.7876, http://www.jbc.org/content/by/year.

Rosenberg, H.F., 1998. The eosinophil ribonucleases. Cellular and Molecular Life Sciences (CMLS) 54 (8), 795–803. Available at https://doi.org/10.1007/s000180050208.

Rosenberg, H.F., 2008a. RNase A ribonucleases and host defense: an evolving story. Journal of Leukocyte Biology 83 (5), 1079–1087. Available at https://doi.org/10.1189/jlb.1107725, http://www.jleukbio.org/cgi/reprint/83/5/1079.

Rosenberg, H.F., 2008b. Eosinophil-derived neurotoxin/RNase 2: connecting the past, the present and the future. Current Pharmaceutical Biotechnology 9 (3), 135–140. Available at https://doi.org/10.2174/138920108784567236, http://www.ingentaconnect.com/content/ben/cpb/2008/00000009/00000003/art00002.

Rosenberg, H., 2015. Eosinophil-derived neurotoxin (EDN/RNase 2) and the mouse eosinophil-associated RNases (mEars): expanding roles in promoting host defense. International Journal of Molecular Sciences 16 (7), 15442–15455. Available at https://doi.org/10.3390/ijms160715442.

Rosenberg, H.F., Dyer, K.D., Foster, P.S., 2013. Eosinophils: changing perspectives in health and disease. Nature Reviews Immunology 13 (1), 9–22. Available at https://doi.org/10.1038/nri3341.

Rugeles, M.T., Trubey, C.M., Bedoya, V.I., Pinto, L.A., Oppenheim, J.J., Rybak, S.M., et al., 2003. Ribonuclease is partly responsible for the HIV-1 inhibitory effect activated by HLA alloantigen recognition. AIDS 17 (4), 481–486. Available at https://doi.org/10.1097/00002030-200303070-00002.

Rutkoski, T.J., Raines, R.T., 2008. Evasion of ribonuclease inhibitor as a determinant of ribonuclease cytotoxicity. Current Pharmaceutical Biotechnology 9 (3), 185–199. Available at https://doi.org/10.2174/138920108784567344, http://www.ingentaconnect.com/content/ben/cpb/2008/00000009/00000003/art00008.

Salazar, V.A., Arranz-Trullén, J., Prats-Ejarque, G., Torrent, M., Andreu, D., Pulido, D., et al., 2019. Insight into the antifungal mechanism of action of human RNase N-terminus derived peptides. International Journal of Molecular Sciences 20 (18), 4558. Available at https://doi.org/10.3390/ijms20184558.

Sandín, D., Valle, J., Chaves-Arquero, B., Prats-Ejarque, G., Larrosa, M.N., González-López, J.J., et al., 2021. Rationally modified antimicrobial peptides from the N-terminal domain of human RNase 3 show exceptional serum stability. Journal of Medicinal Chemistry 64 (15), 11472–11482. Available at https://doi.org/10.1021/acs.jmedchem.1c00795, http://pubs.acs.org/jmc.

Scarborough, R.J., Gatignol, A., 2015. HIV and ribozymes. Advances in Experimental Medicine and Biology 848, 97–116. Available at https://doi.org/10.1007/978-1-4939-2432-5_5, http://www.springer.com/series/5584.

Siddiqui, M.A., Malathi, K., 2012. RNase L induces autophagy via c-Jun N-terminal kinase and double-stranded RNA-dependent protein kinase signaling pathways. Journal of Biological Chemistry 287 (52), 43651–43664. Available at https://doi.org/10.1074/jbc.M112.399964, http://www.jbc.org/content/287/52/43651.full.pdf+html.

Silverman, R.H., 2007. Viral encounters with 2′,5′-oligoadenylate synthetase and RNase L during the interferon antiviral response. Journal of Virology 81 (23), 12720–12729. Available at https://doi.org/10.1128/JVI.01471-07.

Singh, A., Batra, J.K., 2011. Role of unique basic residues in cytotoxic, antibacterial and antiparasitic activities of human eosinophil cationic protein. Biological Chemistry 392 (4), 337–346. Available at https://doi.org/10.1515/BC.2011.037.

Singh, U.P., Ardelt, W., Saxena, S.K., Holloway, D.E., Vidunas, E., Lee, H.S., et al., 2007. Enzymatic and structural characterisation of amphinase, a novel cytotoxic ribonuclease from *Rana pipiens* oocytes. Journal of Molecular Biology 371 (1), 93–111. Available at https://doi.org/10.1016/j.jmb.2007.04.071, https://www.journals.elsevier.com/journal-of-molecular-biology.

Sorrentino, S., 2010. The eight human "canonical" ribonucleases: molecular diversity, catalytic properties, and special biological actions of the enzyme proteins. FEBS Letters 584 (11), 2194–2200. Available at https://doi.org/10.1016/j.febslet.2010.04.018.

Sorrentino, S., Libonati, M., 1994. Human pancreatic-type and nonpancreatic-type ribonucleases: a direct side-by-side comparison of their catalytic properties. Archives of Biochemistry and Biophysics 312 (2), 340–348. Available at https://doi.org/10.1006/abbi.1994.1318.

Sorrentino, S., Libonati, M., 1997. Structure-function relationships in human ribonucleases: main distinctive features of the major RNase types. FEBS Letters 404 (1), 1–5. Available at https://doi.org/10.1016/S0014-5793(97)00086-0.

Suzuki, H.I., Arase, M., Matsuyama, H., Choi, Y.L., Ueno, T., Mano, H., et al., 2011. MCPIP1 ribonuclease antagonizes dicer and terminates microRNA biogenesis through precursor microRNA degradation. Molecular Cell 44 (3), 424–436. Available at https://doi.org/10.1016/j.molcel.2011.09.012.

Tamkovich, N., Koroleva, L., Kovpak, M., Goncharova, E., Silnikov, V., Vlassov, V., et al., 2016. Design, RNA cleavage and antiviral activity of new artificial ribonucleases derived from mono-, di- and tripeptides connected by linkers of different hydrophobicity. Federation Bioorganic and Medicinal Chemistry 24 (6), 1346–1355. Available at https://doi.org/10.1016/j.bmc.2016.02.007, http://www.journals.elsevier.com/bioorganic-and-medicinal-chemistry/.

Thomas, S.P., Kim, E., Kim, J.S., Raines, R.T., 2016. Knockout of the ribonuclease inhibitor gene leaves human cells vulnerable to secretory ribonucleases. Biochemistry 55 (46), 6359–6362. Available at https://doi.org/10.1021/acs.biochem.6b01003, http://pubs.acs.org/journal/bichaw.

Tischendorf, F.W., Brattig, N.W., Büttner, D.W., Pieper, A., Lintzel, M., 1996. Serum levels of eosinophil cationic protein, eosinophil-derived neurotoxin and myeloperoxidase in infections with filariea and schistosomes. Acta Tropica 62 (3), 171–182. Available at https://doi.org/10.1016/s0001-706x(96)00038-1.

Torrent, M., Cuyás, E., Carreras, E., Navarro, S., López, O., De La Maza, A., et al., 2007. Topography studies on the membrane interaction mechanism of the eosinophil cationic protein. Biochemistry 46 (3), 720–733. Available at https://doi.org/10.1021/bi061190e.

Torrent, M., Navarro, S., Moussaoui, M., Nogués, M.V., Boix, E., 2008. Eosinophil cationic protein high-affinity binding to bacteria-wall lipopolysaccharides and peptidoglycans. Biochemistry 47 (11), 3544–3555. Available at https://doi.org/10.1021/bi702065b.

Torrent, G., Ribó, M., Benito, A., Vilanova, M., 2009a. Bactericidal activity engineered on human pancreatic ribonuclease and onconase. Molecular Pharmaceutics 6 (2), 531–542. Available at https://doi.org/10.1021/mp8001914Spain, http://pubs.acs.org/doi/pdfplus/10.1021/mp8001914.

Torrent, M., de la Torre, B.G., Nogués, V.M., Andreu, D., Boix, E., 2009b. Bactericidal and membrane disruption activities of the eosinophil cationic protein are largely retained in an N-terminal fragment. Biochemical Journal 421 (3), 425–434. Available at https://doi.org/10.1042/BJ20082330Spain, http://www.biochemj.org/bj/421/0425/4210425.pdf.

Torrent, M., Sánchez, D., Buzón, V., Nogués, M.V., Cladera, J., Boix, E., 2009c. Comparison of the membrane interaction mechanism of two antimicrobial RNases: RNase 3/ECP and RNase 7. Biochimica et Biophysica Acta - Biomembranes 1788 (5), 1116–1125. Available at https://doi.org/10.1016/j.bbamem.2009.01.013.

Torrent, M., Odorizzi, F., Nogués, M.V., Boix, E., 2010. Eosinophil cationic protein aggregation: identification of an N-terminus amyloid prone region. Biomacromolecules 11 (8), 1983–1990. Available at https://doi.org/10.1021/bm100334u.

Torrent, M., Andreu, D., Nogués, V.M., Boix, E., 2011a. Connecting peptide physicochemical and antimicrobial properties by a rational prediction model. PLoS ONE 6 (2). Available at https://doi.org/10.1371/journal.pone.0016968Spain, http://www.plosone.org/article/fetchObjectAttachment.action?uri=info%3Adoi%2F10.1371%2Fjournal.pone.0016968&representation=PDF.

Torrent, M., Pulido, D., De La Torre, B.G., García-Mayoral, M.F., Nogués, M.V., Bruix, M., et al., 2011b. Refining the eosinophil cationic protein antibacterial pharmacophore by rational structure minimization. Journal of Medicinal Chemistry 54 (14), 5237–5244. Available at https://doi.org/10.1021/jm200701g.

Torrent, M., Pulido, D., Nogués, M.V., Boix, E., 2012. Exploring new biological functions of amyloids: bacteria cell agglutination mediated by host protein aggregation. PLoS Pathogens 8 (11). Available at https://doi.org/10.1371/journal.ppat.1003005Spain, http://www.plospathogens.org/article/fetchObjectAttachment.action?uri=info%3Adoi%2F10.1371%2Fjournal.ppat.1003005&representation=PDF.

Torrent, M., Pulido, D., Valle, J., Nogués, M.V., Andreu, D., Boix, E., 2013. Ribonucleases as a host-defence family: evidence of evolutionarily conserved antimicrobial activity at the N-terminus. Biochemical Journal 456 (1), 99–108. Available at https://doi.org/10.1042/BJ20130123Spain, http://www.biochemj.org/bj/456/0099/4560099.pdf.

Wang, Y.N., Lee, H.H., Chou, C.K., Yang, W.H., Wei, Y., Chen, C.T., et al., 2018. Angiogenin/Ribonuclease 5 Is an EGFR ligand and a serum biomarker for erlotinib sensitivity in pancreatic cancer. Cancer Cell 33 (4), 752. Available at https://doi.org/10.1016/j.ccell.2018.02.012, https://www.journals.elsevier.com/cancer-cell.

Waters, L.S., Taverne, J., Tai, P.C., Spry, C.J., Targett, G.A., Playfair, J.H., 1987. Killing of *Plasmodium falciparum* by eosinophil secretory products. Infection and Immunity 55 (4), 877–881. Available at https://doi.org/10.1128/iai.55.4.877-881.1987.

Wu, Y., Mikulski, S.M., Ardelt, W., Rybak, S.M., Youle, R.J., 1993. A cytotoxic ribonuclease. Study of the mechanism of onconase cytotoxicity. Journal of Biological Chemistry 268 (14), 10686–10693. Available at https://doi.org/10.1016/s0021-9258(18)82252-9.

Wu, N., Nguyen, X.N., Wang, L., Appourchaux, R., Zhang, C., Panthu, B., et al., 2019. The interferon stimulated gene 20 protein (ISG20) is an innate defense antiviral factor that discriminates self versus non-self translation. Public Library of Science, France PLoS Pathogens 15 (10). Available at https://doi.org/10.1371/journal.ppat.1008093, https://journals.plos.org/plospathogens/article/file?id=10.1371/journal.ppat.1008093&type=printable.

Xu, L., Liao, W.L., Lu, Q.J., Li, C.G., Yuan, Y., Xu, Z.Y., et al., 2016. ANG promotes proliferation and invasion of the cell of lung squamous carcinoma by Directly Up-regulating HMGA2. Journal of Cancer 7 (7), 862–871. Available at https://doi.org/10.7150/jca.14440, http://www.jcancer.org/v07p0862.pdf.

Xu, S.Y., Fomenkov, A., Chen, T.H., Yigit, E., 2021. Expression of human ACE2 N-terminal domain, part of the receptor for SARS-CoV-2, in fusion with maltose-binding protein, *E. coli* ribonuclease I and human RNase A. Frontiers in Microbiology 12. Available at https://doi.org/10.3389/fmicb.2021.660149, https://www.frontiersin.org/journals/microbiology#.

Yang, D., Rosenberg, H.F., Chen, Q., Dyer, K.D., Kurosaka, K., Oppenheim, J.J., 2003. Eosinophil-derived neurotoxin (EDN), an antimicrobial protein with chemotactic activities for dendritic cells. Blood 102 (9), 3396–3403. Available at https://doi.org/10.1182/blood-2003-01-0151, https://www.journals.elsevier.com/blood.

Yu, W., Goncalves, K.A., Li, S., Kishikawa, H., Sun, G., Yang, H., et al., 2017. Plexin-B2 mediates physiologic and pathologic functions of angiogenin. Cell 171 (4), 849. Available at https://doi.org/10.1016/j.cell.2017.10.005, https://www.sciencedirect.com/journal/cell.

Yu, A.M., Choi, Y.H., Tu, M.J., 2020. RNA drugs and RNA targets for small molecules: principles, progress, and challenges. Pharmacological Reviews 72 (4), 862–898. Available at https://doi.org/10.1124/pr.120.019554, https://pharmrev.aspetjournals.org/content/72/4/862.

Index

Note: Page numbers followed by "*f*" and "*t*" refer to figures and tables, respectively.